Springer Texts in Business and Economics

Springer Texts in Business and Economics (STBE) delivers high-quality instructional content for undergraduates and graduates in all areas of Business/Management Science and Economics. The series is comprised of self-contained books with a broad and comprehensive coverage that are suitable for class as well as for individual self-study. All texts are authored by established experts in their fields and offer a solid methodological background, often accompanied by problems and exercises.

More information about this series at http://www.springer.com/series/10099

Pak-Sing Choi • Felix Munoz-Garcia

Auction Theory

Introductory Exercises
with Answer Keys

 Springer

Pak-Sing Choi
Graduate Institute of Industrial Economics
National Central University
Taoyuan City, Taiwan

Felix Munoz-Garcia
School of Economic Sciences
Washington State University
Pullman, WA, USA

ISSN 2192-4333 ISSN 2192-4341 (electronic)
Springer Texts in Business and Economics
ISBN 978-3-030-69577-4 ISBN 978-3-030-69575-0 (eBook)
https://doi.org/10.1007/978-3-030-69575-0

© The Editor(s) (if applicable) and The Author(s), under exclusive license to Springer Nature Switzerland AG 2021

This work is subject to copyright. All rights are solely and exclusively licensed by the Publisher, whether the whole or part of the material is concerned, specifically the rights of translation, reprinting, reuse of illustrations, recitation, broadcasting, reproduction on microfilms or in any other physical way, and transmission or information storage and retrieval, electronic adaptation, computer software, or by similar or dissimilar methodology now known or hereafter developed.

The use of general descriptive names, registered names, trademarks, service marks, etc. in this publication does not imply, even in the absence of a specific statement, that such names are exempt from the relevant protective laws and regulations and therefore free for general use.

The publisher, the authors, and the editors are safe to assume that the advice and information in this book are believed to be true and accurate at the date of publication. Neither the publisher nor the authors or the editors give a warranty, expressed or implied, with respect to the material contained herein or for any errors or omissions that may have been made. The publisher remains neutral with regard to jurisdictional claims in published maps and institutional affiliations.

This Springer imprint is published by the registered company Springer Nature Switzerland AG
The registered company address is: Gewerbestrasse 11, 6330 Cham, Switzerland

Preface

This textbook provides a short introduction to Auction Theory by presenting exercises with detailed answer keys. Our presentation should be accessible for undergraduate and masters-level students in Economics and Finance, although some of the more challenging questions can be used as an introduction to these topics for Ph.D. students.

There are, currently, some textbooks on Auction Theory at this level, such as Klemperer (2004), or at the Ph.D. level, such as Menezes and Monteiro (2005) and Krishna (2009). Other contributions are clearly more applied and discuss the design of real-world auctions in different countries and industries, such as Milgrom (2004, 2017), Haeringer (2018), and Mochón and Sáez (2014). Our presentation, however, differs from these textbooks along several dimensions:

1. *Worked-out exercises.* We provide 83 step-by-step exercises with detailed answer keys, so readers can understand how to solve similar questions on their own. We also present the intuition behind each mathematical step and result. The book covers the typical materials required in Auction Theory courses (and the topics on Auction Theory are also presented in other courses, such as Game Theory, Market Design, Economics of Information, or Contract Theory) at the undergraduate level, and by most courses at the master's level.

    In contrast, the above textbooks on Auction Theory tend to focus on theoretical tools, often using a relatively general approach, providing few examples, and rarely offering step-by-step answers to exercises. In our experience, students learn more effectively when they can reproduce some of the mathematical steps on their own, as that improves their ability to apply the theoretical concepts to their own research projects. Our book, then, seeks to complement the theoretical presentation of current textbooks, offering practice exercises that predict bidders' equilibrium behavior in different auction formats, along with the seller's strategic incentives to organize one auction over another.

2. *Tools and algebra support.* We emphasize the game-theoretic tools used in each type of exercise, so that readers can apply similar tools to other auction formats. In addition, we assume little mathematical background in algebra and calculus, providing most algebraic steps and simplifications, helping readers follow every step.

3. *Exercises based on journal articles*. Several exercises are based on published articles, but reducing the model to its main elements and dividing the question into several easy-to-answer parts. This more user-friendly version of the published article helps students to reproduce the main findings of the article on their own. In our experience, we found that this teaching tool is highly effective for students in their senior year and for graduate students, as they can subsequently use the main steps of these published papers in their own research.
4. *Ranked exercises*. Finally, we rank exercises according to their level of difficulty with a letter A next to its title to indicate easy exercises, a B to mark exercises of intermediate difficulty, and a C for more challenging exercises. This ranking should help undergraduate (master's) students start reviewing exercises with an A (B) and, once they feel comfortable with this type of questions, move on to exercises marked with a letter B (C, respectively).

Organization of the Book

We first examine second-price auctions, where every bidder privately observes the valuation for the object being sold and submits a sealed bid. Bids are then ranked, where the bidder submitting the highest bid wins the object and pays the second-highest bid (not his bid, which is the highest bid) for the object. The reason we start with this auction format is pedagogical, as it allows readers to find equilibrium bidding strategies without having to use calculus and requires little algebra, while most other auction formats require more extensive use of algebra and calculus tools.

Chapters 2 and 3 analyze first-price auctions, where the highest-valued bidder still wins the object but must now pay the bid submitted. Chapter 2 introduces the reader to this type of auction, first with only two bidders drawing their valuation for the object from a uniform distribution (which facilitates the mathematical steps), and subsequently considering more bidders who draw their valuations from other distributions. The chapter finishes with several exercises on order statistics, which are commonly used in other chapters in the book. Chapter 3 extends the analysis of first-price auctions to evaluate the seller's expected revenue from the auction, examining how equilibrium bids are affected when bidders are risk averse and allowing bidders to draw their valuations from different distributions. We seek to identify the entry fee or the reservation price that maximizes the seller's expected revenue and investigate how collusion among bidders affects this auction.

Chapter 4 analyzes all-pay auctions, where the winner is still the bidder submitting the highest bid, but all bidders must pay the bid they submitted. As expected, bidders in this setting submit less aggressive bids than in the other two auction formats studied in previous chapters. This chapter also considers a first-price auction, where bidders are asymmetrically informed about each other's valuations; that is, each bidder only observes his or her own valuation, while the rival may observe both bidders' values.

Chapter 5 studies third-price auctions, where the winner pays the third-highest bid and, more generally, kth-price auctions, where the winner pays the kth-highest bid. First- and second-price auctions are, of course, special cases of the kth-price auction, where $k=1$ and $k=2$, respectively, so this auction format can help us predict equilibrium bids in a large array of settings. We also analyze lottery auctions, where bidders submit a bid that they must pay, regardless of whether they win the auction, and their probability of winning is a function of how their bids are compared relative to the sum of bids submitted by all bidders.

In Chap. 6, we evaluate the seller's expected revenue in different auctions and then present the Revenue Equivalence principle, first with a short and then in a long, more technical, proof. Informally, this principle states that if two auction formats assign the object to the same bidder (e.g., the same winner), the expected utility of the bidder with the lowest valuation coincides across both auction formats, and bidders are risk neutral, then these two auctions must generate the same expected revenue for the seller. In other words, the bidder could run either auction and earn the same expected revenue.

Chapter 7 then presents common-value auctions, where bidders receive a signal about the value of the object being sold (e.g., an estimate) but know that the value of the object (e.g., profits to be made) is the same for all bidders. In this auction, bidders may fall prey of the so-called winner's curse, where the winner submits a bid lower than the signal privately observed, but higher than the value of the object, ultimately yielding a loss from winning the auction. We analyze the winner's curse and then study the equilibrium bidding function that players in this auction should use to avoid this. For presentation purposes, we find this equilibrium bidding function in the case of two bidders and then allow for more bidders. We end this chapter with auctions where every bidder's valuation is a function of one's signal and that of the rival (i.e., interdependent values), so that bidders need to observe all signals in order to evaluate the overall value of the object. In this context, we analyze the seller's expected revenue in first- and second-price auctions.

Chapter 8 examines multi-unit auctions, where the seller offers several units of the same good, analyzing first- and second-price auctions and whether selling each unit of the good separately yields a larger expected revenue for the seller than selling them simultaneously as part of the same auction.

In Chap. 9, we take a more general approach, allowing the seller to design any auction format, not only those considered in previous chapters. We first study that, in direct revelation mechanisms where bidders are asked to report their valuation for the object, bidders have no incentives to report their true valuation in the first-price auction, but they do in the second-price auction. We then consider a direct revelation mechanism often used in the literature, the Vickrey–Clarke–Groves (VCG) mechanism, allowing for one and several units being sold.

Finally, Chap. 10 provides exercises on procurement auctions, where the seller (e.g., a municipality) procures a service to firms (such as building a bridge, or managing the water distribution service), and the firm making the lowest bid wins the contract. For generality, we first present a standard procurement auction with perfect monitoring from the seller and no external effects between the participating bidders, relaxing subsequently these assumptions, one at a time, and allowing for external effects between bidders or imperfect monitoring of the service being contracted.

How to Use This Book

Instructors may cover specific exercises in class to illustrate how to extend typical auction formats to other settings, such as allowing for more bidders, risk aversion, or other distribution functions. Alternatively, instructors may assign some exercises as a required reading before class —helping students prepare for the auction formats covered in class—or after class, so that students can use the exercises as a guideline for their homework assignments.

Since exercises are ranked according to their difficulty, instructors at the undergraduate level can assign the reading of A-type exercises, spending class time in more challenging topics. Similarly, instructors at the graduate level can assign the reading of B-type exercises, letting them focus on more technical extensions of basic auction formats.

Table 1 includes a list of suggested exercises for each chapter, for instructors teaching undergraduate and master's courses, such as Auction Theory, Game Theory and Strategy, Market Design, Economics of Information, and senior-level topics in Microeconomic Theory, and for instructors teaching Ph.D.-level courses, such as Auction Theory, Contract Theory, or Topics in Microeconomics.

Table 1 Suggested exercises in each chapter

Chapter	Undergraduate level	Master's level	Ph.D. level
Chapter 1	1.1–1.7,	1.4, 1.8–1.12	1.8–1.12
Chapter 2	2.1–2.4, 2.7, 2.11	2.5, 2.6, 2.8–2.10	2.5, 2.6, 2.8–2.10
Chapter 3	3.1, 3.4, 3.6, 3.8, 3.9	3.1–3.3, 3.5, 3.7, 3.10–3.14	3.2, 3.5, 3.7, 3.10, 3.11, 3.14
Chapter 4	4.2, 4.5, 4.8	4.1, 4.3, 4.4, 4.6–4.10	4.3, 4.6–4.10
Chapter 5	5.1, 5.5,5.6,5.8	5.2, 5.3, 5.4, 5.7	5.2, 5.3, 5.4
Chapter 6	6.1, 6.2, 6.3	6.2, 6.4, 6.5	6.5, 6.6
Chapter 7	7.1, 7.2	7.2–7.6	7.3, 7.4, 7.5
Chapter 8	8.1, 8.2	8.3–8.5	8.4–8.5
Chapter 9	9.1–9.4	9.4, 9.5, 9.6	9.4, 9.5, 9.6
Chapter 10	10.1	10.2	10.2–10.5

Acknowledgments

The exercises in this book were inspired by the courses we taught at Washington State University and the courses that Pak-Sing teaches at the Graduate Institute of Industrial Economics, National Central University (Taiwan).

We are extremely grateful to faculty members and students at Washington State University, for their constant feedback and suggestions, and to our teaching assistants in several courses who, along the years, helped improve the presentation and clarity of many of the exercises in this volume. Felix particularly thanks Ana Espinola-Arredondo, Xueying Ma, Joseph Navelski, Minyi Yang, Magdana Kondaridze, Almira Salimgarieva, Kiriti Kanjilal, Lucy Wang, William King, Jui Sen, GC Apar, Reetwika Basu, and Yenjae Chang at Washington State University; and Pak-Sing Choi thanks Kuan-Cheng Shen, Hong-en Deng, and I-Pin Ku at National Central University. We also thank Lorraine Klimowich, Maria David, Deepika Sam, and the Springer Nature team for all their help and support. Last but not least, we thank our families and friends for their constant motivation.

Taoyuan City, Taiwan Pak-Sing Choi
Pullman, WA, USA Felix Munoz-Garcia

Contents

1 **Second-Price Auctions** 1
　Introduction 1
　Exercise #1.1: Second-Price Auction with Complete Information[A] 3
　Exercise #1.2: Second-Price Auctions Under Incomplete Information[B] .. 4
　Exercise #1.3: Second-Price Auctions Under Incomplete
　Information and Discrete Valuations[B] 7
　Exercise #1.4: Robustness of Equilibrium Bidding in the
　Second-Price Auction[A] 9
　Exercise #1.5: Efficiency in Second-Price Auctions[A] 10
　Exercise #1.6: Bidding Behavior in the English Auction[A] 10
　Exercise #1.7: Expected Revenue in the Second-Price Auction[C] 11
　Exercise #1.8: Second-Price Auctions with Reservation Prices[A] 17
　Exercise #1.9: Second-Price Auctions with Entry Fees[B] 19
　Exercise #1.10: Asymmetric Bidding Equilibria in the
　Second-Price Auction[B] 21
　Exercise #1.11: Collusion in Second-Price Auctions, Based on
　Graham and Marshall (1987)[A] 23
　Exercise #1.12: Second-Price Auctions with Budget Constrained
　Bidders, Based on Che and Gale (1998)[B] 24

2 **First-Price Auctions** 27
　Exercise #2.1: First-Price Auction Under Complete Information[A] 29
　Exercise #2.2: First-Price Auction with Only Two Bidders and
　Uniformly Distributed Valuations[A] 30
　Exercise #2.3: First-Price Auction with $N \geq 2$ Bidders and
　Uniformly Distributed Valuations[A] 32
　Exercise #2.4: First-Price Auction with Generic Distribution of
　Valuations–The Direct Approach[B] 36
　Exercise #2.5: First-Price Auction with Uniformly or
　Exponentially Distributed Values[B] 39
　Exercise #2.6: First-Price Auction with Generic Distribution of
　Valuations–The Envelope Theorem Approach[C] 43
　Exercise #2.7: Efficiency in First-Price Auctions[A] 47
　Exercise #2.8: The First-Order Statistic[B] 48

	Exercise #2.9: The Second-Order Statistic^B	51
	Exercise #2.10: The kth-Order Statistic^C	55
	Exercise #2.11: Bidding Behavior in the Dutch Auction^A	60
3	**First-Price Auctions: Extensions**	**63**
	Introduction	63
	Exercise #3.1: Expected Revenue in the First-Price Auction—Direct Proof^B	64
	Exercise #3.2: Expected Revenue in the First-Price Auction—Proof Using the First-Order Statistic^C	71
	Exercise #3.3: Expected Payment in the First-Price Auction^B	73
	Exercise #3.4: First-Price Auction with Risk-Averse Bidders—An Introduction^A	74
	Exercise #3.5: First-Price Auction with Risk-Averse Bidders—General Setting^B	78
	Exercise #3.6: Efficiency with Risk Aversion^A	82
	Exercise #3.7: First-Price Auction with Asymmetrically Distributed Valuations^B	83
	Exercise #3.8: Sequential Version of the First-Price Auction^A	87
	Exercise #3.9: First-Price Auctions with Reservation Prices—One Bidder^A	88
	Exercise #3.10: First-Price Auctions with Reservation Prices—Several Bidders^C	90
	Exercise #3.11: First-Price Auction with Entry Fees^C	99
	Exercise #3.12: First-Price Auction with Discrete Valuations^B	109
	Exercise #3.13: Collusion in First-Price Auctions, Based on McAfee and McMillan (1992)^B	115
	Exercise #3.14: First-Price Auctions with Budget Constrained Bidders, Based on Che and Gale (1998)^B	118
4	**All-Pay Auctions and Auctions with Asymmetrically Informed Bidders**	**125**
	Introduction	125
	Exercise #4.1: All-Pay Auction Under Complete Information, Based on Baye et al. (1996)^B	126
	Exercise #4.2: Testing a Bidding Function in a First-Price, All-Pay Auction^A	129
	Exercise #4.3: Finding the Equilibrium Bidding Function in the First-Price All-Pay Auction Using the Envelope Theorem Approach^C	130
	Exercise #4.4: Finding the Equilibrium Bidding Function in the First-Price All-Pay Auction Using the Direct Approach^B	137
	Exercise #4.5: Efficiency in All-Pay Auctions^A	139
	Exercise #4.6: Finding the Expected Revenue in the First-Price All-Pay Auction^B	140

Exercise #4.7: Finding Equilibrium Bids in the Second-Price
All-Pay Auction^B 142
Exercise #4.8: War of Attrition^A 146
Exercise #4.9: Asymmetrically Informed Risk-Neutral Bidders,
Based on Kim and Che (2014)^B 150
Exercise #4.10: Asymmetrically Informed Risk-Averse Bidders,
Based on Orozco-Aleman and Munoz-Garcia (2011)^B 156

5 Third-Price Auctions, kth-Price Auctions, and Lotteries 165
Exercise #5.1: Third-Price Auction, A Numerical Example^A 166
Exercise #5.2: Finding the Equilibrium Bidding Function in a
Third-Price Auction^C 167
Exercise #5.3: Equilibrium Bidding Function in a Third-Price
Auction with Uniformly or Exponentially Distributed Values^B 170
Exercise #5.4: kth-Price Auction^C 175
Exercise #5.5: Efficiency in kth-Price Auctions^A 180
Exercise #5.6: Lottery Auction, An Introduction^A 180
Exercise #5.7: Lottery Auction, A More General Approach^B 181
Exercise #5.8: Efficiency in Lottery Auctions^A 185

6 The Revenue Equivalence Principle 187
Exercise #6.1: Revenue Comparison in Four Auction Formats^A 188
Exercise #6.2: Revenue Comparison Between First- and
Second-Price Auctions^B 192
Exercise #6.3: The Revenue Equivalence Principle with Risk
Averse Bidders^A 195
Exercise #6.4: The Revenue Equivalence Principle with Risk
Averse Sellers^B 196
Exercise #6.5: Revenue Equivalence Theorem–Short Proof^B 197
Exercise #6.6: Revenue Equivalence Theorem–Longer Proof^C 201

7 Common-Value Auctions 205
Exercise #7.1: The Winner's Curse in Common-Value
Auctions–Introduction^A 206
Exercise #7.2: Equilibrium Bidding in Common-Value Auctions
with Two Bidders^B 207
Exercise #7.3: Equilibrium Bidding in Common-Value Auctions
with $N \geq 2$ Bidders^C 209
Exercise #7.4: First-Price Auction When Bidders Have
Interdependent Values^B 215
Exercise #7.5: Second-Price Auction When Bidders have
Interdependent Values^B 218
Exercise #7.6: Revenue Comparison in Auctions When Bidders
Have Interdependent Values^B 221

8 Multi-Unit Auctions 225
Introduction 225

Exercise #8.1: First-Price Auction Selling Multiple Units
(Discriminatory Auction)[A] 226
Exercise #8.2: Second-Price Auction Selling Multiple Units
(Uniform-Price Auction)[A] 226
Exercise #8.3: Second-Price Auction Selling Multiple Units,
Allowing for Different Prices[B] 227
Exercise #8.4: Divide Bidders and Conquer in a First-Price Auction[B] ... 234
Exercise #8.5: Divide Bidders and Conquer in a Second-Price Auction[A] .. 235

9 Mechanism Design 237
Exercise #9.1: Incentives to Truthfully Reveal Valuations[A] 238
Exercise #9.2: First-Price Auction as a Direct Revelation Mechanism[B] .. 239
Exercise #9.3: Second-Price Auction as a Direct Revelation
Mechanism[A] 241
Exercise #9.4: VCG Mechanism Selling a Single Unit[B] 243
Exercise #9.5: VCG Mechanism Selling Several Units[B] 245
Exercise #9.6: VCG Mechanism and the Generalized Second-Price
Auction[B] 247

10 Procurement Auctions 251
Exercise #10.1: Procurement Auctions Under Complete Information[A] ... 252
Exercise #10.2: Procurement Auctions Under Incomplete Information[B] . 256
Exercise #10.3: Procurement Auctions with External Effects,
Based on Choi et al. (2018)[C] 261
Exercise #10.4: Procurement Auctions with Perfect Monitoring[B] 267
Exercise #10.5: Procurement Auctions with Imperfect Monitoring[C] 273

A Game Theory Appendix 283
Background 283
Dominated Strategies 284
Nash Equilibrium 285
   Mixed-Strategy Nash Equilibrium 286
Subgame Perfect Equilibrium 288
Bayesian Nash Equilibrium 289

References 291

Index 293

Second-Price Auctions

Keywords

Second-price auction · Second-price auction under complete information · Second price-auction under incomplete information · Discrete valuations · Continuous valuations · Uniform distribution · Ebay · Bing · Google · Yahoo · Radio spectrum · Keyword-based advertising · Bid shading · More bidders · Risk aversion · Risk neutrality · Risk loving · Correlated valuations · Efficiency · English auction · Initial bid · Reservation price · Ascending bids · Expected revenue · Second-order statistic · Third-order statistic · K-th order statistic · Entry fee · Symmetric bids · Asymmetric bids · Collusion · Budget constraints · Affordable bids · Weakly dominant strategy · Rivals' bid profile · Highest bidder · Second highest bidder · Third highest bidder · Overbidding · Incentives to deviate · Payoff from winning · Payoff from losing · Probability of winning · Exponential distribution · Standing bid · Pareto efficiency · Price clock · Raising hands · Second-highest valuation · Third-highest valuation · Kth-highest valuation · Other distribution forms · Rate parameter in exponential distribution · Bayesian Nash equilibrium · Bid equal to valuation

Introduction

We start our analysis of auction formats with the second-price auction, where the winning bidder does not pay the bid he submitted but, instead, the second-highest bid. This auction format is relatively easy to analyze, requiring limited mathematical steps, and thus we believe that it can serve as a first approximation to bidding behavior for non-technical readers. Examples of this auction format include eBay sales, auctioning radio spectrums, and the pricing that search engines, such as Google, Bing, or Yahoo, use to sell keyword-based advertising.

We first show that an equilibrium bid can be sustained where every bidder submits a bid equal to his valuation for the object being sold in the auction. This result, in addition, holds under many settings, such as complete information environments where every bidder observes the valuations of his rivals (as shown in Exercise 1.1), incomplete information contexts where every bidder only observes his valuation for the object (Exercise 1.2), and when bidders' valuations are discrete rather than continuous (Exercise 1.3). In addition, Exercise 1.2 verbally discusses why bidders do not have incentives to shade their bids (submitting a bid lower than their value for the object) in a second-price auction.

Exercise 1.4 then confirms that every bidder submitting a bid equal to his valuation can be sustained as an equilibrium of the second-price auction even if we alter the number of bidders, if they are risk averse or risk lovers, if their valuations are drawn from different cumulative distribution functions, or if their valuations for the object are correlated.

Exercise 1.5 discusses whether second-price auctions are efficient, in the sense that the winner is the bidder with the highest valuation for the object. Otherwise, the seller could reassign the good, potentially increasing his revenue from the auction, or bidders could trade between themselves improving the payoff of at least one of them. We show that the second-price auction is efficient as bids are increasing in bidders' valuations, implying that the bidder submitting the highest bid must also have the highest valuation for the good.

Exercise 1.6 then considers an English auction, where the seller announces an initial bid of zero (or a reservation price) and bidders submit ascending bids. If only one bidder remains at any point during the auction, that bidder is declared the winner. We show that, despite its differences with the second-price auction, both auction formats yield the same equilibrium bidding function, with every bidder submitting a bid equal to his valuation. Exercise 1.7 measures the seller's expected revenue in a second-price auction, where we need to use the concept of second-order statistic in the value distribution; a concept that we will use in future chapters too.

Exercises 1.8 and 1.9 analyze the role of reservation prices and entry fees, respectively, in a second-price auction. Intuitively, a reservation price is the minimum bid that the seller running an auction will consider or, alternatively, bids below the reservation price are ignored by the seller. Entry fees, however, must be paid by all bidders who wish to enter the auction room, independent of the bids they submit once the auction starts. We examine how each of these tools affect the pool of bidders participating in the auction and, for those who participate, how their equilibrium bids change.

In all settings considered in this chapter, the equilibrium bidding function was symmetric (every bidder submits his valuation for the object). In other words, even if submitted bids are different across bidders because their values differ, the bidding function they use is symmetric. Exercise 1.10 examines whether asymmetric bidding strategies can also be sustained as equilibria of the second-price auction, thus showing that bidders have no incentives to deviate.

Finally, Exercise 1.11 considers collusion in second-price auctions, demonstrating that, in this setting too, every bidder has incentives to submit a bid equal to his valuation. Exercise 1.12 allows for bidders to face budget constraints, as in Che and Gale (1998), showing that bidders submit a bid equal to their valuation as long as that valuation is affordable.

Exercise #1.1: Second-Price Auction with Complete Information[A]

1.1 Consider a second-price auction with $N \geq 2$ risk-neutral bidders. As a benchmark, this exercise assumes a complete information setting where every bidder i can observe not only his valuation for the object, v_i, but also that of his rivals, as represented in valuation profile

$$v_{-i} = (v_1, \ldots v_{i-1}, v_{i+1}, \ldots, v_N).$$

Assume, without loss of generality, that bidder valuations are ranked as $v_1 > v_2 > \cdots > v_N$. Show that a strategy profile where every bidder i submits a bid equal to his valuation, $b(v_i) = v_i$, is a weakly dominant strategy.[1]

- We first identify expected payoffs if all bidders submit a bid equal to their valuation, $b(v_i) = v_i$.
  - *Highest-value bidder.* Bidder 1, observing that he has the highest valuation, submits a bid equal to his valuation, $b(v_1) = v_1$, which is the highest bid. This bidder wins the auction and pays a price equal to the second-highest bid, $b(v_2) = v_2$, earning a payoff $v_1 - b(v_2) = v_1 - v_2$.
  - *All other bidders.* Bidder 2 submits a bid equal to his valuation, $b(v_2) = v_2$, which lies below that of bidder 1 since $v_1 > v_2$, entailing that bidder 2 loses the auction and earns a zero payoff. A similar argument applies to all other bidders with lower valuations, such as bidders $3, 4, \ldots, N$.
- We now check that bidders have no incentive to deviate from the above strategy profile:
  - *Highest-value bidder.* If bidder 1 submits a bid above his valuation, he still wins the auction and pays the same price (the second-highest bid), thus earning the same payoff as before deviating, $v_1 - b(v_2) = v_1 - v_2$. If bidder 1, instead, submits a bid lower than his valuation, he either keeps winning the auction and earning the same payoff (since the price has not changed) or loses the auction seeing his payoff drop to zero. Therefore, bidder 1 has no incentive to deviate from submitting a bid $b(v_1) = v_1$.
  - *All other bidders.* If bidder 2 submits a bid above his valuation, he either keeps losing the auction (if $b(v_2) < b(v_1)$) or wins the auction at a loss.

---

[1]Recall that strategy s_i weakly dominates another strategy s'_i if player i's utility satisfies $u_i(s_i, s_{-i}) \geq u_i(s'_i, s_{-i})$ for every strategy profile s_{-i} that his rivals choose and $u_i(s_i, s_{-i}) > u_i(s'_i, s_{-i})$ for at least one strategy profile s_{-i}. For more details, see the Game Theory Appendix at the end of the book, which provides a list of game theoretic tools used throughout the book.

(Indeed, if $b(v_2) > b(v_1)$), bidder 2 wins but earns a payoff $v_2 - b(v_1)$, which is negative since $v_2 - b(v_1) = v_2 - v_1 < 0$. If, instead, bidder 2 submits a bid below his valuation, he still loses the auction and earns the same zero payoff as when he submits a bid equal to his valuation. As a consequence, bidder 2 has no incentive to deviate from submitting a bid $b(v_2) = v_2$. A similar argument applies to bidders $3, 4, \ldots, N$.
- Therefore, every bidder i finds that submitting a bid equal to his valuation, $b(v_i) = v_i$, produces a payoff that cannot be strictly improved by deviating to bids above or below his valuation, implying that bid $b(v_i) = v_i$ is a weakly dominant strategy.

Exercise #1.2: Second-Price Auctions Under Incomplete Information[B]

1.2 Consider a second-price auction with the following rules. Every player i learns his valuation for the object v_i but does not observe his rivals' valuations,

$$v_{-i} = (v_1, \ldots v_{i-1}, v_{i+1}, \ldots, v_N),$$

but he knows that valuation v_j is distributed according to a cumulative distribution function $F_j(v_j)$, where $j \neq i$. Given this information, every bidder i submits his bid, b_i, to the seller. The seller assigns the object to the individual who submitted the highest bid and this bidder must pay the second-highest bid. All other bidders pay zero. If two or more bidders submit the highest bid, then the object is randomly assigned among them with equal probabilities. (Part f of this exercise shows that equilibrium results are unaffected by the tiebreaking rule.)

In this exercise, we seek to show that bidding your valuation, $b_i(v_i) = v_i$, is a weakly dominant strategy for all players. That is, regardless of the valuation you assign to the object, and independent of your opponents' valuations, submitting a bid equal to your valuation, $b_i(v_i) = v_i$, yields an expected payoff equal to or higher than that of submitting any other bid, $b_i(v_i) \neq v_i$.

(a) Find bidder i's expected payoff from submitting a bid that coincides with his own valuation v_i. [*Hint*: Compare bidder i's bid, b_i, against the highest bid, $h_i = \max_{j \neq i}\{b_j\}$.][2]

---

[2] Intuitively, expression $h_i = \max_{j \neq i}\{b_j\}$ finds the highest bid among all bidders different from bidder i, $j \neq i$. Alternatively, h_i can be written more explicitly as $h_i = \max\{b_1, b_2, \ldots, b_{i-1}, b_{i+1}, \ldots, b_N\}$, where we find the highest bid among all N bidders except for bidder i (note that we wrote everyone's bid but i's, b_i).

- If the bidder submits his own valuation, $b_i(v_i) = v_i$, then either of the following situations can arise:
  - If the highest competing bid lies below his bid, $h_i < b_i$, then bidder i wins the auction. In this case, he obtains a net payoff of $v_i - h_i$ because in a second-price auction, the winning bidder does not pay the bid he submitted, but rather the second-highest bid, h_i, and in this case, $b_i > h_i$.
  - If, instead, the highest competing bid lies above his bid, $h_i > b_i$, then he loses the auction, earning zero payoff.
  - We do not consider the case when his bid coincides with the highest competing bid (i.e., $b_i = h_i$), and thus a tie occurs. Ties are normally solved in auctions by randomly assigning the object to the bidders who submitted the highest bids (e.g., if bidders 3 and 4 are tied in the highest bid, the auctioneer can flip a coin to determine whether bidder 3 or 4 will receive the object). As a consequence, bidder i's payoff becomes $v_i - h_i$, but with only $\frac{1}{2}$ probability, i.e., his expected payoff becomes $\frac{1}{2}(v_i - h_i)$.[3] However, because $v_i = h_i$ in this case, the bidder earns a zero expected payoff.

(b) Find bidder i's expected payoff from submitting a bid that shades his valuation (i.e., $b_i < v_i$).

- In this case, we can see three cases emerging (see Fig. 1.1), depending on the ranking between bidder i's bid, b_i, and the highest competing bid, h_i:
  - **Case 2a.** If the highest competing bid h_i lies below his bid (i.e., $h_i < b_i$), then he still wins the auction, obtaining the same net payoff as when he does not shade his bid, $v_i - h_i$.
  - **Case 2b.** If the highest competing bid h_i lies between b_i and v_i, bidder i loses, making zero payoff. Had he submitted a bid equal to his valuation for the object, he would have won the auction, earning a payoff of $v_i - h_i > 0$.
  - **Case 2c.** If the highest competing bid h_i is higher than v_i, bidder i loses the auction, thus yielding the same outcome as when he submits a bid, $b_i = v_i$.
  - Hence, we just showed that when bidder i shades his bid, $b_i < v_i$ in cases 2a–2c, he earns the same or lower payoff than when he submits a bid that coincides with his valuation for the object ($b_i = v_i$). Therefore, he does not have incentives to shade his bid because his payoff would not improve from doing so.

---

[3] More generally, if $K \geq 2$ bidders are tied submitting the highest bid, and the auctioneer randomly assigns the object to any of them, each of these bidders earns an expected payoff of $\frac{1}{K}(v_i - h_i)$.

Fig. 1.1 Cases arising when bidder i shades his bid, $b_i < v_i$

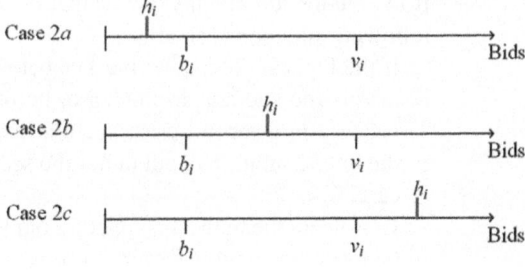

Fig. 1.2 Cases arising when bidder i bids above his value, $b_i > v_i$

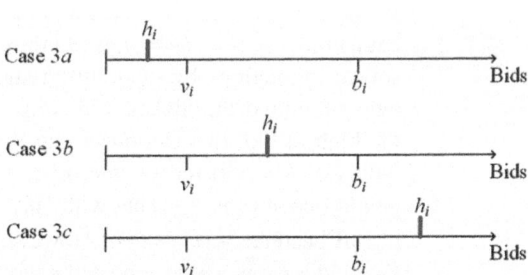

(c) Find bidder i's expected payoff from submitting a bid above his own valuation (i.e., $b_i > v_i$).

- Three cases also arise in this setting (see Fig. 1.2).
    - **Case 3a.** If the highest competing bid h_i lies below bidder i's valuation, v_i, he still wins, earning a payoff of $v_i - h_i$, which coincides with that when he submits his valuation, $b_i = v_i$.
    - **Case 3b.** If the highest competing bid h_i lies between v_i and b_i, bidder i wins the object but earns a negative payoff because $v_i - h_i < 0$. If, instead, bidder i submits a bid $b_i = v_i$, he would have lost the object, earning a higher payoff (zero).
    - **Case 3c.** If the highest competing bid h_i lies above b_i, bidder i loses the auction, thus yielding the same outcome as when he submits a bid, $b_i = v_i$.

(d) Use your results in parts (a)–(c) to argue that every bidder i cannot strictly increase his payoff by submitting a bid different from his valuation of the object.

- As shown in part b (c), every bidder i's payoff from submitting a bid below (above, respectively) his valuation either coincides with his payoff from submitting a bid equal to his value for the object, $b_i(v_i) = v_i$, or becomes strictly lower, thus eliminating his incentives to deviate from his equilibrium bid of $b_i(v_i) = v_i$.

- In summary, there is no bidding strategy that provides a strictly higher payoff than $b_i(v_i) = v_i$ in the second-price auction, and all players bid their own valuation. In other chapters, we show that this result differs from the optimal bidding function in the first-price auction, where players shade their bids, $b_i(v_i) < v_i$, unless the number of bidders, N, satisfies $N \to \infty$.
- **Remark:** Our above discussion focuses on symmetric bidding equilibria in the second-price auction. However, asymmetric bidding equilibria can also be supported where, for instance, a bidder submits a bid equal to the highest valuation $b_i(v_i) = 1$, while all other bidders submit a bid $b_j(v_j) = k$, where $1 > k \geq 0$ for all $j \neq i$. Exercise 1.10 in this chapter shows that this bidding profile is, indeed, an equilibrium in the second-price auction.

(d) Verbally discuss why bid shading is not part of the equilibrium bid in the second-price auction.

- By shading his bid, $b_i(v_i) < v_i$, bidder i lowers the chance that he wins the auction but does not lower the amount he pays when he wins, which is still equal to the highest competing bid. Instead, he lowers the amount the winner would have to pay from winning the auction!

(e) *Other tiebreaking rules.* In previous parts of the exercise, we considered a specific tiebreaking rule, namely, if two bidders i and j submit the highest bid, the object is randomly assigned among them with equal probabilities, $p_i = p_j = 1/2$. Show that if, without loss of generality, these probabilities satisfy $p_i > p_j$, the results in parts (a)–(d) are unaffected.

- Our analysis in part (a) would be unaffected, except for the last case where bidder i's bid coincides with the highest competing bid (i.e., $b_i = h_i$), yielding an expected payoff $p_i(v_i - h_i)$, which is zero since $b_i = v_i$. We then obtain the same results as in part (a).
- Our analysis in parts (b) and (c) is unchanged by the exact probabilities, p_i and p_j, with which the object is allocated among the bidders submitting the highest bid. Therefore, part (d), which just combines our findings in parts (a)–(c), is unaffected, and we can still conclude that in the second-price auction, every bidder submits a bid that coincides with his valuation.

Exercise #1.3: Second-Price Auctions Under Incomplete Information and Discrete Valuations[B]

1.3 Consider a second-price auction with two bidders, each privately observing his valuation for the object, v_i, which is either high (v_H) or low (v_L), where

$1 > v_H > v_L > 0$. The probability of bidder i drawing a high valuation, v_H, is $p \in (0,1)$. If two bidders submit the same bid, assume that the seller randomly assigns the object. In this exercise, we seek to show that every bidder i submitting a bid equal to his valuation $b(v_i) = v_i$ is a weakly dominant strategy.

(a) Find bidder i's expected payoff from submitting a bid that coincides with his valuation, $b(v_i) = v_i$. [*Hint*: Comparing bidder i's and j's bids, three cases can arise.]

- When every bidder i submits a bid that coincides with his valuation, $b(v_i) = v_i$, three cases can arise:
  (i) Bidder i wins the auction because his bid satisfies $b(v_i) > b(v_j)$, which occurs when his valuation satisfies $v_i > v_j$. In this case, his payoff is $v_i - b(v_j) = v_i - v_j > 0$ since bidder i pays the second-highest bid $b(v_j) = v_j$.
  (ii) Bidder i wins the auction because his bid is tied with his rival's, $b(v_i) = b(v_j)$, which in this strategy profile can only occur when their valuations coincide, $v_i = v_j$, and the object is randomly assigned to bidder i with 1/2 probability. In this case, his expected payoff is

  $$\frac{1}{2}(v_i - b(v_j)) = \frac{1}{2}(v_i - v_j) = 0$$

  which is equal to zero since, in this case, valuations must coincide, $v_i = v_j$, for bids to be tied.
  (iii) Bidder i loses the auction because his bid satisfies $b(v_i) < b(v_j)$, which occurs when his valuation satisfies $v_i < v_j$. In this case, his payoff is zero.

(b) Find bidder i's expected payoff from submitting a bid below his valuation, $b(v_i) < v_i$.

- If, instead, bidder i submits a lower bid, $b(v_i) < v_i$, he can no longer win the auction in case (ii), which now yields a payoff of zero rather than $\frac{1}{2}(v_i - v_j)$; and case (iii) is unaffected. If his bid still lies in case (i), he wins the auction, but the price he pays for the object is still the second-highest bid, $b(v_j) = v_j$, thus not changing his expected payoff in this case, $v_i - v_j$. Therefore, bidder i has no incentives to shade his valuation.

(c) Find bidder i's expected payoff from submitting a bid above his valuation, $b(v_i) > v_i$.

- If bidder i submits a higher bid, $b(v_i) > v_i$, case (i) becomes more likely, but the price he pays for the object is unaffected, leaving his expected payoff unchanged. Case (ii) cannot be sustained since $b(v_i)$ has increased, thus breaking the tie between players' bids. Finally, case (iii) is less likely to arise, but if winning, bidder i may suffer a loss since $v_i - b(v_j) = v_i - v_j < 0$, given that $v_i < v_j$ in case (iii).

(d) Use your results from parts (a)–(c) to argue that bidder i cannot strictly increase his payoff by submitting a bid strictly below or above his valuation.

- Overall, every bidder i does not have incentives to deviate from submitting a bid $b(v_i) = v_i$. Therefore, this bidding profile can be sustained as an equilibrium of the second-price auction when bidders draw their valuations both from a discrete and from a continuous distribution.

Exercise #1.4: Robustness of Equilibrium Bidding in the Second-Price Auction[A]

1.4 Consider a second-price auction with $N \geq 2$ bidders competing for an object. Bidders are all risk neutral, and their valuations are drawn from a uniform distribution, $F(v_i) = v_i$ and $v_i \in [0, 1]$. We now analyze how equilibrium bidding behavior in the second-price auction (as described in Exercise 1.2) is affected by the following changes:
(a) The number of bidders increases to N', where $N' > N$.

- The equilibrium bidding results found in Exercise 1.2, where every bidder i behaves according to $b_i(v_i) = v_i$, apply to any number of bidders. Therefore, an increase in the number of competing bidders does not increase or decrease equilibrium bids, remaining at $b_i(v_i) = v_i$ for every bidder i.

(b) All bidders are risk averse. What if they are all risk lovers?

- The equilibrium bidding results found in Exercise 1.2 did not rely on bidders being risk neutral, risk averse, or risk lover. Specifically, for a given value of the highest competing bid, h_i, bidder i's expected payoff from submitting a bid, $b_i(v_i) = v_i$, would still be weakly larger than when deviating to a bidding strategy above his true valuation for the object, $b_i(v_i) > v_i$, or below it, $b_i(v_i) < v_i$. As a consequence, equilibrium bids remain at $b_i(v_i) = v_i$ for every bidder i. Equilibrium payoffs, however, would be affected.

(c) Valuations are not uniformly, but exponentially, distributed, with cumulative distribution function $F(v_i) = 1 - e^{-\lambda v_i}$, where $\lambda > 0$ and $v_i \in [0, +\infty)$.

- Again, the equilibrium bidding results found in Exercise 1.2 did not rely on bidders' valuations being distributed in any specific form, being applicable to any cumulative distribution function $F(v_i)$. As a consequence, equilibrium bids remain at $b_i(v_i) = v_i$ for every bidder i.

(d) Bidders' valuations are positively correlated.

- By the same argument as in previous parts of the exercise, the equilibrium bidding function found in Exercise 1.2 was not affected by whether valuations were independently distributed or correlated. The same argument applies if bidders' valuations are negatively correlated.
- In summary, the equilibrium bidding function in the second-price auction (bidding one's true valuation) is robust to changes in the number of bidders, risk preferences, the distribution function, and correlation.

Exercise #1.5: Efficiency in Second-Price Auctions[A]

1.5 Consider the second-price auction in Exercise 1.2. Argue that the object, in equilibrium, is assigned to the bidder with the highest valuation. When this occurs, we say that the auction is "efficient," because there is no potential reassignment of the object to another bidder (or a trade between bidders) that can improve the payoff of at least one individual without reducing the payoff of any other bidder.

- Since every bidder i uses the bidding function $b_i(v_i) = v_i$, the individual submitting the highest bid, h, satisfies $b_h(v_h) > b_j(v_j)$ for every bidder $j \neq h$, which implies $v_h > v_j$. Therefore, the bidder who wins the auction, h, also has the highest valuation for the object, entailing that the second-price auction is efficient. (The same result applies if, in the case of a tie, the object is randomly assigned to the bidder with the highest valuation, thus still satisfying Pareto efficiency.)

Exercise #1.6: Bidding Behavior in the English Auction[A]

1.6 Consider an ascending-price auction (also known as the English auction) where the seller starts announcing a zero price (or a reservation price if the auction has one) and then allows bidders to submit increasingly higher bids. The highest bid at any given moment is the "standing bid." If another bidder submits a higher bid, then that becomes the standing bid. If no competing bidder displaces the standing bid (often after a few seconds as determined by the seller), the standing bid is declared the winner of the auction, and the bidder who made this bid receives the object and pays the price where the auction was stopped.

Exercise #1.7: Expected Revenue in the Second-Price Auction^C

Like the Dutch auction analyzed in Chap. 2, the English auction is an open auction since every bidder can observe other bidders' bidding behavior (by seeing whether they are raising their hand or not), as opposed to the second-price auction we considered in previous exercises of this chapter which is known as a closed (or sealed bid) auction given that bidders cannot observe other bidders' strategies until the end of the auction.

A variation of the English auction is the so-called Japanese auction, where the seller uses a price clock, starting from zero (or a reservation price) which continuously increases. Bidders do not raise their hands but, instead, decide if, given the price at any given moment, they want to leave the auction. When a single bidder remains, the clock is stopped, the remaining bidder is declared the winner, and he pays the price where the clock stopped.

Show that the English auction is strategically equivalent to the second-price auction, implying that every bidder uses the same equilibrium bidding function in both auction formats.

- Consider bidder i with privately observed valuation v_i. He has incentives to submit increasingly higher bids until the standing bid coincides with his valuation, $b = v_i$.
    - If the standing bid b satisfies $b < v_i$, he still has incentives to submit a bid b' (stay in the auction) such that $v_i > b' > b$, so he can win the auction and earn a positive surplus $v_i - b' > 0$ (if he is the only bidder remaining in the room).
    - If, instead, the standing bid b lies above his valuation, $b > v_i$, he has no incentive to submit a bid above b (stay in the auction), since that would entail either a zero utility if he loses the auction or a negative payoff if he wins the auction, earning $v_i - b < 0$.
- Therefore, every bidder i has incentives to submit a bid that coincides with his valuation, $b(v_i) = v_i$, as in the second-price auction, where the seller earns the same expected revenue in the English (or Japanese) auction and in the second-price auction.

Exercise #1.7: Expected Revenue in the Second-Price Auction^C

1.7 Consider again the second-price auction in Exercise 1.2. In this exercise, we examine the second-order statistic and how to use it to express the seller's expected revenue in the second-price auction.

(a) Find the cumulative distribution function of the second-order statistic $F^2(x)$, that is, the probability that the second-highest valuation is lower than x.

- The cumulative distribution function $F^2(x)$ describes the probability that the second-highest valuation is lower than x, that is, $\Pr\{v^2 \leq x\}$, which can occur if either (but not both) of these two events occurs:

- The valuations of all N bidders are below x or, formally, $v_i \leq x$ for every bidder i. This event happens with probability

$$\Pr\{v_1 \leq x\} \times \cdots \times \Pr\{v_N \leq x\} = \underbrace{F(x) \times \cdots \times F(x)}_{N \text{ times}} = [F(x)]^N$$

- The valuations of $N-1$ bidders are below x, $v_i \leq x$, but that of only one bidder j is above x, $v_j > x$. This event can occur in N different ways:
  - (1) $v_1 > x$ for bidder 1, but $v_i \leq x$ for every bidder $i \neq 1$;
  - (2) $v_2 > x$ for bidder 2, but $v_i \leq x$ for every bidder $i \neq 2$; and so on up to N, where $v_N > x$ for bidder N, but $v_i \leq x$ for every bidder $i \neq N$.
- Each of these N cases happens with probability

$$(1 - F(x))[F(x)]^{N-1},$$

where $(1 - F(x))$ denotes the probability that $v_i > x$ for a given bidder i, while $[F(x)]^{N-1}$ represents the probability that $v_j \leq x$ for all other bidders $j \neq i$.

Summing over N, we find that this event happens with probability

$$\sum_{i=1}^{N}(1 - F(x))[F(x)]^{N-1} = N(1 - F(x))[F(x)]^{N-1}$$

- Summarizing the cumulative distribution function of the second-order statistic $F^{[2]}(x)$ is

$$F^{[2]}(x) = [F(x)]^N + N(1 - F(x))[F(x)]^{N-1}.$$

Rearranging the above expression simplifies to

$$F^{[2]}(x) = N[F(x)]^{N-1} - (N-1)[F(x)]^N.$$

(b) Find the density of the second-order statistic $f^{[2]}(x)$.

- Differentiating the expression of $F^{[2]}(x)$ that we found in part (a) with respect to x yields

$$\begin{aligned} f^{[2]}(x) &= N(N-1)F(x)^{N-2}f(x) - N(N-1)F(x)^{N-1}f(x) \\ &= N(N-1)F(x)^{N-2}[1 - F(x)]f(x) \end{aligned}$$

Exercise #1.7: Expected Revenue in the Second-Price Auction[C]

(c) Use the density of the second-order statistic $f^{[2]}(x)$ to obtain the seller's expected revenue in the second-price auction, R^2.

- The seller's expected revenue is equal to the expected price that the winner pays in this auction, namely, the second-highest bid, that is,

$$R^2 = \int x f^{[2]}(x) dx$$

$$= \int x \underbrace{N(N-1) F(x)^{N-2} [1 - F(x)] f(x)}_{f^{[2]}(x)} dx.$$

(d) *Uniformly distributed valuation.* Assume that all bidders' valuations are uniformly distributed, $F(v_i) = v_i$, and $v_i \in [0, 1]$. Find the seller's expected revenue in this context.

- Since valuations are uniformly distributed, $F(x) = x$ and $f(x) = 1$, implying that the seller's expected revenue found in part (c) becomes

$$R^2 = \int_0^1 x N(N-1) x^{N-2} [1-x] dx$$

$$= (N-1) \int_0^1 N x^{N-1} dx - N(N-1) \int_0^1 x^N dx$$

$$= (N-1) x^N \Big|_0^1 - \frac{N(N-1) x^{N+1}}{N+1} \Big|_0^1$$

$$= (N-1) - \frac{N(N-1)}{N+1}$$

$$= \frac{N-1}{N+1}.$$

The expected revenue in the second-price auction coincides with that we will find for the first-price auction in Exercises 3.1–3.3. We return to this coincidence in expected revenues, under more general auction formats, in Chap. 6.

- Figure 1.3 plots this expected revenue, R^2, as a function of the number of bidders, N, showing that it is increasing in N. This property can also be checked by finding the derivative

$$\frac{\partial R^2}{\partial N} = \frac{2}{(N+1)^2},$$

which is clearly positive.

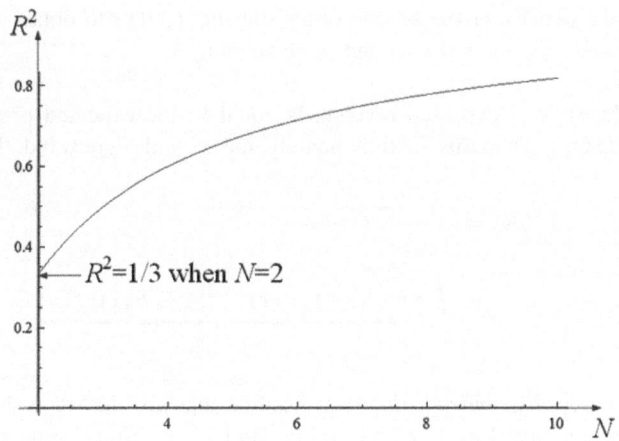

Fig. 1.3 Expected revenue in the second-price auction—uniform distribution

- In addition, when $N = 2$ bidders compete for the object, expected revenue is $R^2 = \frac{2-1}{2+1} = \frac{1}{3}$, as depicted in the vertical intercept of Fig. 1.3. While expected revenue increases in N, it does so at a decreasing rate, since

$$\frac{\partial^2 R^2}{\partial N^2} = -\frac{4}{(N+1)^3} < 0,$$

so R^2 has a concave shape. In the limit, when infinitely many bidders compete, the expected revenue approaches 1, that is,

$$\lim_{N \to +\infty} \frac{N-1}{N+1} = 1.$$

Intuitively, as infinitely many bidders participate in the auction, the valuation of the second-highest bidder is extremely close to 1.

(e) *Exponentially distributed valuations.* Consider now that individual valuations are drawn from an exponential distribution, $F(v_i) = 1 - \exp(-\lambda v_i)$, where $v_i \in [0, \infty)$, and there are $N = 2$ bidders. Find the seller's expected revenue in this context. How does expected revenue change with parameter λ? Interpret your results.

- We first insert $F(v_i)^{2-2} = F(v_i)^0 = 1$ into the expression of the expected revenue found in part (c), yielding

Exercise #1.7: Expected Revenue in the Second-Price Auction[C]

$$R^2 = \int_0^\infty xN(N-1)F(x)^{N-2}[1-F(x)]f(x)dx.$$

$$= \int_0^\infty 2(2-1)xF(x)^{2-2}[1-F(x)]f(x)dx$$

$$= 2\int_0^\infty x[1-F(x)]f(x)dx$$

In addition, $1 - F(v_i) = 1 - (1 - \exp(-\lambda v_i)) = \exp(-\lambda v_i)$ and $f(v_i) = F'(v_i) = \lambda \exp(-\lambda v_i)$, so that

$$R^2 = 2\int_0^\infty x\underbrace{\exp(-\lambda x)}_{1-F(x)}\underbrace{\lambda \exp(-\lambda x)}_{f(x)}dx$$

which we can rearrange as follows:

$$R^2 = -\int_0^\infty (-2\lambda x)\exp(-2\lambda x)\,dx$$

$$= -\int_0^\infty xd\left[\exp(-2\lambda x)\right]$$

Applying integration by parts yields

$$R^2 = -\left[x\exp(-2\lambda x)\right]_0^\infty + \int_0^\infty \exp(-2\lambda x)\,dx.$$

Applying the L'Hôpital rule, we find

$$R^2 = \lim_{x\to\infty}\frac{\frac{\partial x}{\partial x}}{\frac{\partial \exp(2\lambda x)}{\partial x}} - \frac{1}{2\lambda}\left[\exp(-2\lambda x)\right]_0^\infty$$

$$= \lim_{x\to\infty}\frac{1}{2\lambda \exp(2\lambda x)} - \frac{1}{2\lambda}\left[\lim_{x\to\infty}\exp(-2\lambda x) - 1\right]$$

$$= \frac{1}{2\lambda}$$

- Formally, R^2 is monotonically decreasing in λ since

$$\frac{dR^2}{d\lambda} = -\frac{1}{2\lambda^2} < 0,$$

indicating that the seller's expected revenue in the second-price auction decreases as bidders' valuations are more likely to be low (captured by a

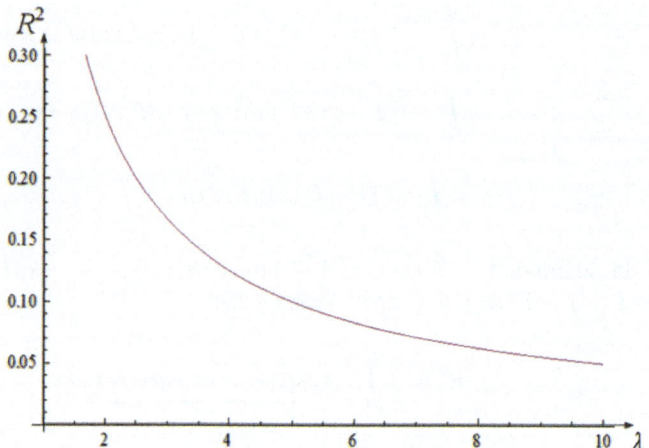

Fig. 1.4 Expected revenue in the second-price auction—exponential distribution

high λ).[4] For instance, when $\lambda = 1$, the expected revenue is $R^2 = 1/2$, when $\lambda = 2$, this revenue decreases to $R^2 = 1/4$, and when $\lambda = 3$, it further decreases to $R^2 = 1/6$, as illustrated in Fig. 1.4, which depicts R^2 as a function of λ.

(f) *Other distribution forms.* Consider the following distribution function:

$$F(x) = (1 + \alpha) x - \alpha x^2$$

where $x \in [0, 1]$, and parameter α satisfies $\alpha \in [-1, 1]$. When $\alpha = 0$, this function collapses to the uniform distribution, $F(x) = x$; when $\alpha > 0$, it becomes concave, thus putting more probability weight on low valuations; and when $\alpha < 0$, it is convex, assigning more probability weight on high valuations. Find the seller's expected revenue in the setting of $N = 2$ bidders. How does the expected revenue change with parameter α? Interpret your results.

- Since $F(x) = (1 + \alpha) x - \alpha x^2$ and $f(x) = F'(x) = 1 + \alpha - 2\alpha x$, we obtain that

---

[4] Recall that, in the exponential distribution, λ represents the rate parameter, describing how quickly the decay of the function is. In addition, the expected value of the exponential distribution, $E[x] = \frac{1}{\lambda}$, and its variance, $Var[x] = \frac{1}{\lambda^2}$, are both decreasing in λ, implying that bidders' expected valuation is more concentrated around the lower bound.

$$R^2 = 2\int_0^1 x\left[1 - F(x)\right] f(x)\, dx$$

$$= 2\int_0^1 x\left[1 - \underbrace{\left[(1+\alpha)x - \alpha x^2\right]}_{F(x)}\right]\underbrace{[1+\alpha - 2\alpha x]}_{f(x)}\, dx$$

$$= 2\int_0^1 x\left[1 + \alpha - \left(1 + 4\alpha + \alpha^2\right)x + 3\alpha(1+\alpha)x^2 - 2\alpha^2 x^3\right] dx$$

$$= \frac{1}{30}\left[30(1+\alpha) - 20\left(1 + 4\alpha + \alpha^2\right) + 45\alpha(1+\alpha) - 24\alpha^2\right]$$

$$= \frac{\alpha^2 - 5\alpha + 10}{30}.$$

- Differentiating R^2 with respect to α, we find that

$$\frac{\partial R^2}{\partial \alpha} = \frac{2\alpha - 5}{30}$$

which is positive if and only if $2\alpha > 5$ or $\alpha > 2.5$. Since parameter α satisfies $\alpha \in [-1, 1]$ by definition, we obtain that R^2 is decreasing in α. Intuitively, as the probability of low valuations increases (higher α), the seller's expected revenue decreases.
- As a remark, note that when $\alpha = 0$ (uniform distribution), the expected revenue simplifies to $R^2 = \frac{1}{3}$, as shown in part (d) of this exercise where, for $N = 2$ bidders, we found that $R^2 = \frac{2-1}{2+1} = \frac{1}{3}$. Figure 1.5 depicts R^2 as a function of α, illustrating the above results.

Exercise #1.8: Second-Price Auctions with Reservation Prices[A]

1.8 Repeat the analysis in Exercise 1.2, now assuming that bidders interact in a second-price auction during the second stage of the game. In the first stage, the seller sets a reservation price r, where $r < 1$ (that is, r is lower than the highest valuation), so only bids above r are considered by the seller (intuitively, r can be understood as the minimum admissible bid).

(a) *Second stage.* Starting from the second stage, find the optimal bidding function for bidder i, $b_i(v_i)$.

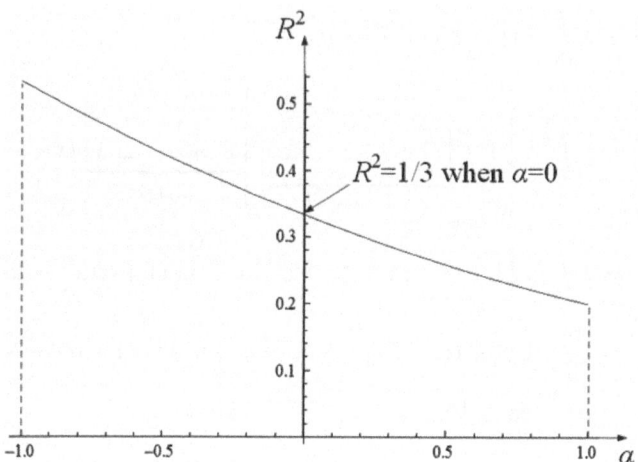

Fig. 1.5 Expected revenue in the second-price auction—other distributions

- As we know, every bidder i bidding according to his valuation, $b_i(v_i) = v_i$, is a weakly dominant strategy in a second-price auction and constitutes the Bayesian Nash equilibrium of the game.[5]

(b) How does the optimal bidding function, $b_i(v_i)$, change in the reservation price r?

- The optimal bidding function, $b_i(v_i) = v_i$, is independent of the reservation price r, so the bid of participating players (i.e., those with valuations above the reservation price, $v_i \geq r$) is unaffected by the reservation price r. However, bidders with valuations below the reservation price, $v_i < r$, do not participate in the auction.
- This bidding behavior, however, is sometimes not observed in controlled experiments where subjects tend to reduce their bids when the reservation price is increased; see, for instance, Sayman and Akçay (2020).

(c) *First stage.* Anticipating the optimal bidding function, $b_i(v_i)$, found in part (a), what is the optimal reservation price r^* that the seller sets in the first stage to maximize his expected revenue from the auction?

---

[5]Recall that a strategy profile is a Bayesian Nash Equilibrium if every player chooses a best response (in expectation) given his rivals' strategies, s_{-i}, and given his type. For a more formal definition, see the Game Theory Appendix at the end of this book. For a more detailed presentation, see Tadelis (2013, Chapter 12), and for examples of how to apply this game-theory tool to several games, see Munoz-Garcia and Toro-Gonzalez (2020, Chapter 7).

- Unlike in the first-price auction, the introduction of a positive reservation price r^* does not induce more aggressive bidding from participating bidders. Therefore, the reservation price does not produce a positive effect in the seller's expected revenue in second-price auctions, and only a negative effect (limiting the number of bidders submitting bids) remains. As a result, the seller has incentives to set a zero reservation price when conducting a second-price auction, $r^* = 0$.

Exercise #1.9: Second-Price Auctions with Entry Fees[B]

1.9 Consider a second-price auction where the seller announces an entry fee, E, that all bidders must pay in order to enter the auction room. Entry fees differ from reservation prices in the sense that they must be paid by all participating bidders, whether they win or not, whereas reservation prices are just the lowest bid that the seller considers from participating bidders. Every bidder i's valuation is independently distributed according to $F(v_i)$.
  (a) Find the optimal bidding function for bidder i, $b_i(v_i, E)$. How is the number of bidders participating in the auction affected by the entry fee E?

- Every bidder i bidding according to his valuation, $b_i(v_i) = v_i$, is a weakly dominant strategy in a second-price auction and constitutes the Bayesian Nash equilibrium of the game. This equilibrium result is unaffected by the entry fee, which only affects the type of bidders who participate in the auction. In particular, a bidder with valuation v_0 is indifferent between participating and not participating if

$$v_0 F(v_0)^{N-1} - E = 0$$

The first term represents his expected utility upon winning, while the second term denotes the entry fee he must pay both when winning and losing the auction. Rearranging this expression, we obtain

$$v_0 F(v_0)^{N-1} = E.$$

Therefore, every bidder i with valuation $v_i < v_0$ does not participate in this second-price auction, while all other bidders with valuations $v_i \geq v_0$ pay entry fee E to participate in the auction.
- To investigate how participation in the auction is affected by a marginal increase in the entry fee, E, note that the left term in the above expression, $v_0 F(v_0)^{N-1}$, is unaffected by E, while the right term increases in E. Therefore, a marginal increase in E reduces the pool of bidders with valuations $v_i \geq v_0$, shrinking the number of participants in the auction.

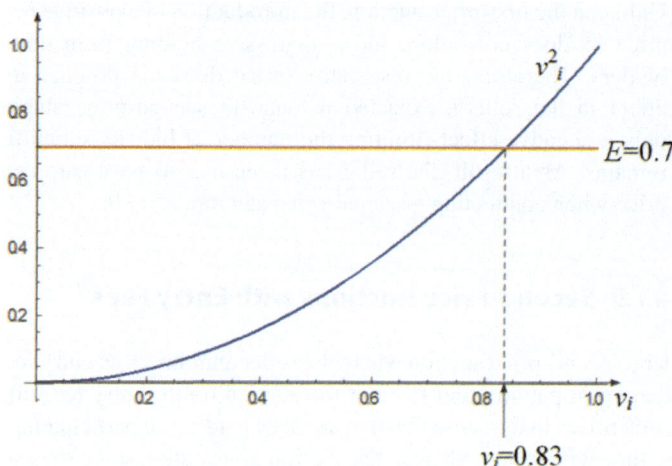

Fig. 1.6 Entry fee and participating bidders

- (b) Assume that valuations are uniformly distributed in [0, 1]. Which bidders participate in the auction, and what are their bidding functions?

    - In a setting where valuations are uniformly distributed, expression
    $$v_0 F(v_0)^{N-1} = E$$
    simplifies to $v_0^N = E$, so the cutoff valuation v_0 is
    $$v_0 = E^{\frac{1}{N}}.$$
    Differentiating v_0 with respect to E, we obtain
    $$\frac{\partial v_0}{\partial E} = \frac{1}{N} E^{-\frac{N-1}{N}} > 0$$
    Therefore, an increase in the entry fee E increases cutoff valuation v_0, thus shrinking the pool of bidders participating in the auction, i.e., those with valuations $v_i \geq v_0$. These bidders submit a bid equal to their valuations, $b_i(v_i) = v_i$. Figure 1.6 illustrates this result. Assuming an entry fee $E = 0.7$ and $N = 2$ bidders, only those bidders with valuations $v_i \geq 0.84$ submit a bid equal to their valuations. Graphically, an increase in entry fee E shifts the horizontal line representing E in Fig. 1.6, producing a rightward shift in the crossing point between E and v_i^N, limiting the pool of bidders who participate in the auction.
    - Next, differentiating v_0 with respect to N, we obtain

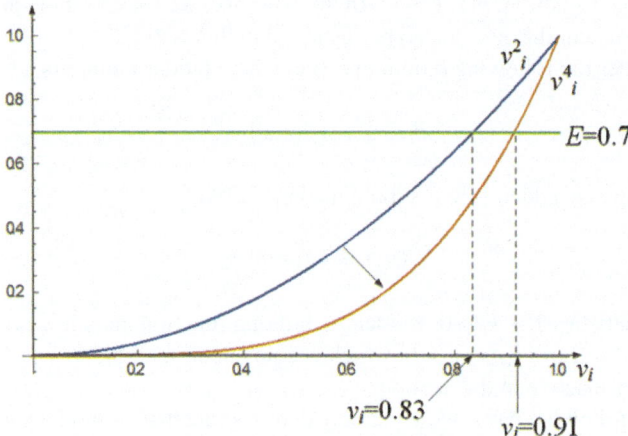

Fig. 1.7 Entry fee and participating bidders—more bidders

$$\frac{\partial v_0}{\partial N} = \underbrace{\frac{-\log E}{N^2}}_{>0} \underbrace{E^{\frac{1}{N}}}_{>0} > 0$$

Figure 1.7 examines how the crossing point between E and v_i^N is affected by the number of bidders. As N increases from $N=2$ to $N=4$, for a given entry fee E, the auction becomes more competitive, as it is more likely that other bidders have a higher valuation for the object, and fewer bidders choose to participate.

- This bidding behavior does not always coincide with that observed in controlled experiments where subjects often decrease their bids when an entry fee is imposed, or when the entry fee increases. Once a bidder has paid the entry fee, however, this cost is sunk, implying that his bidding behavior should not be affected by the entry fee. For this reason, this bidding behavior is commonly cited as an example of the "sunk cost fallacy." For more details, see Augenblick (2016) and references therein.

Exercise #1.10: Asymmetric Bidding Equilibria in the Second-Price Auction[B]

1.10 Consider again the second-price auction in Exercise 1.2 and, for simplicity, assume that every bidder i independently draws his valuation from the same cumulative distribution function $F(v_i)$, where $v_i \in [0, 1]$. In previous exercises, we found a symmetric bidding equilibrium in which every bidder submits his valuation, $b(v_i) = v_i$. In this exercise, we seek to show that other, *asymmetric*, equilibria can also be sustained in this setting.

Consider the following bidding profile, where bidder i submits a bid

$$b_i(v_i) = 1 \text{ for all } v_i,$$

while all other bidders $j \neq i$ submit a constant bid,

$$b_j(v_j) = k \text{ for all } v_j,$$

where k satisfies $1 > k \geq 0$. Bidder j's bidding function includes, as a special case, a bid of zero, $b_j(v_j) = 0$, which is equivalent to all bidders $j \neq i$ being essentially inactive in the auction.

In the following parts, we separately demonstrate that no bidder can strictly increase the payoff he earns in this strategy profile, making it a weakly dominant strategy in the second-price auction.

(a) Show that bidder i does not have incentives to deviate from his bid in this strategy profile. [*Hint*: Start identifying who wins the auction and the price he pays for the object.]

- If bidder i behaves as prescribed, he submits the highest bid $b_i(v_i) = 1$, while all his rivals submit $b_j(v_j) = k$. Therefore, bidder i wins the auction, paying the second-highest bid, which is k. His payoff from winning is, then, $v_i - k$.
- *Deviations:*
  - If, instead, bidder i lowers his bid (e.g., $b_i(v_i) = 1 - \varepsilon$, where $\varepsilon \to 0$), he still wins the auction and still pays a price of k, earning the same payoff as before deviating, $v_i - k$.
  - If bidder i lowers his bid all the way to k, $b_i(v_i) = k$, then there is a tie, and the object is assigned to him with probability $\frac{1}{N}$ since all N bidders submit the same bid k. In that setting, his payoff is $\frac{v_i - k}{N}$, which is strictly lower than his payoff from submitting bid $b_i(v_i) = 1$, $v_i - k$.
  - Furthermore, if bidder i lowers his bid to below k, he loses the auction, entailing a zero payoff, which falls below that of submitting a bid $b_i(v_i) = k$.
- Therefore, bidder i does not strictly improve his payoff from submitting a bid other than $b_i(v_i) = 1$, implying that it is a weakly dominant strategy for him.

(b) Show that bidder j does not have incentives to deviate from his bid in this strategy profile.

- If bidder j behaves as prescribed, he submits one of the lowest bids $b_j(v_j) = k$, along with the other $N - 2$ losing bidders, while bidder i submits $b_i(v_i) = 1$. Therefore, bidder j loses the auction, earning a zero payoff.
- *Deviations:*
  - If, instead, bidder j unilaterally increases his bid (e.g., $b_j(v_j) = k + \varepsilon$, where $\varepsilon \to 0$), he still loses the auction. His payoff from this deviation is, of course, still zero as before he deviated. (The same argument applies if bidder j unilaterally decreases his bid to $b_j(v_j) = k - \varepsilon$, as he would still lose the auction.)
  - If bidder j unilaterally increases his bid all the way to 1, $b_j(v_j) = 1$, then there is a tie with bidders i and j submitting a bid of 1, and the object is assigned to bidder j with probability $\frac{1}{2}$. In that setting, his payoff is $\frac{1}{2}(v_j - 1)$, which is negative for all $v_j < 1$ and zero when $v_j = 1$ (that is, his deviating payoff is, at most, zero).
- Therefore, bidder j does not strictly improve his payoff from submitting a bid other than $b_j(v_j) = k$, implying that it is a weakly dominant strategy for him too. As a consequence, no bidder has incentives to deviate from this asymmetric bidding profile, implying that it constitutes an equilibrium of the second-price auction.
- **Remark:** There are several other asymmetric bidding equilibria in the second-price auction. For a complete characterization, see Blume and Heidhues (2004).

Exercise #1.11: Collusion in Second-Price Auctions, Based on Graham and Marshall (1987)[A]

1.11 Consider a seller running a second-price auction with $N \geq 2$ risk-neutral bidders. Every bidder i independently draws his valuation for the object from a uniform distribution, $F(v_i) = v_i$, where $v_i \in [0, 1]$. Assume that all N bidders met before the auction started and shared their private valuations for the object, asking the bidder with the highest valuation to submit a bid according to his valuation, while all other bidders submit a bid of zero. Show that every bidder has incentives to respect this agreement when the auction starts. (A verbal discussion suffices.)

- *Winning bidder.* If the winning bidder respects the agreement, bidding $b(v_i) = v_i$, he wins the auction, as his bid is larger than all other (zero) bids, and pays for the object the second-highest bid, which is zero, thus earning a net payoff $v_i - 0 = v_i$. Let us now check his potential deviations.
  - If, instead, he submits a higher bid, $b(v_i) > v_i$, he still wins the auction, not affecting the price he pays for the object, thus earning the same net payoff v_i.

- If, instead, he submits a lower bid, $b(v_i) < v_i$, he still wins the auction, affecting neither his price for the object nor his net payoff.

  As a consequence, the bidder with the highest valuation has no incentives to deviate from the collusive agreement.
- *Losing bidder.* If one of the losing bidders respects the agreement, bidding zero, he loses the auction and earns zero. If, instead, he submits a positive bid, he only wins the auction if $b(v_j) > b(v_i)$, or $b(v_j) > v_i$. In that case, he pays the second-highest bid, v_i, which is higher than his own valuation, v_j, since bidder i was selected in the pre-auction stage because he had the highest valuation.

  Therefore, every bidder $j \neq i$ has no incentives to deviate from the collusive agreement.
- We analyze collusion in other auction formats in future chapters. For an accessible literature review on collusion in different auction formats, see Hendricks and Porter (1989).

Exercise #1.12: Second-Price Auctions with Budget Constrained Bidders, Based on Che and Gale (1998)[B]

1.12 Consider again a second-price auction with $N \geq 2$ bidders, but assume now that every bidder privately observes his valuation for the object, v_i, and his budget, w_i. Bidder i's type in this context is, then, a pair (v_i, w_i), where both v_i and w_i are independently drawn from the [0, 1] interval, that is, $(v_i, w_i) \in [0, 1]^2$. For simplicity, assume that if a bidder wins the auction and the winning price is above his budget, w_i, he cannot afford to pay this price, and the seller imposes a fine on the buyer for having to renege.

   (a) Show that every bidder i finds it dominated to bid above his budget, $b_i > w_i$.

   - Let us consider a bidder who submits a bid above his budget, $b_i > w_i$, and wins the auction, paying the second-highest bid $h_i = \max_{j \neq i}\{b_j\}$, which denotes the highest bid among all losing bidders $j \neq i$. Two cases can arise:
     - If $h_i \leq w_i$, then he also wins submitting a bid $b_i = w_i$, which entails a price h_i too. Therefore, he would have incentives to deviate to a bid equal to his budget, $b_i = w_i$.
     - If $h_i > w_i$, he cannot afford to pay the winning price h_i, does not win the object, and, in addition, receives a fine from the seller. In this context, he would have been better off submitting any bid equal or lower than his budget, $b_i \leq w_i$.

     In summary, we found that bidder i is weakly or strictly better off submitting a bid $b_i = w_i$ than bidding strictly above his budget, $b_i >$

Exercise #1.12: Second-Price Auctions with Budget Constrained Bidders,...

w_i. In other words, bidding strictly above his budget, $b_i > w_i$, is a dominated strategy.

(b) If bidder i's valuation, v_i, satisfies $v_i \leq w_i$ (i.e., his budget constraint does not bind), show that bidding according to his valuation, $b_i = v_i$ (as in Exercise 1.2), is still a weakly dominant strategy in the second-price auction.

- In this case, the budget constraint does not bind. Therefore, we can follow the same argument as in Exercise 1.2 to show that bidding according to his valuation, $b_i = v_i$, is a weakly dominant strategy in the second-price auction.

(c) If bidder i's valuation, v_i, satisfies $v_i > w_i$ (i.e., his budget constraint binds), show that submitting a bid equal to his budget, $b_i = w_i$, is a weakly dominant strategy.

- If $v_i > w_i$, it is easy to show that bidder i finds that $b_i = w_i$ is a weakly dominant strategy. Two cases can arise depending on how h_i ranks relative to w_i:
  - If $h_i \leq w_i$, he wins the object, paying h_i, which he can afford. If he decreases his bid, he still pays h_i if he wins the object, but he may lose the auction if his bid satisfies $b_i < h_i$, so he has no strict incentives to deviate from $b_i = w_i$.
  - If $h_i > w_i$, he either loses the auction because the highest bid among his rivals, h_i, exceeds his bid, $b_i = w_i$, or wins the auction, which is not affordable for bidder i in this case. Recall that if he bids above his budget and wins, where $b_i > h_i > w_i$, he would not be able to pay the price, losing the auction and receiving a penalty from the seller after reneging. Therefore, deviations from bidding $b_i = w_i$ do not strictly increase bidder i's payoff.

  Overall, we showed that bidding $b_i = w_i$ yields a payoff that bidder i cannot strictly increase by submitting a different bid.

(d) Combine your results from parts (b) and (c) to describe the equilibrium bidding function in the second-price auction with budget constraints, $b_i(v_i, w_i)$. Depict it as a function of v_i.

- We found that every bidder i submits a bid equal to his valuation, v_i, when his budget constraint is not binding (as in part b) and a bid equal to his budget, w_i, otherwise (as in part c). More compactly,

$$b_i(v_i, w_i) = \begin{cases} v_i & \text{if } v_i \leq w_i \\ w_i & \text{otherwise.} \end{cases}$$

Fig. 1.8 Equilibrium bids in a second-price auction with budget constraints

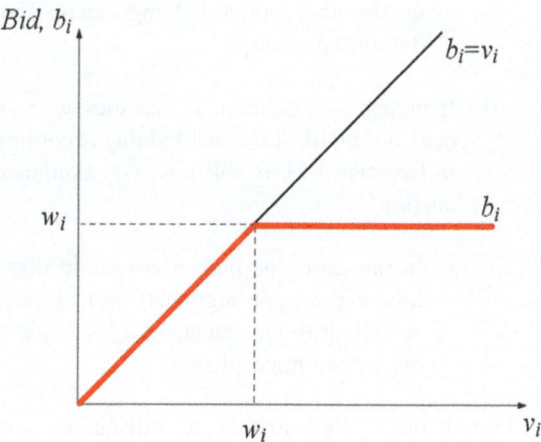

As depicted in Fig. 1.8, this bid coincides with the 45°-line, $b_i = v_i$, when his valuation satisfies $v_i \leq w_i$, but it becomes $b_i = w_i$ otherwise (see the flat segment). Graphically, when the budget constraint does not bind, $v_i \leq w_i$, the bidder behaves as in a standard second-price auction, where $b_i = v_i$, but otherwise submits a bid equal to his budget w_i. This bidding function is often presented as

$$b_i(v_i, w_i) = \min\{v_i, w_i\},$$

since bidder i's bid is the lowest of his valuation for the object (when such valuation is lower than his budget, w_i) and his budget (when his valuation exceeds his budget).

First-Price Auctions

Keywords

First-price auction · Sealed envelope · Marginal cost · Marginal benefit · First-price auction under complete information · First-price auction under incomplete information · Equilibrium bidding function · Direct approach · Envelope theorem approach · Dutch auction · Descending auction · Babylon · Tulip fever · Government securities · Independent and identically distributed values · Expected utility maximization problem · Generic distribution · Integration by parts · Chain rule · Probability weight · Density function · Cumulative distribution function · Expected value · Variance

Introduction

In this chapter, we analyze first-price auctions, where every bidder privately observes his valuation for the object and submits his bid to the seller (e.g., in a sealed envelope). The seller ranks all bids, selects as a winner the bidder who submitted the highest bid, and this bidder pays the bid he submitted. All other bidders pay zero and not receive the object.

As a benchmark, we start the chapter discussing equilibrium bids in a complete information setting where bidders can observe each other's valuations for the object. Exercise 2.2 moves to the more realistic incomplete information environment and begins, for simplicity, in a stylized setting with only two bidders, each of them drawing his valuation from a uniform distribution. This setting helps us understand the main steps at solving similar auction formats in subsequent exercises and chapters. We show that the equilibrium bidding function in the first-price auction has bidders shading their bids, as opposed to their behavior in second-price auctions where bidders submit bids equal to their valuations.

Exercise 2.3 extends the setting of Exercise 2.2 to allow for more than two bidders, but still assuming that valuations are uniformly distributed, and examining how an increase in the number of bidders reduces bid shading.

Exercises 2.4–2.6 further extend our above analysis to allow every bidder i to draw his valuation v_i from a common, but not necessarily uniform, cumulative distribution function $F(v_i)$. (Chapter 3 considers first-price auctions where every bidder draws his valuation from a different distribution.) First, Exercise 2.4 to finds the equilibrium bidding function in this setting using the so-called direct approach, where we differentiate with respect to bidder i's bid, b_i; and Exercise 2.5, as an illustration, evaluates the equilibrium bidding function in the case that bidders' valuations are uniformly or exponentially distributed. In contrast, Exercise 2.6 follows the "Envelope Theorem" approach, where we differentiate with respect to bidder i's valuation and then show that he has no incentives to bid according to a valuation different from his real value for the object. At the end of Exercise 2.6, we evaluate our equilibrium bids in two special settings, one assuming that bidders' valuations are uniformly distributed (where we confirm the results in Exercises 2.2 and 2.3) and another assuming that bidders' valuations are exponentially distributed. Exercise 2.7 investigates the efficiency properties of equilibrium bids in first-price auctions. We show that, as with second-price auctions in Exercise 1.5, the object is assigned to the bidder with the highest valuation, leaving no room for profitable renegotiations between bidders.

Exercises 2.8–2.10 present three statistical concepts that are often used in subsequent chapters: the first-, second-, and (generally) the kth-order statistic, which denotes, respectively, the highest value, the second-highest value, and the kth-highest value, among all bidders' valuations. These valuations can, of course, be treated as random variables, so we can find its cumulative distribution function, its density function, its expected value and variance, and evaluate each of them in the special case where bidders' valuations are, for instance, drawn from a uniform distribution.

Finally, Exercise 2.11 considers a Dutch, or descending, auction, where the seller announces an extremely high price for the object. If no bidder accepts the standing price, the seller decreases it, until one bidder raises his hand accepting that price, who is declared the winner. We show that, despite its differences, this auction format yields the same equilibrium bids as the first-price auction, explaining why the literature regards them as equivalent. Dutch auctions have been common in history since Babylon, in the Netherlands during the "tulip fever" of 1636–1637 (hence the auction's name), and are still used nowadays by the US Department of Treasury to sell government securities to banks and brokers.[1]

---

[1] For a more detailed historical account of Dutch auctions, among other auction formats, see Lucking-Reiley (2000) and Karpowicz (2010).

Exercise #2.1: First-Price Auction Under Complete Information[A]

2.1 Consider a first-price auction with $N \geq 2$ risk-neutral bidders. As a benchmark, this exercise assumes a complete information setting where every bidder i can observe not only his valuation for the object, v_i, but also that of his rivals, $v_{-i} = (v_1, \ldots v_{i-1}, v_{i+1}, \ldots, v_N)$. Find the equilibrium bidding function in this first-price auction under complete information.

- Assume, without loss of generality, that bidder valuations are ranked as $v_1 > v_2 > \ldots > v_N$.
- *Highest-value bidder.* Bidder 1, observing that he has the highest valuation, bids slightly above the second-highest valuation; that is, $b(v_1) = v_2 + \varepsilon$, where $\varepsilon \to 0$, which guarantees that the bidder with the second-highest valuation cannot submit a higher bid. In addition, bidder 1 makes $\varepsilon > 0$ as small as possible (e.g., one cent) to maximize his payoff upon winning the auction, $v_1 - b(v_1) = v_1 - v_2 - \varepsilon$.
- *Second-highest-value bidder.* Bidder 2 submits a bid equal to his valuation, $b(v_2) = v_2$, which guarantees that he loses the auction, given that bidder 1 submits a higher bid, earning a zero payoff. If, instead, bidder 2 lowers his bid, he still loses the auction and earns a zero payoff. Suppose otherwise bidder 2 increases his bid above his valuation, he can win the auction if

$$b(v_2) > b(v_1) = v_2 + \varepsilon,$$

but suffers a loss of

$$v_2 - b(v_2) = v_2 - (v_2 + \varepsilon) = -\varepsilon < 0.$$

Overall, bidder 2 has no strict incentives to deviate from submitting a bid equal to his valuation.
- *All other bidders.* Bidders with the third-highest valuation and below can submit a bid equal to their valuation, or any other bid $b(v_i) \in [0, v_i]$ that guarantees losing the auction and earning a zero payoff. Suppose bidder i bids above the highest bid, that is, $b(v_i) = v_2 + \varepsilon$ for $i = \{3, \ldots, N\}$, then this bidder wins the object but suffers a loss of

$$v_i - b(v_i) = v_i - (v_2 + \varepsilon) = -(v_2 - v_i) - \varepsilon < 0.$$

Therefore, deviations from the true valuation v_i do not strictly increase bidder i's payoff.
- Intuitively, this bidding equilibrium resembles a Bertrand game of price competition with asymmetric firms, where the most competitive firm (lowest production cost) sets a price equal to the marginal cost of the second most

competitive firm (less one cent), capturing all market sales. All other firms set a price equal to their marginal cost (or higher) and make no sales.

Exercise #2.2: First-Price Auction with Only Two Bidders and Uniformly Distributed Valuations[A]

2.2 Consider a first-price auction with two participants, each of them privately observing his valuation for the object: Person A observes his valuation v_A while person B observes his valuation v_B. It is common knowledge that valuations are drawn from a uniform distribution in $[0, 1]$.

For simplicity, you can assume every bidder uses a bidding function

$$b_i = a \times v_i,$$

where parameter $a \in (0, 1)$ measures bid shading. Intuitively, when a approaches one (zero), bidder i submits a bid that is close to (far from) his true valuation for the object.

(a) If the seller runs a first-price auction, who wins the auction?

- *Person A.* We know that person A's expected utility from choosing bid b_A is

$$EU_A(b_A|v_A) = \Pr(win) \times (v_A - b_A) + \Pr(lose) \times 0$$

Since every bidder uses a bidding function of the form $b_i = a \times v_i$, the probability of winning is

$$\Pr(win) = \Pr\{b_A > b_B\}$$
$$= \Pr\{a \times v_A > a \times v_B\}$$
$$= \Pr\{v_A > v_B\}$$

and, given that $v_A = \frac{b_A}{a}$, we can express this probability as

$$\Pr(win) = \Pr\left\{\frac{b_A}{a} > v_B\right\}$$
$$= \frac{b_A}{a}$$

where the last equality comes from the fact that bidder B's valuation, v_B, is uniformly distributed in $[0, 1]$. Thus, person A must choose his bid b_A that maximizes his expected utility,

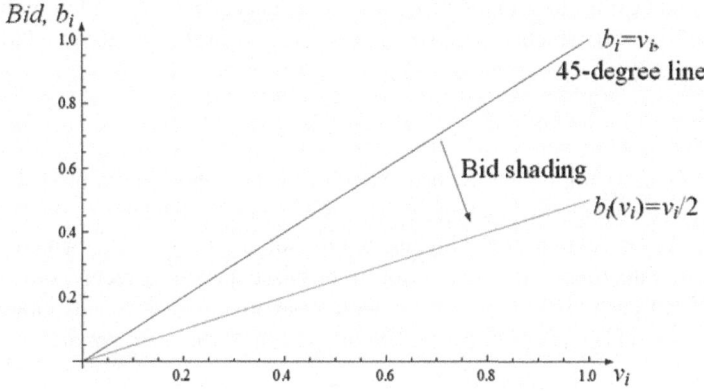

Fig. 2.1 Equilibrium bidding function in the first-price auction with two bidders

$$\max_{b_A} \ EU_A(b_A|v_A) = \underbrace{\frac{b_A}{a}}_{\Pr(win)} (v_A - b_A)$$

Differentiating with respect to b_A yields,

$$\frac{\partial EU_A(b_A|v_A)}{\partial b_A} = \frac{v_A - 2b_A}{a} = 0$$

Since a is alone in the denominator, it cancels out. We are left with $v_A - 2b_A = 0$, which we can solve to obtain person A's equilibrium bid,

$$b_A = \frac{v_A}{2}$$

This implies that person A bids half of his valuation. Figure 2.1 illustrates this bidding function where, as a reference, we include the 45° line where $b_A = v_A$.

- *Person B.* Similar to person A, person B must choose his bid b_B that maximizes his expected utility, as follows:

$$\max_{b_B} \ EU_B(b_B|v_B) = \underbrace{\frac{b_B}{a}}_{\Pr(win)} (v_B - b_B)$$

Differentiating with respect to b_B yields,

$$\frac{\partial EU_B(b_B|v_B)}{\partial b_B} = \frac{v_B - 2b_B}{a} = 0$$

Since a is alone in the denominator, it cancels out. We are left with $v_B - 2b_B = 0$, which we can solve to obtain person B's equilibrium bid,

$$b_B = \frac{v_B}{2}$$

Once again, person B shades half of his bid, choosing to bid half of his true valuation when there exists only one other bidder.

- Person A submits the highest bid if and only if, $\frac{v_A}{2} > \frac{v_B}{2}$, which occurs if valuations satisfy $v_A > v_B$. In this case, person A receives the object and pays his bid, $b_A = \frac{v_A}{2}$, with a payoff $v_A - \frac{v_A}{2} = \frac{v_A}{2}$. Otherwise, person B wins the auction, leaving person A with zero payoff.

(b) Suppose now that Person A was able to observe Person B's private valuation prior to the auction. Would Person A change his bid? If so, how? If not, why?

- If person A could observe person B's private valuation, he would know that person B would enter a bid of $\frac{v_B}{2}$ for the object. Thus, if person A has a higher valuation than person B, that is, $v_A > v_B$, person A could simply bid $\frac{v_B}{2} + \varepsilon$, where $\varepsilon \to 0$, and win the object. This is a profitable bid for bidder A since

$$v_A - \frac{v_B}{2} > v_A - \frac{v_A}{2} > 0$$

In this case, bidder A obtains a utility higher than that of bidding at $\frac{v_A}{2}$.

Exercise #2.3: First-Price Auction with $N \geq 2$ Bidders and Uniformly Distributed Valuations[A]

2.3 Consider a first-price auction with $N \geq 2$ bidders. Every bidder i's valuation for the object, v_i, is drawn from a uniform distribution function, $F(v_i) = v_i$, where $v_i \in [0, 1]$. Valuations are independent and identically distributed. Bidders submit their bids, the winner is the bidder who submitted the highest bid, receives the object, and must pay a price equal to his own bid. As in Exercise 2.2, you can assume every bidder i uses a bidding function $b_i = a \times v_i$, where parameter $a \in (0, 1)$ measures bid shading.

(a) Write bidder i's expected utility maximization problem when he submits a bid b_i and his valuation for the object is v_i.

- *Writing expected utility.* We can write bidder i's expected utility maximization problem (UMP) as follows:

Exercise #2.3: First-Price Auction with $N \geq 2$ Bidders and Uniformly...

$$\max_{b_i \geq 0} \Pr(win) \times (v_i - b_i)$$

which denotes the probability of winning the object times bidder i's net payoff from winning, $v_i - b_i$, because he values the object at v_i and pays his bid b_i for it.

- *Finding the probability of winning*. At this point, we need to write the probability of winning, $\Pr(win)$, as a function of bidder i's bid, b_i. Since bidding functions are *ex-ante* symmetric across bidders, we can say that bidder i wins the auction when his bid exceeds that of bidder j,

$$b_i = b(v_i) > b(v_j) = b_j$$

which is equivalent to his valuation exceeding that of bidder j, $v_i > v_j$. We can express this probability as

$$\Pr(v_j < v_i) = F(v_i)$$

Therefore, when bidder i faces $N-1$ rivals, his probability of winning the auction is the probability that his valuation exceeds that of all other $N-1$ bidders. Because valuations are independently distributed, we can write this probability as the following product:

$$\Pr(v_1 < v_i) \times \ldots \times \Pr(v_{i-1} < v_i) \times \Pr(v_{i+1} < v_i) \times \ldots \times \Pr(v_N < v_i)$$

$$= \underbrace{F(v_i) \times F(v_i) \times \ldots \times F(v_i)}_{N-1 \text{ times}}$$

$$= F(v_i)^{N-1}$$

Now, we can express bidder i's expected utility maximization problem as follows:

$$\max_{b_i \geq 0} F(v_i)^{N-1}(v_i - b_i)$$

Since valuations are uniformly distributed, $F(v_i) = v_i$, this problem becomes

$$\max_{b_i \geq 0} \underbrace{v_i^{N-1}}_{F(v_i)^{N-1}} (v_i - b_i)$$

Since the bidding function satisfies $b_i = a \times v_i$ by assumption, where a is the proportion bidder i shades his bid, we can solve for v_i to obtain $v_i = \frac{b_i}{a}$. Therefore, we can rewrite the above maximization problem as follows:

$$\max_{b_i \geq 0} \left(\frac{b_i}{a}\right)^{N-1} (v_i - b_i)$$

(b) Differentiate with respect to b_i, and rearrange, to find the equilibrium bidding function $b_i(v_i)$.

- Differentiating the above expected utility maximization problem with respect to b_i, we find

$$\frac{1}{a^{N-1}} \left[(N-1) b_i^{N-2} v_i - N b_i^{N-1} \right] = 0$$

which we can rearrange to yield

$$(N-1) b_i^{N-2} v_i = N b_i^{N-1}$$

Simplifying, we obtain

$$\frac{b_i^{N-1}}{b_i^{N-2}} = \frac{N-1}{N} v_i$$

Solving for b_i, we find the equilibrium bidding function when bidders' valuations are uniformly distributed,

$$b_i(v_i) = \frac{N-1}{N} v_i$$

(c) *Comparative statics.* How is the equilibrium bidding function $b_i(v_i)$ affected by the number of bidders? Interpret.

- When only two bidders compete for the object, $N = 2$, this bidding function simplifies to $b_i(v_i) = \frac{v_i}{2}$, which coincides with the result in Exercise 2.2. When $N = 3$, equilibrium bids increase to $b_i(v_i) = \frac{2v_i}{3}$, and a similar result occurs when $N = 4$ bidders compete for the object, where $b_i(v_i) = \frac{3v_i}{4}$. More generally, the derivative of bidding function $b_i(v_i) = v_i \left(\frac{N-1}{N}\right)$ with respect to the number of bidders N yields

$$\frac{\partial b_i(v_i)}{\partial N} = \frac{1}{N^2} v_i$$

which is clearly positive.

As depicted in Fig. 2.2, as more bidders participate in the auction, every bidder i submits more aggressive bids because he faces a higher probability that another bidder j has a higher valuation for the object

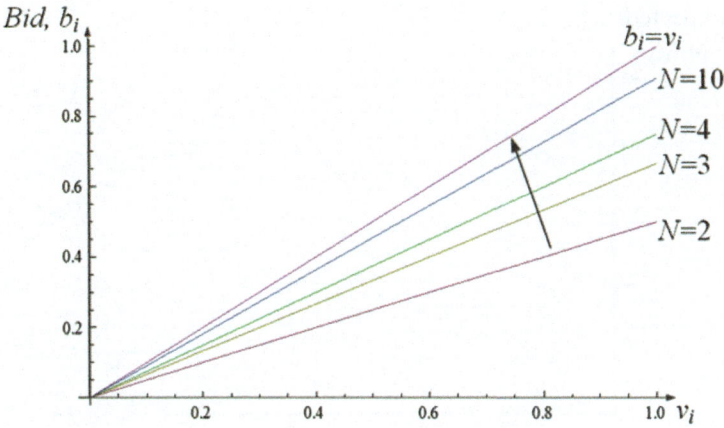

Fig. 2.2 Equilibrium bidding function in the first-price auction—More bidders

than he has. As a reference, the figure also includes the 45°-line where bids coincide with valuations, $b_i = v_i$. When the number of bidders is sufficiently high, $N \to +\infty$, equilibrium bids approach $b_i = v_i$, which entails no bid shading.

(d) *Equilibrium payoff.* Find the equilibrium payoff that every bidder i obtains from participating in this auction. How is this expected payoff affected by the number of bidders competing for the object?

- Every bidder i uses bidding function $b_i(v_i) = \frac{N-1}{N} v_i$, implying that, in equilibrium, his expected utility is

$$EU(b_i(v_i), v_i) = v_i^{N-1} \left(v_i - \frac{N-1}{N} v_i \right)$$

$$= \frac{v_i^N}{N}$$

Differentiating the expected utility with respect to N, we obtain

$$\frac{\partial EU(b_i(v_i), v_i)}{\partial N} = \frac{(N \log v_i - 1) v_i^N}{N^2}$$

which is negative since $\log v_i < 0$ for $0 \leq v_i \leq 1$, so that every bidder i obtains a lower expected utility when more bidders compete for the object.
- Figure 2.3 evaluates this expected utility at three different values of N. At $N = 2$, the expected utility becomes $\frac{v_i^2}{2}$; at $N = 3$, this expected

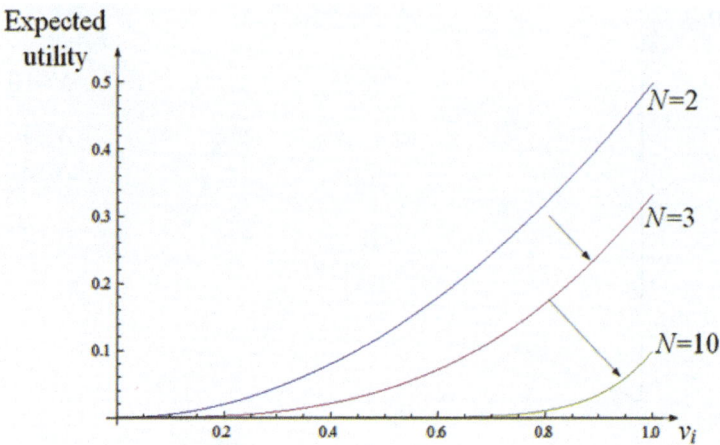

Fig. 2.3 Expected utility from participating in the first-price auction

utility shifts downward to $\frac{v_i^3}{3}$; and a similar argument applies when the number of bidders increases to $N = 10$, where the expected utility becomes $\frac{v_i^{10}}{10}$. When the number of bidders satisfies $N \to +\infty$, expected utility converges to zero, graphically coinciding with the horizontal axis in Fig. 2.3.

Exercise #2.4: First-Price Auction with Generic Distribution of Valuations–The Direct Approach[B]

2.4 Consider again the first-price auction with $N \geq 2$ bidders in Exercise 2.3, but now allow for valuations to be drawn from more general cumulative distribution functions, $F(v_i)$, with positive density in all its support–that is, $f(v_i) > 0$ for all $v_i \in [0, 1]$.

(a) Write bidder i's expected utility maximization problem when he submits a bid b_i and his valuation for the object is v_i.

- *Writing expected utility.* We can write bidder i's expected utility maximization problem (UMP) as follows:

$$\max_{b_i \geq 0} \ \Pr(win) \times (v_i - b_i)$$

which denotes the probability of winning the object times bidder i's net payoff from winning, $v_i - b_i$, because he values the object at v_i and pays his bid b_i for it.

Exercise #2.4: First-Price Auction with Generic Distribution of Valuations–The...

- *Finding the probability of winning.* At this point, we need to write the probability of winning, Pr(*win*), as a function of bidder i's bid, b_i. Bidder i wins the auction when his bid exceeds that of bidder j, $b_i > b_j$, which is equivalent to his valuation exceeding that of bidder j, $v_i > v_j$. We can express this probability as

$$\Pr(v_j < v_i) = F(v_i)$$

Therefore, when bidder i faces $N - 1$ rivals, his probability of winning the auction is the probability that his valuation exceeds that of all other $N - 1$ bidders. Because valuations are independently distributed, we can write this probability as the following product:

$$\Pr(v_1 < v_i) \times \ldots \times \Pr(v_{i-1} < v_i) \times \Pr(v_{i+1} < v_i) \times \ldots \times \Pr(v_N < v_i)$$
$$= \underbrace{F(v_i) \times F(v_i) \times \ldots \times F(v_i)}_{N-1 \text{ times}}$$
$$= F(v_i)^{N-1}$$

Now we can express the expected utility maximization problem as follows:

$$\max_{b_i \geq 0} F(v_i)^{N-1}(v_i - b_i)$$

Using this bidding function, we can write $b_i(v_i) = x_i$, where $x_i > 0$ represents bidder i's bid when his valuation is v_i. We claim that the bidding function $b_i(v_i)$ is increasing in v_i, which we show below, so we can apply the inverse $b^{-1}(\cdot)$ on both sides of $b_i(v_i) = x_i$, yielding $v_i = b_i^{-1}(x_i)$. This result helps us rewrite the probability of winning, $F(v_i)^{N-1}$, as $F(b_i^{-1}(x_i))^{N-1}$, so the above maximization problem becomes

$$\max_{x_i \geq 0} F(b_i^{-1}(x_i))^{N-1}(v_i - x_i)$$

where we expressed the bid as x_i in the last term because $b_i(v_i) = x_i$.

(b) Differentiate with respect to b_i, and rearrange, to find the equilibrium bidding function $b_i(v_i)$.

- Now that the probability of winning is written as a function of bidder i's bid, x_i, we are ready to differentiate with respect to x_i to find the equilibrium bidding function. Differentiating with respect to x_i, yields

$$-\left[F(b_i^{-1}(x_i))^{N-1}\right] + (N-1)F(b_i^{-1}(x_i))^{N-2} f\left(b_i^{-1}(x_i)\right)$$
$$\times \frac{\partial b_i^{-1}(x_i)}{\partial x_i}(v_i - x_i) = 0.$$

Since $b_i^{-1}(x_i) = v_i$ and $\frac{\partial b_i^{-1}(x_i)}{\partial x_i} = \frac{1}{b'\left(b_i^{-1}(x_i)\right)}$, this expression simplifies to[2]

$$-\left[F(v_i)^{N-1}\right] + (N-1)F(v_i)^{N-2} f(v_i) \frac{1}{b'(v_i)}(v_i - x_i) = 0$$

Further rearranging the above expression, we obtain

$$(N-1)F(v_i)^{N-2} f(v_i) v_i - (N-1)F(v_i)^{N-2} f(v_i) x_i = F(v_i)^{N-1} b'(v_i)$$

that is simplified into

$$F(v_i)^{N-1} b'(v_i) + (N-1)F(v_i)^{N-2} f(v_i) x_i = (N-1)F(v_i)^{N-2} f(v_i) v_i$$

The left side is $\frac{\partial \left[F(v_i)^{N-1} b(v_i)\right]}{\partial v_i}$, which helps us write this expression as

$$\frac{\partial \left[F(v_i)^{N-1} b(v_i)\right]}{\partial v_i} = (N-1)F(v_i)^{N-2} f(v_i) v_i$$

Integrating both sides yields

$$F(v_i)^{N-1} b_i(v_i) = \int_0^{v_i} (N-1)F(x)^{N-2} f(x) x \, dx$$

Applying integration by parts on the right side,[3] we find

---

[2]Recall that $b_i\left(b_i^{-1}(x_i)\right) = x_i$. Differentiating both sides with respect to x_i, and applying the Chain rule, we obtain $\frac{\partial b_i\left(b_i^{-1}(x_i)\right)}{\partial b_i^{-1}(x_i)} \frac{\partial b_i^{-1}(x_i)}{\partial x_i} = 1$, which we can rearrange to yield $\frac{\partial b_i^{-1}(x_i)}{\partial x_i} = \frac{1}{b_i'\left(b_i^{-1}(x_i)\right)}$.

[3]Recall that, when integrating by parts, we consider two functions $g(x)$ and $h(x)$, such that $(gh)' = g'h + gh'$. Integrating both sides yields $g(x)h(x) = \int g'(x)h(x)dx + \int g(x)h'(x)dx$. Reordering this expression, we find $\int g'(x)h(x)dx = g(x)h(x) - \int g(x)h'(x)dx$. At this point, we can apply integration by parts in our auction setting by defining $g'(x) \equiv (N-1)F(x)^{N-2} f(x)$ and $h(x) \equiv x$, so we obtain the given results.

Exercise #2.5: First-Price Auction with Uniformly or Exponentially Distributed...

$$\int_0^{v_i} (N-1)F(x)^{N-2} f(x) x\, dx = F(v_i)^{N-1} v_i - \int_0^{v_i} F(x)^{N-1} dx$$

so we can write our above first-order condition as

$$F(v_i)^{N-1} b_i(v_i) = F(v_i)^{N-1} v_i - \int_0^{v_i} F(x)^{N-1} dx$$

We can now solve for the equilibrium bidding function $b_i(v_i)$ that we seek to find. Dividing both sides by $F(v_i)^{N-1}$ yields

$$b_i(v_i) = v_i - \underbrace{\frac{\int_0^{v_i} F(x)^{N-1} dx}{F(v_i)^{N-1}}}_{\text{bid shading}}$$

Intuitively, bidder i submits a bid equal to his valuation for the object, v_i, minus an amount captured by the second term in this expression, which is referred to as his "bid shading." We can then claim that the bidding function $b_i(v_i)$ constitutes the Bayesian Nash equilibrium of the first-price auction when bidders' valuations are distributed according to $F(v_i)$.

Exercise #2.5: First-Price Auction with Uniformly or Exponentially Distributed Values[B]

2.5 The equilibrium bidding function found in Exercise 2.4 is

$$b_i(v_i) = v_i - \frac{\int_0^{v_i} F(x)^{N-1} dx}{F(v_i)^{N-1}}$$

for a generic distribution function $F(v_i)$ where $v_i > 0$. In this exercise, we evaluate this bidding function in some commonly used settings, such as the uniform and exponential distribution.

(a) *Uniformly distributed valuations.* Consider that individual valuations are uniformly distributed, $F(v_i) = v_i$, where $v_i \in [0, 1]$. Find the equilibrium bidding function $b_i(v_i)$ in this context.

- In this scenario, we obtain $F(v_i)^{N-1} = v_i^{N-1}$ and

$$\int_0^{v_i} F(x)^{N-1} dx = \frac{1}{N} v_i^N$$

generating a bidding function of

$$b_i(v_i) = v_i - \frac{\frac{1}{N} v_i^N}{v_i^{N-1}}$$

$$= v_i - \frac{v_i^N}{N v_i^{N-1}}$$

$$= \frac{N-1}{N} v_i$$

which coincides with the result in Exercise 2.3.

- When only two bidders compete for the object, $N = 2$, this bidding function simplifies to $b_i(v_i) = \frac{v_i}{2}$, which coincides with the result in Exercise 2.2. When $N = 3$, equilibrium bids increase to $b_i(v_i) = \frac{2v_i}{3}$, and a similar result is obtained when $N = 4$ bidders compete for the object, where $b_i(v_i) = \frac{3v_i}{4}$. More generally, the derivative of bidding function $b_i(v_i) = \frac{N-1}{N} v_i$ with respect to the number of bidders N yields

$$\frac{\partial b_i(v_i)}{\partial N} = \frac{1}{N^2} v_i$$

which is clearly positive.

- Informally, as more bidders participate in the auction, every bidder i submits more aggressive bids because he faces a higher probability that another bidder j has a higher valuation for the object than he has.

(b) *Exponentially distributed valuations.* Consider now that individual valuations are drawn from an exponential distribution, $F(v_i) = 1 - \exp(-\lambda v_i)$ where $v_i \in [0, +\infty)$, and $N = 2$ bidders. Find the equilibrium bidding function $b_i(v_i)$ in this context. Evaluate your results at different values of λ, and interpret your results.

- In this case, we obtain $F(v_i)^{N-1} = F(v_i)^{2-1} = 1 - \exp(-\lambda v_i)$, and

$$\int_0^{v_i} [1 - \exp(-\lambda x)] \, dx = \left[x + \frac{1}{\lambda} \exp(-\lambda x) \right]_0^{v_i}$$

$$= \left(v_i + \frac{1}{\lambda} \exp(-\lambda v_i) \right) - \frac{1}{\lambda}$$

since $\exp(0) = 1$. Simplifying, we obtain

$$v_i - \frac{1}{\lambda} [1 - \exp(-\lambda v_i)]$$

In this context, the bidding function becomes

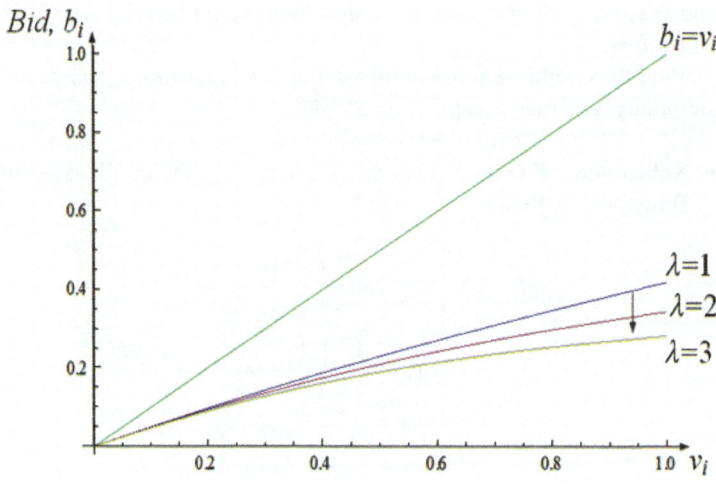

Fig. 2.4 Equilibrium bidding function with exponentially distributed valuations

$$b_i(v_i) = v_i - \frac{v_i - \frac{1}{\lambda}\left[1 - \exp(-\lambda v_i)\right]}{1 - \exp(-\lambda v_i)}$$

$$= \frac{1}{\lambda} + v_i \left(1 - \frac{1}{1 - \exp(-\lambda v_i)}\right)$$

$$= \frac{1}{\lambda} + v_i \left(1 + \frac{\exp(\lambda v_i)}{1 - \exp(\lambda v_i)}\right)$$

$$= \frac{1}{\lambda} + \frac{v_i}{1 - \exp(\lambda v_i)}$$

- Evaluating this bidding function at $\lambda = 1$, we obtain $b_i(v_i) = 1 + \frac{v_i}{1 - \exp(v_i)}$; at $\lambda = 2$, the bidding function is $b_i(v_i) = \frac{1}{2} + \frac{v_i}{1 - \exp(2v_i)}$; and, similarly, at $\lambda = 3$, we find a bidding function of $b_i(v_i) = \frac{1}{3} + \frac{v_i}{1 - \exp(3v_i)}$. Figure 2.4 plots these bidding functions, showing that, as the exponential distribution puts a larger probability weight on small valuations (higher λ), every bidder submits more conservative bids (i.e., bidding function $b_i(v_i)$ shifts downward in λ), enlarging the bid shading.

(c) *Other distribution forms.* Consider the following distribution function,

$$F(v_i) = (1 + \alpha) v_i - \alpha v_i^2$$

where $v_i \in [0, 1]$, and parameter α satisfies $\alpha \in [-1, 1]$. When $\alpha = 0$, this function collapses to the uniform distribution, $F(v_i) = v_i$; when $\alpha > 0$, it becomes concave, thus putting more probability weight on low valuations;

and when $\alpha < 0$, it is convex, assigning more probability weight on high valuations.

Find the equilibrium bid in this setting, and explain its change in α. For simplicity, you may assume $N = 2$ bidders.

- Substituting $F(v_i) = (1+\alpha) v_i - \alpha v_i^2$ into the equilibrium bidding function, we obtain

$$b_i(v_i) = v_i - \frac{\int_0^{v_i} F(x)\, dx}{F(v_i)}$$

$$= v_i - \frac{\int_0^{v_i} \left[(1+\alpha) x - \alpha x^2\right] dx}{(1+\alpha) v_i - \alpha v_i^2}$$

$$= v_i - \frac{\frac{1+\alpha}{2} x^2 \big|_0^{v_i} - \frac{\alpha}{3} x^3 \big|_0^{v_i}}{(1+\alpha) v_i - \alpha v_i^2}$$

$$= v_i - \frac{\frac{1+\alpha}{2} v_i^2 - \frac{\alpha}{3} v_i^3}{(1+\alpha) v_i - \alpha v_i^2}$$

$$= \frac{v_i \left[3(1+\alpha) - 4\alpha v_i\right]}{6(1+\alpha - \alpha v_i)}$$

- Differentiating the equilibrium bid with respect to α, we obtain

$$\frac{\partial b_i(v_i)}{\partial \alpha} = \frac{v_i}{6} \frac{(3 - 4v_i)\left[1 + \alpha - \alpha v_i - \alpha(1 - v_i)\right] - 3(1 - v_i)}{(1 + \alpha - \alpha v_i)^2}$$

$$= -\frac{v_i^2}{6(1+\alpha - \alpha v_i)^2} < 0$$

As α increases, the distribution function becomes more concave. Therefore, bidders put a larger probability weight on low valuations, submitting less aggressive bids. (Alternatively, the effect of α can be understood by finding the density function in this case, $f(v_i) = 1 + \alpha(1 - 2v_i)$. When $\alpha = 1$, this density becomes $f(v_i) = 2 - 2v_i$, assigning most probability weight on low valuations; when $\alpha = 0$, this density collapses to $f(v_i) = 1$, as in the uniform distribution; and when $\alpha = -1$, this density simplifies to $f(v_i) = 2v_i$, which assigns most probability weight on high valuations.)

- Figure 2.5 depicts this bidding function, evaluated at three different values of α:
  - When $\alpha = 0$, the equilibrium bid simplifies to $b_i(v_i) = \frac{v_i}{2}$, thus coinciding with our results when valuations are uniformly distributed (part a).

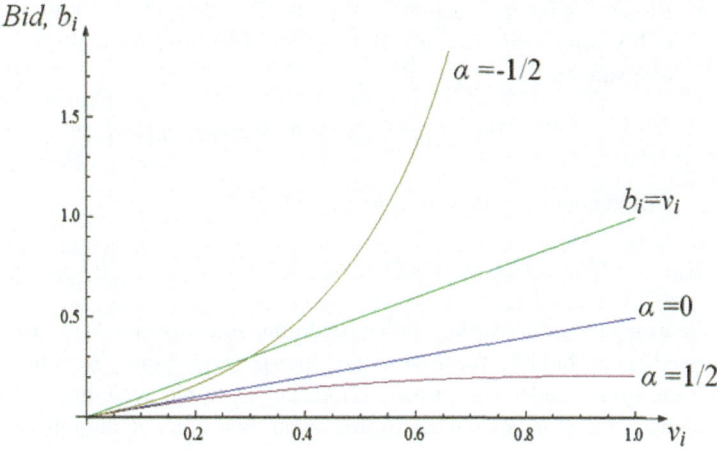

Fig. 2.5 Equilibrium bidding function—Other cumulative distribution

- When $\alpha = \frac{1}{2}$, the equilibrium bid becomes $b_i(v_i) = \frac{v_i(9-4v_i)}{6(3-v_i)}$, being located below $b_i(v_i) = \frac{v_i}{2}$ since bidders submit less aggressive bids.
- Finally, when $\alpha = -\frac{1}{2}$, the equilibrium bid becomes $b_i(v_i) = \frac{v_i(3+4v_i)}{6(1+v_i)}$, which lies above other bidding functions in Fig. 2.5. Intuitively, when $\alpha < 0$, the cumulative distribution function is convex, putting more probability weight on high valuations, which induces every bidder i to submit more aggressive bids as he expects that his rivals' valuations are likely to be high.

Exercise #2.6: First-Price Auction with Generic Distribution of Valuations–The Envelope Theorem Approach[C]

2.6 Consider a first-price auction with $N \geq 2$ risk-neutral bidders, where every bidder i privately observes his valuation v_i for the object drawn from cumulative distribution function $F(v_i)$. In this exercise, we seek to show that the equilibrium bidding function in Exercise 2.4 can be found by directly differentiating with respect to valuation v_i, as opposed to differentiating with respect to the bid b.

Assume a symmetric bidding function, $b(v_i)$, and take bidder i who, rather than bidding according to his true valuation v_i, uses the bidding function $b(\cdot)$ but bids according to a different valuation $z_i \neq v_i$.

(a) Find bidder i's expected utility from participating in the first-price auction and differentiate with respect to valuation z_i.

- Bidder i wins the auction with probability $F(z_i)^{N-1}$, so his expected utility from participating in the auction, submitting a bid $b(z_i)$ when his true valuation is v_i, is

$$EU(z_i|v_i) = F(z_i)^{N-1} \times [v_i - b(z_i)]$$

Differentiating with respect to z_i, we find

$$(N-1)F(z_i)^{N-2} f(z_i)[v_i - b(z_i)] - F(z_i)^{N-1} b'(z_i) = 0$$

(b) Rearrange your above first-order condition, and solve for $b(v_i)$, to find an equilibrium bidding function in this first-price auction. [*Hint*: In equilibrium, every bidder i must have no incentives to bid according a different valuation $z_i \neq v_i$, so you can evaluate your first-order conditions at his real valuation, $z_i = v_i$].

- In equilibrium, $z_i = v_i$ and $b(z_i) = b(v_i)$. Inserting this property in the above first-order condition, we obtain

$$(N-1)F(v_i)^{N-2} f(v_i)[v_i - b(v_i)] - F(v_i)^{N-1} b'(v_i) = 0$$

Rearranging, yields

$$(N-1)F(v_i)^{N-2} f(v_i) b(v_i) + F(v_i)^{N-1} b'(v_i) = (N-1)F(v_i)^{N-2} f(v_i) v_i$$

The left-hand side can be alternatively represented as the derivative $\frac{d[F(v_i)^{N-1} \times b(v_i)]}{dv_i}$, so the above first-order condition can be more compactly expressed as

$$\frac{d\left[F(v_i)^{N-1} \times b(v_i)\right]}{dv_i} = (N-1)F(v_i)^{N-2} f(v_i) v_i$$

At this point, recall that we seek to solve for the equilibrium bidding function $b(v_i)$, which is only in the derivative on the left-hand side. To eliminate of the derivative, we integrate both sides of the above equality to obtain

$$F(v_i)^{N-1} \times b(v_i) = \int_0^{v_i} (N-1) F(x)^{N-2} f(x) x \, dx$$

Solving for $b(v_i)$, we find that the equilibrium bidding function in the first-price auction is

$$b(v_i) = \frac{1}{F(v_i)^{N-1}} \int_0^{v_i} (N-1)F(x)^{N-2} f(x) x \, dx$$

Since $(N-1)F(x)^{N-2}f(x) = \frac{dF(x)^{N-1}}{dx}$, this bidding function is often expressed more compactly as

$$b(v_i) = \frac{1}{F(v_i)^{N-1}} \int_0^{v_i} x \, dF(x)^{N-1}$$

(c) Show that the equilibrium bidding function that you obtained using the Envelope Theorem approach in part (b) coincides with that found using the direct approach in Exercise 2.4(b).

- The equilibrium bidding function using the direct approach that we found in Exercise 2.4(b) is

$$b_i(v_i) = v_i - \frac{\int_0^{v_i} F(x)^{N-1} dx}{F(v_i)^{N-1}}$$

which we can rewrite as follows:

$$b_i(v_i) = \frac{1}{F(v_i)^{N-1}} \left(v_i F(v_i)^{N-1} - \int_0^{v_i} F(x)^{N-1} dx \right)$$

$$= \frac{1}{F(v_i)^{N-1}} \int_0^{v_i} x(N-1) F(x)^{N-2} f(x) dx$$

where we apply the inverse of integration by parts in the second line. Furthermore, since $\frac{\partial F(x)^{N-1}}{\partial x} = (N-1)F(x)^{N-2} f(x)$, this expression can be presented more compactly as

$$b_i(v_i) = \frac{1}{F(v_i)^{N-1}} \int_0^{v_i} x \, dF(x)^{N-1}$$

implying that, as expected, both approaches produce the same equilibrium bidding function.

(d) *Uniform distribution.* Assume that bidders draw their valuation from a uniform distribution in [0, 1]. Evaluate the equilibrium bidding function you found in part (b). Interpret.

- If valuations are uniformly distributed, $F(v_i) = v_i$ and $f(v_i) = 1$, so the equilibrium bidding function of part (b) becomes

$$b(v_i) = \frac{1}{v_i^{N-1}} \int_0^{v_i} (N-1)x^{N-2} x \, dx$$

$$= \frac{1}{v_i^{N-1}} \int_0^{v_i} (N-1)x^{N-1} \, dx$$

Further rearranging, we obtain

$$b(v_i) = \frac{N-1}{v_i^{N-1}} \int_0^{v_i} x^{N-1} \, dx$$

$$= \frac{N-1}{v_i^{N-1}} \left[\frac{x^N}{N} \right]_0^{v_i}$$

$$= \frac{N-1}{v_i^{N-1}} \frac{v_i^N}{N}$$

$$= \frac{N-1}{N} v_i$$

This bidding function, as expected, coincides with that in part (b) of Exercise 2.3.

(e) *Exponentially distributed valuations.* Consider now that individual valuations are drawn from an exponential distribution, $F(v_i) = 1 - \exp(-\lambda v_i)$ where $v_i \in [0, +\infty)$, and $N = 2$ bidders. Find the equilibrium bidding function $b_i(v_i)$ in this context.

- In this case, we obtain $F(v_i)^{N-1} = F(v_i)^{2-1} = 1 - \exp(-\lambda v_i)$, and

$$\int_0^{v_i} x \, dF(x)^{N-1} = \int_0^{v_i} x \, d\overbrace{\left[1 - \exp(-\lambda x) \right]}^{F(x)^{2-1}}$$

$$= -\int_0^{v_i} x \, d\left[\exp(-\lambda x) \right]$$

$$= -\left[x \exp(-\lambda x) \right]_0^{v_i} + \int_0^{v_i} \exp(-\lambda x) \, dx$$

$$= -v_i \exp(-\lambda v_i) - \frac{1}{\lambda}\left[\exp(-\lambda x)\right]_0^{v_i}$$

$$= \frac{1}{\lambda}\left[1 - \exp(-\lambda v_i)\right] - v_i \exp(-\lambda v_i)$$

In this context, the bidding function $b_i(v_i) = \frac{1}{F(v_i)^{N-1}} \int_0^{v_i} x\, dF(x)^{N-1}$ becomes

$$b_i(v_i) = \frac{1}{1 - \exp(-\lambda v_i)} \left[\frac{1}{\lambda}\left[1 - \exp(-\lambda v_i)\right] - v_i \exp(-\lambda v_i)\right]$$

$$= \frac{1}{\lambda} - \frac{\frac{v_i}{\exp(\lambda v_i)}}{1 - \frac{1}{\exp(\lambda v_i)}}$$

$$= \frac{1}{\lambda} + \frac{v_i}{1 - \exp(\lambda v_i)}$$

and this bidding function, as expected, coincides with that in part (b) of Exercise 2.5.

Exercise #2.7: Efficiency in First-Price Auctions[A]

2.7 Consider the first-price auction in Exercise 2.4. Argue that the object, in equilibrium, is assigned to the bidder with the highest valuation.

- In this auction format, every bidder i uses a bidding function

$$b_i(v_i) = v_i - \frac{\int_0^{v_i} F(x)^{N-1} dx}{F(v_i)^{N-1}}$$

which is monotonically increasing in bidder i's valuation, v_i, since

$$\frac{db_i(v_i)}{dv_i} = 1 - \frac{F(v_i)^{2(N-1)} - (N-1)F(v_i)^{N-2} f(v_i) \int_0^{v_i} F(x)^{N-1} dx}{F(v_i)^{2(N-1)}}$$

$$= \frac{(N-1)f(v_i) \int_0^{v_i} F(x)^{N-1} dx}{F(v_i)^N} > 0$$

Therefore, individuals with higher valuations submit higher bids. In other words, the individual submitting the highest bid, h, who satisfies $b_h(v_h) > b_j(v_j)$ for every bidder $j \neq h$, assigns the highest value to the object, $v_h >$

v_j. In this context, the bidder who wins the auction, h, has the highest value for the object, entailing that the first-price auction is efficient.
- From Exercise 1.5 (analyzing efficiency in second-price auctions), we know that, when the object goes to the individual who assigns the highest value, we say that the auction is "efficient," because the seller cannot reassign the object to another bidder and increase his expected revenue. Similarly, allowing for trade between bidders once the auction is over would not improve the payoff of at least one bidder without reducing the payoff of any other bidder.

Exercise #2.8: The First-Order Statistic[B]

2.8 Consider an auction where every bidder i privately observes his valuations for the good, v_i, according to the same cumulative distribution function $F(v_i)$, where $v_i \in [\underline{v}, \overline{v}]$ and $0 \leq \underline{v} < \overline{v}$. Assume that valuations are independently distributed, and let $(v_1, v_2, \ldots, v_N)$ denote the (randomly drawn) valuation profile for $N \geq 2$ bidders.

(a) Find the probability that a valuation v lies weakly above the highest element in valuation profile $(v_1, v_2, \ldots, v_N)$. The highest element in the valuation profile is often known as the "first-order statistic," and denoted as $v^{[1]}$.

- Since $v^{[1]} \equiv \max\{v_1, v_2, \ldots, v_N\}$, for $v^{[1]} \leq v$ to hold, we need that $v_1 < v$, $v_2 < v$, and similarly for the valuations of every individual. Therefore, the probability that $v^{[1]} < v$ is

$$\Pr\{v^{[1]} \leq v\} = \Pr\{v_1 \leq v\} \times \Pr\{v_2 \leq v\} \times \ldots \times \Pr\{v_N \leq v\}$$

and since $\Pr\{v_i \leq v\} = F(v)$ for any bidder i, we obtain that

$$\Pr\{v^{[1]} \leq v\} = \underbrace{F(v) \times F(v) \times \ldots \times F(v)}_{N \text{ times}} = F(v)^N$$

This is, formally, referred to as the cumulative distribution function of the first-order statistic, $v^{[1]}$, and denoted as $F^{[1]}(v)$.

(b) Find the density function of the first-order statistic.

- Differentiating $F^{[1]}(v) = F(v)^N$ with respect to v, we find the density function

$$f^{[1]}(v) = N F(v)^{N-1} f(v)$$

Exercise #2.8: The First-Order Statistic[B]

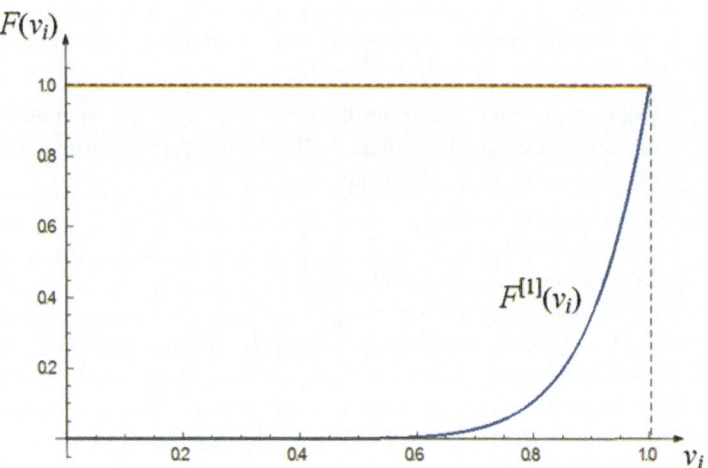

Fig. 2.6 First-order statistic of a uniform distribution

(c) If every bidder's valuation is uniformly distributed, $F(v_i) = v_i$ and $v_i \in [0, 1]$, find the cumulative distribution function and the density of the first-order statistic.

- Evaluating our results from part (a) at $F(v_i) = v_i$ and $v_i \in [0, 1]$, yields a cumulative distribution function

$$F^{[1]}(v) = F(v)^N$$
$$= v^N$$

And evaluating our results from part (b) at a uniform distribution, and given that $f(v_i) = F'(v_i) = 1$, we obtain

$$f^{[1]}(v) = NF(v)^{N-1}f(v)$$
$$= Nv^{N-1}$$

Figure 2.6 depicts the cumulative distribution function of a first-order statistic where, for illustration purpose, we consider $N = 10$ bidders.

(d) *Expected value.* Still assuming that valuations are uniformly distributed, find the expected value of the first-order statistic.

- The expected value of the first-order statistic, for any cumulative distribution $F(v_i)$ where $v_i \in [0, 1]$, is

$$E[v^{[1]}] = \frac{N!}{(N-1)!} \int_0^1 x[1-F(x)]^{N-1} f(x)dx$$

where the first term indicates the number of ways in which we can have all valuations being lower than v_i. If valuations are uniformly distributed, $F(v_i) = v_i$, this expected value simplifies to

$$E\left[v^{[1]}\right] = \frac{1}{B(1,N)} \int_0^1 x(1-x)^{N-1} dx$$

$$= -\frac{N!}{N(N-1)!} \int_0^1 xd\left[(1-x)^N\right]$$

$$= -\left[x(1-x)^N\right]_0^1 + \int_0^1 (1-x)^N dx$$

$$= -\frac{1}{N+1}\left[(1-x)^{N+1}\right]_0^1$$

$$= \frac{1}{N+1}$$

where $B(r, s)$ is the Beta function with parameters r and s. This could, of course, be found more directly, by using the cumulative distribution function of the first-order statistic found in part (c), $F^{[1]}(v) = v^N$, and depicted in Fig. 2.6. Its expected value is the area under the curve $F^{[1]}(v)$ in Fig. 2.6, that is,

$$E[v^{[1]}] = \int_0^1 x^N dx = \left[\frac{x^{N+1}}{N+1}\right]_0^1 = \frac{1}{N+1}$$

(e) *Variance.* Still assuming that valuations are uniformly distributed, find the variance of the first-order statistic.

- To find the variance of the first-order statistic, we need $Var(x) = E[x^2] - E[x]^2$. From part (d) we know $E[x]$, so we only need to calculate $E[x^2]$, as follows:

$$E[x^2] = \frac{1}{B(1,N)} \underbrace{\int_0^1 x^2(1-x)^{N-1} dx}_{B(3,N)}$$

$$= \frac{B(3,N)}{B(1,N)}$$

$$= \frac{\frac{2!(N-1)!}{(N+2)!}}{\frac{(N-1)!}{N!}}$$

$$= \frac{2}{(N+1)(N+2)}$$

- Therefore, the variance of the first-order statistic is

$$Var(x) = E[x^2] - E[x]^2$$

$$= \frac{2}{(N+1)(N+2)} - \left(\frac{1}{N+1}\right)^2$$

$$= \frac{2N+2-N-2}{(N+1)^2(N+2)}$$

$$= \frac{N}{(N+1)^2(N+2)}$$

Exercise #2.9: The Second-Order Statistic[B]

2.9 Consider the same setting as in Exercise 2.8. In this exercise, we focus on the second-highest element in valuation profile $(v_1, v_2, \ldots, v_N)$, which is often known as the "second-order statistic," and denoted as $v^{[2]}$.

(a) Find the cumulative distribution function of the second-order statistic, $F^{[2]}(v)$.

- A given valuation v lying weakly above the second-highest valuation, $v^{[2]}$, can occur in either (but not both) of these two events:
  - *First event.* The valuations of all N bidders are below v or, formally, $v_i \leq v$ for every bidder i. This event happens with probability $[F(v)]^N$, as shown in part (a) of Exercise 2.8.
  - *Second event.* The valuations of $N-1$ bidders are below v, $v_i \leq v$, but that of only one bidder j is above v, $v_j > v$. This event can occur in N different ways: (1) $v_1 > v$ for bidder 1 but $v_i \leq v$ for every bidder $i \neq 1$; (2) $v_2 > v$ for bidder 2 but $v_i \leq v$ for every bidder $i \neq 2$; and so on up to N, where $v_N > v$ for bidder N but $v_i \leq v$ for every bidder $i \neq N$. Each of these N cases happens with probability $[1 - F(v)][F(v)]^{N-1}$, where $[1 - F(v)]$ denotes the probability that $v_j > v$ for a given bidder j, while $[F(v)]^{N-1}$ represents the probability that $v_i \leq v$ for all other bidders $i \neq j$. Summing over N, we find that this event happens with probability

$$\sum_{i=1}^{N}[1-F(v)][F(v)]^{N-1} = N[1-F(v)][F(v)]^{N-1}$$

since probabilities are independent draws from the cumulative distribution function.

- Summarizing, the cumulative distribution function of the second-order statistic, $F^{[2]}(x)$, is

$$F^{[2]}(v) = \underbrace{[F(v)]^N}_{\text{First event}} + \underbrace{N[1-F(v)][F(v)]^{N-1}}_{\text{Second event}}$$

Rearranging, this expression simplifies to

$$F^{[2]}(v) = N[F(v)]^{N-1} - (N-1)[F(v)]^N$$

(b) Find the density function of the second-order statistic.

- Differentiating the expression of $F^{[2]}(v)$ that we found in part (a) with respect to v, yields

$$f^{[2]}(v) = N(N-1)F(v)^{N-2}f(v) - N(N-1)F(v)^{N-1}f(v)$$
$$= N(N-1)F(v)^{N-2}[1-F(v)]f(v)$$

(c) *Uniform distribution.* Assuming that every bidder's valuation is uniformly distributed, $F(v_i) = v_i$ and $v_i \in [0,1]$, find the cumulative distribution function and the density of the second-order statistic.

- Evaluating our result from part (a) at $F(v_i) = v_i$ and $v_i \in [0,1]$, yields a cumulative distribution function

$$F^{[2]}(v) = N[F(v)]^{N-1} - (N-1)[F(v)]^N$$
$$= Nv^{N-1} - (N-1)v^N$$
$$= v^{N-1}(N+v-Nv)$$

And evaluating our results from part (b) at a uniform distribution, we obtain

$$f^{[2]}(v) = N(N-1)F(v)^{N-2}[1-F(v)]f(v)$$
$$= N(N-1)v^{N-2}(1-v)$$

since $f(v) = 1$ when valuations are uniformly distributed. For illustration purposes, Fig. 2.7 depicts the cumulative distribution function of

Exercise #2.9: The Second-Order Statistic[B]

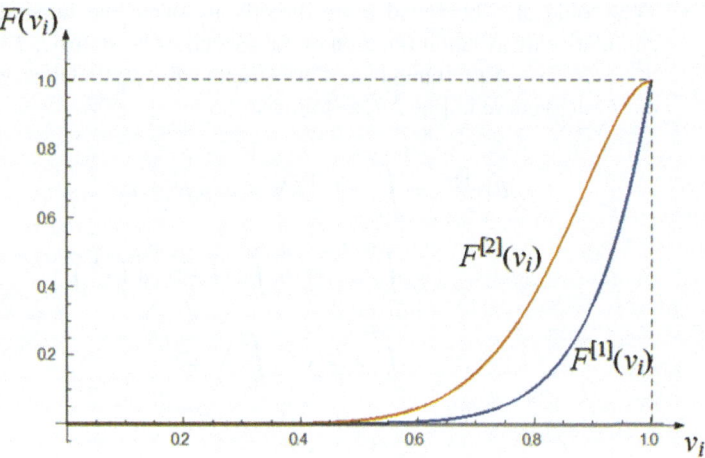

Fig. 2.7 First- and second-order statistics of the uniform distribution

the first- and second-order statistic of a uniform distribution, evaluated at $N = 10$ bidders. Intuitively, the cumulative probability that a given valuation v lies above the second-highest valuation is higher than the cumulative probability that v lies above the highest valuation.

(d) *Expected value.* Still assuming that valuations are uniformly distributed, find the expected value of the second-order statistic.

- The expected value of the second-order statistic, for any cumulative distribution $F(v_i)$ where $v_i \in [0, 1]$, is

$$E[v^{[2]}] = \frac{1}{B(2, N-1)} \int_0^1 x F(x) [1 - F(x)]^{N-2} f(x) dx$$

and if valuations are uniformly distributed, $F(v_i) = v_i$, this expected value simplifies to

$$E[v^{[2]}] = \frac{N!}{(N-2)!} \int_0^1 x^2 (1-x)^{N-2} dx$$

$$= -\frac{N(N-1)}{N-1} \int_0^1 x^2 d\left[(1-x)^{N-1}\right]$$

$$= -N \left[x^2 (1-x)^{N-1}\right]_0^1 + 2N \int_0^1 x(1-x)^{N-1} dx$$

$$= \frac{2}{N+1}$$

- This could also be found more directly by using the integral of the cumulative distribution function of the second-order statistic, $F^{[2]}(v) = v^{N-1}(N+v-Nv)$, found in part (c). Graphically, we find the area under the $F^{[2]}(v)$ curve in Fig. 2.7, as follows:

$$E[v^{[2]}] = \int_0^1 x^{N-1}(N+x-Nx)\,dx$$

$$= \int_0^1 x^N dx + \int_0^1 (1-x)\,d\left(x^N\right)$$

$$= \int_0^1 x^N dx + \int_0^1 x^N dx$$

$$= 2\left[\frac{x^{N+1}}{N+1}\right]_0^1$$

$$= \frac{2}{N+1}$$

(e) *Variance.* Still assuming that valuations are uniformly distributed, find the variance of the second-order statistic.

- To find the variance of the second-order statistic, we need $Var(x) = E[x^2] - E[x]^2$. We already found $E[x]$ in part (d), so we only need to calculate $E[x^2]$, as follows:

$$E[x^2] = \frac{1}{B(2, N-1)} \int_0^1 x^2 \left[x(1-x)^{N-2}\right] dx$$

$$= \frac{1}{B(2, N-1)} \underbrace{\int_0^1 x^3 (1-x)^{N-2}\,dx}_{B(4,N-1)}$$

$$= \frac{B(4, N-1)}{B(2, N-1)}$$

$$= \frac{\frac{3!(N-2)!}{(N+2)!}}{\frac{1!(N-2)!}{N!}}$$

$$= \frac{6}{(N+1)(N+2)}$$

- Therefore, the variance of the second-order statistic is

$$Var(x) = E[x^2] - E[x]^2$$

$$= \frac{6}{(N+1)(N+2)} - \left(\frac{2}{N+1}\right)^2$$

$$= \frac{6N+6-4N-8}{(N+1)^2(N+2)}$$

$$= \frac{2(N-1)}{(N+1)^2(N+2)}$$

Comparing this variance against that of the first-order statistic found in Exercise 2.8, $\frac{N}{(N+1)^2(N+2)}$, we find that

$$\frac{N}{(N+1)^2(N+2)} < \frac{2(N-1)}{(N+1)^2(N+2)}$$

simplifies to $N < 2(N-1)$, which holds for all $N > 2$, indicating that the second-order statistic is more dispersed than the first-order statistic.

Exercise #2.10: The kth-Order Statistic^C

2.10 Consider again the auction in Exercise 2.3. In this exercise, we extend the notion of first- and second-order statistics to the kth-highest element in a list of randomly drawn valuations, $(v_1, v_2, \ldots, v_N)$, which is known as the "kth-order statistic," and denoted as $v^{[k]}$.
(a) Find the cumulative distribution function of the kth-order statistic.

- A given valuation v being weakly above the kth-highest valuation, $v^{[k]}$, can occur in any of the following disjoint events:
  - *First event.* The valuations of all N bidders are below v or, formally, $v_i \leq v$ for every bidder i. This event happens with probability $[F(v)]^N$, as shown in part (a) of Exercise 2.8.
  - *Second event.* The valuations of $N-1$ bidders are below v, $v_i \leq v$, but that of only one bidder j is above v, $v_j > v$. This event occurs with probability $N[1-F(v)][F(v)]^{N-1}$, as shown in part (a) of Exercise 2.9.
  - *Third event.* The valuations of $N-2$ bidders are below v, $v_i \leq v$, but those of the other two bidders j and l are above v, $v_j > v$ and $v_l > v$. This event occurs with probability

$$\binom{N}{2}[1-F(v)]^2[F(v)]^{N-2} = \frac{N!}{2!(N-2)!}[1-F(v)]^2[F(v)]^{N-2}$$

since there are $\frac{N!}{2!(N-2)!}$ ways in which the valuations of two bidders (out of N) lie above v. For instance, in an auction with four bidders, it is easy to show that there are $\frac{4!}{2!(4-2)!} = 6$ ways in which we can have the valuations of two bidders above the third-highest valuation.

- ...
- *kth event.* The valuations of $N - (k-1) = N - k + 1$ bidders are below v, $v_i \leq v$, but those of the remaining $N - (N - k + 1) = k - 1$ bidders are above v, $v_j > v$. This event occurs with probability

$$\binom{N}{k-1} [1 - F(v)]^{k-1} [F(v)]^{N-k+1}$$

$$= \frac{N!}{(k-1)!(N-k+1)!} [1 - F(v)]^{k-1} [F(v)]^{N-k+1}$$

- Summarizing, the cumulative distribution function of the *k*th-order statistic is

$$F^{[k]}(v) = \sum_{j=k}^{N} \binom{N}{j} [1 - F(v)]^{N-j} [F(v)]^{j}$$

Differentiating $F^{[k]}(v)$ with respect to v, we obtain

$$\frac{\partial F^{[k]}(v)}{\partial v} = \sum_{j=k}^{N} \binom{N}{j} j [1 - F(v)]^{N-j} [F(v)]^{j-1} f(v)$$

$$- \sum_{j=k}^{N-1} \binom{N}{j} (N-j) [1 - F(v)]^{N-j-1} [F(v)]^{j} f(v)$$

Rearranging the first term of the above first-order condition, we obtain

$$\sum_{j=k}^{N} \binom{N}{j} j [1 - F(v)]^{N-j} [F(v)]^{j-1} f(v)$$

$$= N \sum_{j=k}^{N} \binom{N-1}{j-1} [1 - F(v)]^{N-j} [F(v)]^{j-1} f(v)$$

$$= Nf(v) \left[\binom{N-1}{k-1} [1 - F(v)]^{N-k} [F(v)]^{k-1} \right.$$

$$\left. + \sum_{j=k+1}^{N} \binom{N-1}{j-1} [1 - F(v)]^{N-j} [F(v)]^{j-1} \right]$$

Exercise #2.10: The kth-Order Statistic^C

Rearranging the second term of the above first-order condition, we obtain

$$\sum_{j=k}^{N-1} \binom{N}{j} (N-j)[1-F(v)]^{N-j-1} [F(v)]^j f(v)$$

$$= Nf(v) \left[\sum_{j=k}^{N-1} \binom{N-1}{j} [1-F(v)]^{N-j-1} [F(v)]^j \right]$$

$$= Nf(v) \left[\sum_{j=k+1}^{N} \binom{N-1}{j-1} [1-F(v)]^{N-j} [F(v)]^{j-1} \right]$$

Subtracting the first and second term, we obtain the density function of the kth-order statistic, as follows:

$$f^{[k]}(v) = N \binom{N-1}{k-1} [1-F(v)]^{N-k} [F(v)]^{k-1} f(v)$$

(b) *Uniform distribution.* Assuming that every bidder's valuation is uniformly distributed, $F(v_i) = v_i$ and $v_i \in [0, 1]$, find the cumulative distribution function and the density function of the kth-order statistic.

- Evaluating $F^{[k]}(v)$ found in part (a) at a uniform distribution, we find

$$F^{[k]}(v) = \sum_{j=k}^{N} \binom{N}{j} (1-v)^{N-j} v^j$$

- Evaluating $f^{[k]}(v)$ also found in part (a), at a uniform distribution, we obtain

$$f^{[k]}(v) = \frac{N!}{(k-1)!(N-k)!} (1-v)^{N-k} v^{k-1}$$

(c) *Expected value.* Still assuming that valuations are uniformly distributed, find the expected value of the kth-order statistic. How is it affected by an increase in k? Interpret.

- The expected value of the kth-order statistic for any distribution $F(v_i)$, where $v_i \in [0, 1]$, is

$$E[v^{[k]}] = \int_0^1 x f^{[k]}(x) dx$$

$$= \frac{N!}{(k-1)!(N-k)!} \int_0^1 x[1-F(x)]^{N-k} F(x)^{k-1} f(x) dx$$

When valuations are uniformly distributed, $F(v_i) = v_i$, this expected value simplifies to

$$E[v^{[k]}] = \frac{N!}{(k-1)!(N-k)!} \int_0^1 x(1-x)^{N-k} x^{k-1} dx$$

$$= \frac{N!}{(k-1)!(N-k)!} \int_0^1 (1-x)^{N-k} x^k dx$$

where $\int_0^1 [1-x]^{N-k} x^k dx$ is the expected value of Beta distribution, $x \sim B(k+1, N-k+1)$, which is

$$\frac{k!(N-k)!}{[(k+1)+(N-k+1)-1]!} = \frac{k!(N-k)!}{(N+1)!}$$

Therefore, the above expected value further simplifies to

$$E[v^{[k]}] = \frac{N!}{(k-1)!(N-k)!} \frac{k!(N-k)!}{(N+1)!}$$

$$= \frac{N!}{(N+1)!} \frac{k!}{(k-1)!}$$

$$= \frac{k}{N+1}$$

- Evaluating this result at $k = 1$, we find that the expected value of the first-order statistic is

$$E[v^{[1]}] = \frac{1}{N+1}$$

while that of the second-order statistic ($k = 2$) is

$$E[v^{[2]}] = \frac{2}{N+1}$$

both of which coincide with the expected values of the first- and second-order statistics found in Exercises 2.8 and 2.9, respectively.
- The expected value of the kth-order statistic is, of course, increasing in k. In other words, when we consider lower ranked values in a distribution, its probability increases, producing an overall increase in its expected value.

(d) *Variance.* Still assuming that valuations are uniformly distributed, find the variance of the kth-order statistic. How is it affected by an increase in k? Interpret.

- To find the variance of the kth-order statistic, where $x \sim B(k, N - k + 1)$, we need $Var(x) = E[x^2] - E[x]^2$. From part (c) of this exercise, we found $E[x] = \frac{k}{N+1}$, so we only need to calculate $E[x^2]$, as follows:

$$E[x^2] = \frac{1}{B(k, N-k+1)} \int_0^1 x^2 [(1-x)^{N-k} x^{k-1}] dx$$

$$= \frac{1}{B(k, N-k+1)} \underbrace{\int_0^1 (1-x)^{N-k} x^{k+1} dx}_{B(k+2, N-k+1)}$$

$$= \frac{B(k+2, N-k+1)}{B(k, N-k+1)}$$

$$= \frac{\frac{(k+1)!(N-k)!}{(N+2)!}}{\frac{(k-1)!(N-k)!}{N!}}$$

$$= \frac{k(k+1)}{(N+1)(N+2)}$$

- Therefore, the variance of the kth-order statistic is

$$Var(x) = E[x^2] - E[x]^2$$

$$= \frac{k(k+1)}{(N+1)(N+2)} - \left(\frac{k}{N+1}\right)^2$$

$$= \frac{k(N-k+1)}{(N+1)^2(N+2)}$$

This variance is increasing in k if

$$\frac{\partial Var(x)}{\partial k} = \frac{N - 2k + 1}{(N+1)^2(N+2)} > 0$$

which holds for all $N > 2k - 1$. In other words, for sufficiently high ranked statistic, where $k > \frac{N+1}{2}$, a lower order statistic is more dispersed than a higher order statistic.

Exercise #2.11: Bidding Behavior in the Dutch Auction[A]

2.11 Consider a descending-price auction, also known as Dutch auction, as they were often used in Holland to sell paintings and real estates during the seventeenth century. In this auction, the seller announces a relatively high price for the object, where no bidder is willing to buy, and subsequently decreases this price until one bidder shouts, agreeing to purchase the object at this price. This is usually known as an "open" auction, since bidders observe the bidding behavior of their rivals, as opposed to the first-price auction analyzed in previous exercises, which is referred to as a "closed" (or sealed bid) auction given that bidders cannot observe their rivals' bidding behavior until the auction is over and bids are revealed.

In this exercise, we seek to show that Dutch auctions are strategically equivalent to first-price auctions, thus producing the same equilibrium bids. For simplicity, consider a setting with $N \geq 2$ bidders, each of them privately observing his valuation v_i, drawn from a uniform distribution in $[0, 1]$. Our results can, nonetheless, be extended to Dutch auctions where every bidder i draws his valuation from a generic cumulative distribution function $F(v_i)$.

(a) Consider a bidder with valuation v_i who intends to raise his hand when the seller announces a bid $b = a \times v_i$, where parameter $a \in [0, 1]$ denotes his bid shading. Assume that the seller announces b but bidder i does not raise his hand, nor do any other bidders (e.g., bidder i waits for the announced price to decrease by one cent). What is bidder i's increase in expected benefit from waiting?

- The expected benefit of waiting one more second is the probability of winning the auction

$$\Pr\{win\} = \overbrace{\Pr\{b > b_1\} \times \ldots \times \Pr\{b > b_N\}}^{N-1 \text{ times}}$$

where, for any bidder $j \neq i$, we have that

$$\Pr\{b > b_j\} = \Pr\{a \times v_i > a \times v_j\}$$
$$= \Pr\{v_i > v_j\}$$
$$= \Pr\left\{\frac{b}{a} > v_j\right\}$$
$$= \frac{b}{a}$$

given that $b = a \times v_i$ entails $\frac{b}{a} = v_i$, and valuations are uniformly distributed. Therefore, the probability of winning by waiting one more second is

Exercise #2.11: Bidding Behavior in the Dutch Auction[A]

$$\Pr\{win\} = \frac{b}{a} \times \ldots \times \frac{b}{a}$$

$$= \left(\frac{b}{a}\right)^{N-1}$$

(b) What is bidder i's net benefit from waiting one more second?

- By waiting one more second, bidder i experiences an increase in expected utility because of marginally decreasing b. Specifically, if $EU(b|v_i) = \left(\frac{b}{a}\right)^{N-1}(v_i - b)$, then

$$\frac{\partial EU}{\partial b} = (N-1)\left(\frac{b}{a}\right)^{N-2}\frac{1}{a}(v_i - b) - \left(\frac{b}{a}\right)^{N-1}$$

where the first term represents the marginal cost of other bidders winning the object, while the second term denotes the marginal benefit of waiting one more second (which is equivalent to decreasing b), as identified in part (a).

(c) If bidder i raises his hand, it must be that the expected benefit and cost from waiting one more second cancel each other. Use your results from part (b) to find the equilibrium bidding function in the Dutch auction.

- If bidder i agrees with the announced price of the seller (raising his hand), it must be that his expected benefit and cost from waiting one more second are equal to each other, that is, $\frac{\partial EU}{\partial b} = 0$, yielding

$$\left(\frac{b}{a}\right)^{N-1} = (N-1)\left(\frac{b}{a}\right)^{N-2}\frac{1}{a}(v_i - b)$$

Rearranging, we obtain

$$b = (N-1)(v_i - b)$$

Solving for b, the equilibrium bidding function in the Dutch auction becomes

$$b(v_i) = \frac{N-1}{N}v_i$$

which, as expected, coincides with that in the first-price auction with $N \geq 2$ bidders and uniformly distributed valuations. Therefore, the expected revenue for the seller also coincides in the first-price auction and in the Dutch auction.

First-Price Auctions: Extensions 3

Keywords

Aggressive bids · Marginal gain · Marginal loss · Sequential first-price auction · Profit margin · Sales volume · Participation · Concave utility function · Mixed strategy bidding profile · Expected payment · Parametric example · L'Hopital's rule · Asymmetrically distributed valuations · Backward induction · Subgame · Subgame perfect equilibrium · First-order stochastic dominance · Unconstrained bidder · Profitable deviation

Introduction

This chapter applies the analysis of equilibrium bidding behavior in first-price auctions from Chap. 2 to answer different questions. In Exercise 3.1, we measure the seller's expected revenue in this auction format when bidders draw their valuations from a generic distribution function, and then evaluate this expected revenue in the case that bidders' valuations are uniformly distributed.

Exercises 3.2 and 3.3 also find the expected revenue in the first-price auction, but using the first-order statistic concept, which allows for a shorter proof.

In Exercise 3.4, we consider a first-price auction with risk-averse, rather than risk-neutral, bidders and examine how their bidding behavior is affected by risk aversion. We show that bidders submit more aggressive bids when they become more risk averse since, intuitively, the marginal loss of losing the auction offsets their marginal gain from obtaining the object at a lower price. Exercise 3.4 assumes, for simplicity, that bidders draw their valuations from a uniform distribution, and Exercise 3.5 extends our analysis to settings where valuations are drawn from a generic distribution function. Exercise 3.6 then explores whether the allocation of the object in a first-price auction with risk-averse bidders can be inefficient, so the individual winning the object is not necessarily who values it the most.

While all previous exercises assumed that bidders draw their valuations from the same distribution function, Exercise 3.7 allows for each bidder to draw his value from a different distribution. We then find their equilibrium bidding functions in this context and analyze which bidder submits a higher bid.

Exercise 3.8 extends our analysis of first-price auctions to a sequential setting, where a bidder submits his bid first (the leader) and, observing this bid, another bidder responds submitting his own (which we can understand as the follower). We show that the leader behaves as in the simultaneous version of this auction, but the follower responds offering a cent more than the leader's bid, if such a bid is lower than his own valuation. In this context, we also demonstrate that the sequential version of the first-price auction, unlike its simultaneous version, is inefficient, as it does not guarantee that the object in equilibrium goes to the individual with the highest valuation.

Exercises 3.9 and 3.10 then explore the role of reservation prices in first-price auctions, where bidders must submit bids equal to or above a minimum dollar amount for bids to be considered by the seller. For presentation purposes, Exercise 3.9 considers a single bidder, which helps us understand the seller's trade-off between extracting a larger surplus from the bidder and decreasing the pool of participating bidders, which is analogous to the monopolist's trade-off between profit margin and sales volume. In Exercise 3.10, we extend this analysis to a setting with $N \geq 2$ bidders, showing that, essentially, the seller faces a similar problem in both contexts. Exercise 3.11 studies the seller's decision to set entry fees, which all participating bidders must pay, and how his trade-off resembles that of setting a reservation price.

Exercise 3.12 examines a first-price auction with discrete, rather than continuous, valuations for the object. We first show that a pure-strategy bidding profile cannot be sustained in equilibrium, where bidders use an equilibrium bidding function with certainty. However, a mixed-strategy bidding profile can be supported. We analyze how the support of this bidding randomization is affected by the distribution and valuations of the bidders.

Exercise 3.13 studies how collusion among bidders can be sustained in the first-price auction, evaluates each bidder's expected utility from colluding, and shows that it is larger than what he earns by independently submitting his bid. Finally, Exercise 3.14 analyzes equilibrium bids in a first-price auction where bidders face budget constraints, based on Che and Gale (1998), and compares this bidding function to that in a second-price auction (as found in Exercise 1.12).

Exercise #3.1: Expected Revenue in the First-Price Auction—Direct Proof[B]

3.1 Consider again the first-price auction in Exercise 2.4.
    (a) Assuming that every bidder's valuation is distributed according to a generic cumulative distribution function $F(v_i)$, find the seller's expected revenue from the auction.

Exercise #3.1: Expected Revenue in the First-Price Auction—Direct Proof[B]

- For compactness, let us define $G(x) \equiv (F[x])^{N-1}$ to be the joint cumulative distribution function for $N-1$ bidders, where valuation x satisfies $x \in [0, 1]$. Then the optimal bidding function in the first-price auction (from Exercise 2.4) can be rewritten as

$$b_i(v_i) = v_i - \underbrace{\frac{\int_0^{v_i} F(x)^{N-1} dx}{F(v_i)^{N-1}}}_{\text{bid shading}}$$

$$= v_i - \frac{1}{G(v_i)} \int_0^{v_i} G(x)\, dx$$

We now apply integration by parts to further simplify this expression. First, recall the integration by parts formula:

$$\int_0^{v_i} u\, dv + \int_0^{v_i} v\, du = uv\Big|_0^{v_i},$$

or after rearranging,

$$\int_0^{v_i} u\, dv = uv\Big|_0^{v_i} - \int_0^{v_i} v\, du.$$

Second, let $u = x$ and $v = G(x)$, so that each of the three terms in the formula become

$$\int_0^{v_i} u\, dv = \int_0^{v_i} x\, dG(x),$$

$$uv\Big|_0^{v_i} - xG(x)\Big|_0^{v_i} = v_i G(v_i), \text{ and}$$

$$\int_0^{v_i} v\, du = \int_0^{v_i} G(x) dx.$$

Combining them, we obtain that

$$\int_0^{v_i} x\, dG(x) = v_i G(v_i) - \int_0^{v_i} G(x) dx.$$

Third, we solve for $\int_0^{v_i} G(x) dx$ since this is the last term of the bidding function which we seek to simplify. Solving for $\int_0^{v_i} G(x) dx$, we find

$$\int_0^{v_i} G(x) dx = v_i G(v_i) - \int_0^{v_i} x\, dG(x).$$

Finally, we insert this result in the last term of the bidding function, to obtain

$$b_i(v_i) = v_i - \frac{1}{G(v_i)} \left[v_i G(v_i) - \int_0^{v_i} x\, dG(x) \right],$$

which simplifies to

$$b_i(v_i) = v_i - v_i + \frac{1}{G(v_i)} \int_0^{v_i} x\, dG(x)$$

$$= \frac{1}{G(v_i)} \int_0^{v_i} x\, dG(x)$$

$$= \frac{1}{G(v_i)} \int_0^{v_i} x g(x)\, dx.$$

In summary, we can express the equilibrium bidding function in the first-price auction as

$$b_i(v_i) = \frac{1}{G(v_i)} \int_0^{v_i} x g(x)\, dx.$$

From an *ex-ante* point of view (before observing his own valuation for the object), bidder i's expected payment to the seller is given by the probability of winning the auction times the bid he pays for the object upon winning, that is,

$$m(v_i) = \Pr(win) \times b_i(v_i)$$

$$= (F[v_i])^{N-1} \times b_i(v_i)$$

$$= G(v_i) \times b_i(v_i)$$

$$= G(v_i) \times \frac{1}{G(v_i)} \int_0^{v_i} x g(x)\, dx$$

$$= \int_0^{v_i} x g(x)\, dx$$

where the second line indicates that bidder i wins the auction if his valuation v_i is higher than everyone else's, that is, $v_i \geq v_j$ for every bidder $j \neq i$. The probability of his valuation exceeding that of every other bidder is given by $(F[v_i])^{N-1}$, which we represented more compactly in the third line as $G(v_i) \equiv (F[v_i])^{N-1}$. In the fourth line, we just insert the equilibrium bidding function found above, $b_i(v_i)$, and in the last line we simplify the expression.

- Since the seller cannot observe bidders' values, he finds the expected payment from each bidder i, $E[\pi_i(v_i)]$, and then sums up for all N bidders, $\sum_{i=1}^{N} E[\pi_i(v_i)]$, which gives us the seller's revenue from the auction (this is, of course, understood from an *ex-ante* perspective since the seller does not observe bidders' valuations). We find the seller's revenue as follows:

Exercise #3.1: Expected Revenue in the First-Price Auction—Direct Proof[B]

$$R^1 = \sum_{i=1}^{N} E[\pi_i(v_i)] = N \int_0^1 \underbrace{\left[\int_0^{v_i} xg(x)\,dx\right]}_{\text{Expected payment, } \pi_i(v_i)} f(v_i)\,dv_i$$

(b) *Uniformly distributed valuations.* If every bidder's valuation is uniformly distributed, $F(v_i) = v_i$, where $v_i \in [0,1]$, what is the seller's expected revenue from this auction?

- When valuations are uniformly distributed, we obtain that $G(v_i) = v_i^{N-1}$, which implies that $g(v_i) = G'(v_i) = (N-1)v_i^{N-2}$ and that $f(v_i) = F'(v_i) = 1$. Therefore, the seller's expected revenue is

$$R^1 = N \int_0^1 \left[\int_0^{v_i} xg(x)\,dx\right] f(v_i)\,dv_i$$

$$= N \int_0^1 \left[\int_0^{v_i} \underbrace{(N-1)x^{N-1}}_{xg(x)}\,dx\right] \underbrace{1}_{f(v_i)}\,dv_i$$

$$= N(N-1) \int_0^1 \left[\frac{x^N}{N}\right]_0^{v_i} dv_i$$

$$= (N-1) \int_0^1 v_i^N \, dv_i$$

$$= (N-1) \left[\frac{v_i^{N+1}}{N+1}\right]_0^1$$

$$= (N-1)\left(\frac{1}{N+1} - 0\right)$$

$$= \frac{N-1}{N+1}.$$

- This revenue, $R^1 = \frac{N-1}{N+1}$, coincides with the expected revenue in the second-price auction when valuations are uniformly distributed, $R^2 = \frac{N-1}{N+1}$, found in part (d) of Exercise 1.7.

(c) Does the seller's expected revenue found in part (b) increase or decrease in the number of bidders? What is the seller's expected revenue when $N \to +\infty$?

- The expected revenue is increasing in the number of bidders, N, but at a decreasing rate, since

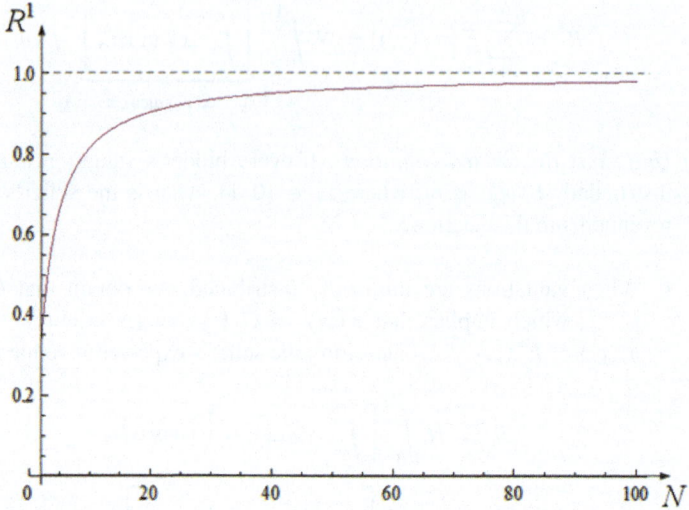

Fig. 3.1 Seller's expected revenue in the first-price auction, R^1, as a function of N

$$\frac{\partial R^1}{\partial N} = \frac{2}{(N+1)^2} > 0, \text{ and}$$

$$\frac{\partial^2 R^1}{\partial N^2} = -\frac{4}{(N+1)^3} < 0.$$

Intuitively, as more bidders participate in the auction, they submit more aggressive bids. As a result, the expected winning bid that the seller collects is higher. For example, when there are $N = 2$ bidders, the seller generates a revenue of $R^1 = \frac{2-1}{2+1} = \frac{1}{3}$ and increases to $R^1 = \frac{3-1}{3+1} = \frac{1}{2}$ when there are $N = 3$ bidders competing for the object. Figure 3.1 depicts the expected revenue in the first-price auction, R^1, as a function of the number of bidders, N.

- Furthermore, when the number of bidders grows to infinity, note that

$$\lim_{N \to \infty} R^1 = \lim_{N \to \infty} \left(1 - \frac{2}{N+1}\right) = 1$$

since the ratio inside the parenthesis converges to zero. Therefore, the seller earns a revenue of $1, which originates from the bid paid by the bidder with the highest valuation, becoming more likely as N grows.

(d) *Exponentially distributed valuations.* Consider now that individual valuations are drawn from an exponential distribution, $F(v_i) = 1 - \exp(-\lambda v_i)$, where $v_i \in [0, +\infty)$, and there are $N = 2$ bidders. Find the seller's

Exercise #3.1: Expected Revenue in the First-Price Auction—Direct Proof[B]

expected revenue in this context. How does expected revenue change with the parameter λ? Interpret your results.

- The seller's expected revenue, with $N = 2$ bidders, is

$$R^1 = 2 \int_0^\infty \left[\int_0^{v_i} x g(x) \, dx \right] f(v_i) \, dv_i$$

Since $G(v_i) = F(v_i)^{2-1} = F(v_i) = 1 - \exp(-\lambda v_i)$, we obtain that $g(v_i) = G'(v_i) = \lambda \exp(-\lambda v_i)$. In addition, $f(v_i) = F'(v_i) = \lambda \exp(-\lambda v_i)$, so the above expected revenue becomes

$$R^1 = 2 \int_0^\infty \left[\int_0^{v_i} \underbrace{x \lambda \exp(-\lambda x) dx}_{g(x)} \right] f(v_i) \, dv_i$$

$$= -2 \int_0^\infty \left[\int_0^{v_i} x \, d \exp(-\lambda x) \right] f(v_i) \, dv_i$$

Integrating by parts yields

$$R^1 = -2 \int_0^\infty \left\{ [x \exp(-\lambda x)]_0^{v_i} - \int_0^{v_i} \exp(-\lambda x) \, dx \right\} f(v_i) \, dv_i$$

$$= -2 \int_0^\infty \left\{ v_i \exp(-\lambda v_i) + \frac{1}{\lambda} [\exp(-\lambda x)]_0^{v_i} \right\} f(v_i) \, dv_i$$

$$= -2 \int_0^\infty \left\{ v_i \exp(-\lambda v_i) + \frac{1}{\lambda} [\exp(-\lambda v_i) - 1] \right\} \underbrace{\lambda \exp(-\lambda v_i) dv_i}_{f(v_i)}$$

$$= -2 \int_0^\infty \lambda v_i \exp(-2\lambda v_i) \, dv_i - 2 \int_0^\infty \exp(-2\lambda v_i) \, dv_i$$

$$+ 2 \int_0^\infty \exp(-\lambda v_i) \, dv_i$$

$$= \int_0^\infty v_i \, d \exp(-2\lambda v_i) - 2 \int_0^\infty \exp(-2\lambda v_i) \, dv_i$$

$$- \frac{2}{\lambda} \int_0^\infty d \exp(-\lambda v_i)$$

Solving each integral and rearranging, we obtain

$$R^1 = [v_i \exp(-2\lambda v_i)]_0^\infty - 3 \int_0^\infty \exp(-2\lambda v_i) \, dv_i - \frac{2}{\lambda} [\exp(-\lambda v_i)]_0^\infty.$$

Finally, applying the L'Hôpital rule, we find

$$R^1 = \lim_{v_i \to \infty} \frac{\frac{\partial v_i}{\partial v_i}}{\frac{\partial \exp(2\lambda v_i)}{\partial v_i}} + \frac{3}{2\lambda}\left[\exp(-2\lambda v_i)\right]_0^\infty + \frac{2}{\lambda}$$

$$= \lim_{v_i \to \infty} \frac{1}{2\lambda \exp(2\lambda v_i)} - \frac{3}{2\lambda} + \frac{2}{\lambda}$$

$$= \frac{1}{2\lambda}.$$

- This revenue, $R^1 = \frac{1}{2\lambda}$, coincides with the expected revenue in the second-price auction when valuations are exponentially distributed, $R^2 = \frac{1}{2\lambda}$, found in part (e) of Exercise 1.7. The same comparative statics as in that exercise apply: an increase in parameter λ implies that the exponential distribution assigns a larger probability weight on low valuations producing a decrease in the seller's expected revenue.

(e) *Other distribution forms.* Consider the following distribution function:

$$F(v_i) = (1+\alpha)v_i - \alpha v_i^2$$

where $v_i \in [0,1]$, and parameter α satisfies $\alpha \in [-1,1]$. When $\alpha = 0$, this function collapses to the uniform distribution, $F(v_i) = v_i$; when $\alpha > 0$, it becomes concave, thus putting more probability weight on low valuations; and when $\alpha < 0$, it is convex, assigning more probability weight on high valuations. Find the seller's expected revenue in the setting of $N = 2$ bidders, and compare it to the seller's revenue under second-price auction. How does it change in α? Interpret your results.

- Since $F(v_i) = (1+\alpha)v_i - \alpha v_i^2$, its associated density function is $f(v_i) = F'(v_i) = 1 + \alpha - 2\alpha v_i$. As we only have $N = 2$ bidders in this context, $g(v_i) = f(v_i)$, we obtain that

$$R^1 = 2\int_0^1 \left[\int_0^{v_i} x g(x)\, dx\right] f(v_i)\, dv_i$$

$$= 2\int_0^1 \left[\int_0^{v_i} \underbrace{x(1+\alpha - 2\alpha x)}_{g(x)} dx\right] f(v_i)\, dv_i$$

$$= 2\int_0^1 \left[\frac{(1+\alpha)v_i^2}{2} - \frac{2\alpha v_i^3}{3}\right] \underbrace{(1+\alpha - 2\alpha v_i)}_{f(v_i)}\, dv_i$$

$$= \frac{1}{3}\int_0^1 \left[3(1+\alpha)^2 v_i^2 - 10\alpha(1+\alpha) v_i^3 + 8\alpha^2 v_i^4\right] dv_i$$

$$= \frac{1}{3}\left[(1+\alpha)^2 - \frac{5\alpha(1+\alpha)}{2} + \frac{8\alpha^2}{5}\right]$$

$$= \frac{\alpha^2 - 5\alpha + 10}{30}.$$

This expected revenue coincides with the seller's expected revenue under second-price auction, R^2, that we found in part (f) of Exercise 1.7. (See our discussion in that exercise for more details about the comparative statics of the expected revenue with respect to parameter α and its interpretation.)

Exercise #3.2: Expected Revenue in the First-Price Auction—Proof Using the First-Order Statistic[C]

3.2 Consider again the first-price auction in Exercise 2.4.
  (a) Assuming that every bidder's valuation is distributed according to a generic cumulative distribution function $F(v_i)$, find the seller's expected revenue from the auction. Use the concept of first-order statistic of valuation profile $v \equiv (v_1, v_2, \ldots, v_N)$. To understand this point, first note that the seller's expected revenue coincides with the expected bid of the bidder who submits the highest bid, which we denote as $E[b_N^1]$, where b_N^1 represents the highest bid among N bidders.

  - To find the highest bid, b_N^1, we need to calculate the first-order statistic of N valuation draws, that is, the probability that N draws are lower than or equal to x. This probability can be computed as follows:

  $$F^{[1]}(x) = \Pr\{\max\{b_1, b_2, \ldots, b_N\} \leq x\}$$
  $$= \Pr\{b_1 \leq x\} \times \ldots \Pr\{b_N \leq x\}$$
  $$= [F(x)]^N$$

  since valuations are independently drawn. Therefore, we can find the density function of the first-order statistic by differentiating with respect to x, yielding

  $$f^{[1]}(x) = N[F(x)]^{N-1} f(x).$$

  Finally, the expected revenue can be written as

  $$E[b_N^1] = \int_0^1 x f^{[1]}(x) dx.$$

(b) *Uniform distribution.* If every bidder's valuation is uniformly distributed, $F(v_i) = v_i$, what is the seller's expected revenue from this auction?

- Since valuations are uniformly drawn from [0, 1], the bid of each bidder i, $b(v_i) = \frac{N-1}{N} v_i$, is uniformly distributed in $\left[0, \frac{N-1}{N}\right]$, where $\frac{N-1}{N}$ is the bid of the individual with the highest valuation, $v_i = 1$. As a consequence,

$$F(b_i) = \frac{b_i}{\frac{N-1}{N} - 0} = \frac{N}{N-1} b_i,$$

implying that the cumulative distribution of the first-order statistic is

$$F^{[1]}(x) = [F(x)]^N$$
$$= \left(\frac{Nx}{N-1}\right)^N \quad \text{for all } 0 \leq x \leq \frac{N-1}{N}$$

We can now obtain the density function of the first-order statistic by differentiating with respect to x, yielding

$$f^{[1]}(x) = N \frac{N}{N-1} \left(\frac{Nx}{N-1}\right)^{N-1}$$
$$= \frac{N}{x} \left(\frac{Nx}{N-1}\right)^N.$$

We can finally insert this density function in the seller's expected revenue, R^1, as follows:

$$R^1 = \int_0^{\frac{N-1}{N}} x f^{[1]}(x) dx$$

$$= \int_0^{\frac{N-1}{N}} x \overbrace{\left[\frac{N}{x} \left(\frac{Nx}{N-1}\right)^N\right]}^{f^{[1]}(x)} dx$$

$$= \int_0^{\frac{N-1}{N}} N \left(\frac{Nx}{N-1}\right)^N dx$$

$$= \frac{N}{N+1} \left(\frac{N}{N-1}\right)^N x^{N+1} \Big|_0^{\frac{N-1}{N}}$$

$$= \frac{N-1}{N+1}$$

which, of course, coincides with the expected revenue found in Exercise 3.1 using a direct proof.

Exercise #3.3: Expected Payment in the First-Price Auction[B]

3.3 Consider again the first-price auction in Exercise 2.4, and, for simplicity, assume that every bidder's valuation is uniformly distributed in [0, 1].
   (a) Find bidder i's expected payment when winning the auction (i.e., the price he pays if winning). As in Exercise 2.8, use the concept of first-order statistic of valuation profile $v \equiv (v_1, v_2, \ldots, v_N)$.

   - Bidder i wins when his bid satisfies $b_i(v_i) > b_j(v_j)$ for every bidder $j \neq i$. Since $b_i(v_i) = \frac{N-1}{N} v_i$, bidder i wins the auction if

   $$\frac{N-1}{N} v_i > \frac{N-1}{N} v_j \text{ for every bidder } j \neq i$$

   or, after rearranging, when $v_i > v_j$ for every bidder $j \neq i$. In other words, bidder i wins the auction when his valuation for the object is higher than that of all other $N - 1$ bidders. This occurs, in particular, when

   $$\underbrace{\Pr\{v_1 \leq v_i\} \times \cdots \times \Pr\{v_N \leq v_i\}}_{N-1 \text{ times}} = F(v_i)^{N-1},$$

   or, more compactly, $G(v_i) \equiv F(v_i)^{N-1}$.
   - Therefore, the expected payment from bidder i is just given by the expected value of v_i, conditional on him winning the auction (which occurs with probability $G(v_i)$); that is,

   $$m(v_i) = \int_0^{v_i} y g(y) dy.$$

   Since $G(v_i) \equiv F(v_i)^{N-1}$ and valuations are uniformly distributed, we find that $F(v_i)^{N-1} = v_i^{N-1}$, and $g(v_i) = (N-1)v_i^{N-2}$, so that the expected payment simplifies to

   $$m(v_i) = \int_0^{v_i} y \underbrace{\left[(N-1)y^{N-2}\right]}_{g(y)} dy$$

   $$= (N-1) \int_0^{v_i} y^{N-1} dy$$

$$= (N-1)\left[\frac{y^N}{N}\right]_0^{v_i}$$

$$= \frac{N-1}{N}v_i^N.$$

(b) Use bidder i's expected payment to find the seller's expected revenue from the auction. Confirm that your results coincide with those in Exercise 2.8.

- The seller's expected revenue is just N times this expected payment, taking expectation over all possible realizations of valuation v_i (which bidder i can observe when computing his expected payment if winning, but the seller cannot observe when calculating the expected revenue from the auction). That is,

$$R^1 = N \int_0^1 m(x)dx$$

$$= N \int_0^1 \underbrace{\frac{N-1}{N}x^N}_{m_i(x)} dx$$

$$= (N-1)\int_0^1 x^N dx$$

$$= (N-1)\left[\frac{x^{N+1}}{N+1}\right]_0^1$$

$$= \frac{N-1}{N+1}.$$

This expression coincides, of course, with the expected revenue found in Exercises 3.1 and 3.2.

Exercise #3.4: First-Price Auction with Risk-Averse Bidders—An Introduction[A]

3.4 Consider a first-price auction where every bidder's valuation is drawn from a uniform distribution, $F(v_i) = v_i$, and $v_i \in [0, 1]$. Every bidder i is risk averse and exhibits a constant relative risk-aversion (CRRA) utility function $u(x) = x^\alpha$, where x is the bidder's income, and $0 < \alpha < 1$. The Arrow–Pratt coefficient of relative risk aversion is

$$r_R = -x\frac{u''(x)}{u'(x)} = -x\frac{\alpha(\alpha-1)x^{\alpha-2}}{\alpha x^{\alpha-1}} = 1 - \alpha.$$

Exercise #3.4: First-Price Auction with Risk-Averse Bidders—An Introduction[A]

Graphically, when bidder's utility function becomes more concave, α decreases (or $1 - \alpha$ increases), which happens when the bidder is more risk averse. For a more detailed presentation of risk aversion, see (Munoz-Garcia 2017, chapter 5).

(a) *Two bidders.* For presentation purposes, consider first a setting with only two bidders. Write every bidder i's expected utility maximization problem. [*Hint*: Assume that every bidder uses a symmetric bidding strategy $b_i = a \times v_i$, where $a \in (0, 1)$ denotes bid shading.]

- Every bidder i's expected utility maximization problem is

$$\max_{b_i \geq 0} \Pr(win) \times (v_i - b_i)^\alpha.$$

If every bidder uses bidding strategy $b_i = a \times v_i$, then $v_i = \frac{b_i}{a}$. Therefore, the probability of winning is $\Pr\{v_j < v_i\} = \frac{b_i}{a}$ because valuations are uniformly distributed in the interval $[0, 1]$. Therefore, the above maximization problem becomes

$$\max_{b_i \geq 0} \frac{b_i}{a} \times (v_i - b_i)^\alpha.$$

(b) Take first-order conditions to find bidder i's bidding function.

- Differentiating the above expected utility maximization problem with respect to b_i yields

$$\frac{1}{a}(v_i - b_i)^\alpha - \frac{b_i}{a}\alpha(v_i - b_i)^{\alpha-1} = 0.$$

Rearranging the above expression, we obtain

$$v_i - b_i = \alpha b_i$$

Solving for b_i, we find bidder i's equilibrium bidding function

$$b_i(v_i) = \frac{1}{1+\alpha} v_i$$

which coincides with that of a first-price auction with two risk-neutral bidders when evaluated at $\alpha = 1$, that is, $b_i(v_i) = \frac{1}{2} v_i$.

(c) How is the equilibrium bidding function you found in part (b) affected by parameter α? Interpret.

- When the bidder becomes more risk averse (α decreases), he submits more aggressive bids because

$$\frac{\partial b_i(v_i)}{\partial \alpha} = -\frac{1}{(1+\alpha)^2} v_i < 0.$$

That is, α and $b_i(v_i)$ move in different directions. Figure 3.2 evaluates the equilibrium bidding function $b_i(v_i)$ at $N = 2$ and then considers different values of the parameter α to illustrate this result. Graphically, the bidding function rotates upward, indicating that bidders submit more aggressive bids. This occurs because the risk-averse bidder seeks to minimize the probability of losing the auction. To understand this point, consider that bidder i reduces his bid from b_i to $b_i - \varepsilon$:

- If he wins the auction, he obtains an additional profit of ε, since he has to pay a lower price for the object.
- However, by lowering his bid, he increases the probability of losing the auction. Importantly, for a risk-averse bidder, the positive effect of getting the object at a cheaper price is offset by the negative effect of increasing the probability of losing the auction.

In other words, for a risk-averse individual, the disutility he suffers from the downside of a lottery is larger than the utility he experiences from the upside of a lottery. Overall, the risk-averse bidder does not have incentives to reduce his bid, but rather to increase it, relative to a risk-neutral bidder. As a result, the more risk averse the bidder becomes, the smaller his bid shading is. When $\alpha = 0$, the bidder is infinitely risk averse so that he does not shade his bid, that is, $b(v_i) = v_i$, and the bidding function coincides with the 45°-line as depicted in Fig. 3.2.

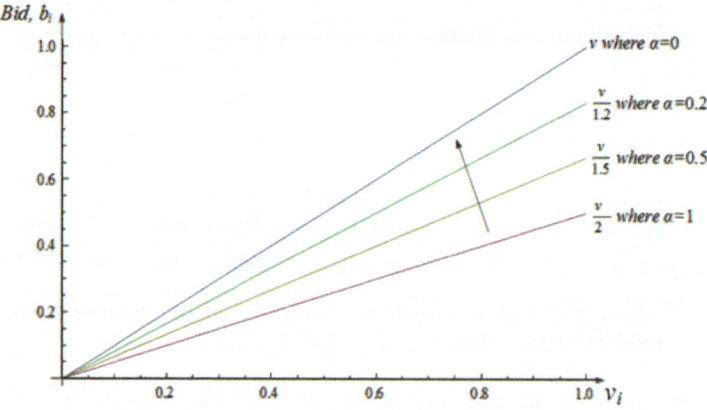

Fig. 3.2 Equilibrium bidding function—changes in α

Exercise #3.4: First-Price Auction with Risk-Averse Bidders—An Introduction[A]

(d) $N \geq 2$ *bidders.* Extend the analysis of parts (a)–(c) to a setting with $N \geq 2$ bidders.

- Bidder i's expected utility maximization problem in this setting becomes

$$\max_{b_i \geq 0} \left(\frac{b_i}{a}\right)^{N-1} \times (v_i - b_i)^\alpha,$$

where $\left(\frac{b_i}{a}\right)^{N-1}$ denotes the probability of winning (i.e., the probability that bidder i's valuation, v_i, exceeds that of his $N-1$ rivals).

- Differentiating the above expression with respect to b_i yields

$$\alpha^{-(N-1)} \left[(N-1) b_i^{N-2} (v_i - b_i)^\alpha - \alpha b_i^{N-1} (v_i - b_i)^{\alpha-1}\right] = 0,$$

which is rearranged to yield

$$\frac{b_i^{N-2} (v_i - b_i)^{\alpha-1}}{\alpha^{N-1}} [(N-1)(v_i - b_i) - \alpha b_i] = 0.$$

Solving for b_i, we find bidder i's equilibrium bidding function

$$b_i(v_i) = \frac{N-1}{N-1+\alpha} v_i$$

which coincides with that of a first-price auction with risk-neutral bidders when evaluated at $\alpha = 1$, that is, $b_i(v_i) = \frac{N-1}{N} v_i$.

- *Comparative statics.* Differentiating this bidding function with respect to α, we find that

$$\frac{\partial b_i(v_i)}{\partial \alpha} = -\frac{N-1}{(N-1+\alpha)^2} v_i < 0$$

thus confirming our above results for the case of two bidders, where bidders submit more (less) aggressive bids when they are more (less) risk averse.

Differentiating the bidding function with respect to N, we obtain

$$\frac{\partial b_i(v_i)}{\partial N} = \frac{\alpha}{(N-1+\alpha)^2} v_i > 0.$$

As depicted in Fig. 3.3, this result indicates that bidders submit more aggressive bids as they face more competitors bidding for the object. (For simplicity, the figure considers $\alpha = 1/2$ and evaluates the equilibrium

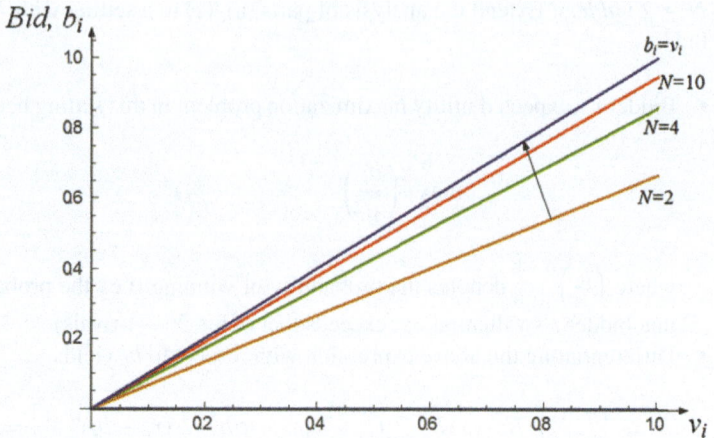

Fig. 3.3 Equilibrium bidding function—changes in N

bidding function at $N=2$, then at $N=4$, and also at $N=10$ bidders.) In particular, when the number of bidders approaches infinity, note that

$$\lim_{N\to\infty} b_i(v_i) = \lim_{N\to\infty} \left(1 - \frac{\alpha}{N-1+\alpha}\right) v_i = v_i$$

since the ratio inside the parenthesis converges to zero. Therefore, risk-averse bidders do not shade their bids when there are many bidders competing for the object. This is because when shading their bids by a small amount, they will lose the object to another bidder who does not shade his bid.

Exercise #3.5: First-Price Auction with Risk-Averse Bidders—General Setting[B]

3.5 Consider a first-price auction where every bidder i's valuation is drawn from a cumulative distribution function, $F(v_i)$, with positive density in all its support, i.e., $f(v_i) > 0$ for all $v_i \in [0,1]$. Every bidder i's utility function is $u : \mathbb{R}_+ \to \mathbb{R}$, where $u(0) = 0$, $u' > 0$, and $u'' < 0$, thus indicating that the bidder is risk averse.
(a) Write every bidder i's expected utility maximization problem.

- We can write bidder i's expected utility maximization problem as follows:

Exercise #3.5: First-Price Auction with Risk-Averse Bidders—General Setting[B]

$$\max_{b_i \geq 0} \Pr(win) \times u(v_i - b_i)$$

which denotes the probability of winning the object times bidder i's net payoff from winning, $u(v_i - b_i)$.

Alternatively, we can use the so-called "indirect approach" to write the above expected utility maximization problem as follows:

$$\max_{z \geq 0} G(z) \times u(v_i - b(z))$$

where z denotes the valuation that bidder i considers, which, for generality, can coincide with his true valuation v_i or not at this point. Inserting this valuation z into the bidding function $b_i : [0, 1] \to \mathbb{R}_+$, he submits bid $b(z)$ and wins the auction with probability $G(z) \equiv F^{N-1}(z)$.

(b) Take first-order conditions to characterize the implicit function that describes bidder i's bidding function.

- Taking first-order conditions with respect to z, we obtain

$$g(z)u(v_i - b(z)) - G(z)u'(v_i - b(z))b'(z) = 0$$

In a symmetric equilibrium, it must be optimal to choose a valuation $z = v_i$, so bidder i does not have incentives to bid according to a valuation z different from his real valuation for the object, v_i. Inserting $z = v_i$ in the above expression yields

$$g(v_i)u(v_i - b(v_i)) - G(v_i)u'(v_i - b(v_i))b'(v_i) = 0$$

which can be rearranged as

$$b'_{RA}(v_i) = \frac{u(v_i - b(v_i))}{u'(v_i - b(v_i))} \frac{g(v_i)}{G(v_i)}$$

where the subscript RA denotes "risk-averse" bidder.

(c) Evaluate your results in a setting where all bidders are risk neutral, so that $u(x) = x$, yielding $u'(x) = 1$.

- Since $u(x) = x$ and $u'(x) = 1$, the above result reduces to

$$b'_{RN}(v_i) = (v_i - b(v_i)) \frac{g(v_i)}{G(v_i)}$$

where the subscript RN denotes "risk-neutral" bidder.

(d) Compare your results in parts (b) and (c). What can you conclude about the bidding behavior when bidders are risk averse? Are the bidders bidding more or less aggressive than when they are risk neutral? Interpret your results.

- Comparing the two expressions, we obtain

$$b'_{RA}(v_i) = \frac{u(v_i - b(v_i))}{u'(v_i - b(v_i))} \frac{g(v_i)}{G(v_i)} > (v_i - b(v_i)) \frac{g(v_i)}{G(v_i)} = b'_{RN}(v_i) \tag{3.1}$$

Since $u(0) = 0$ and $u''(x) < 0$, we obtain that $\frac{u(x)-u(0)}{x-0} > u'(x)$ for all x. Graphically, the slope of the utility function at x is flatter than the ray connecting the origin and x. Setting $x = v_i - b(v_i)$ for $\frac{u(x)}{u'(x)} > x$, we obtain

$$\frac{u(v_i - b(v_i))}{u'(v_i - b(v_i))} > v_i - b(v_i).$$

Therefore, a given increase in valuation v_i increases more significantly the bid that a risk-averse bidder submits, $b'_{RA}(v_i)$, than that of a risk-neutral bidder, $b'_{RN}(v_i)$.

- Since, in addition, at a valuation $v_i = 0$, both bids are zero, $b_{RN}(0) = b_{RA}(0) = 0$, we must have that

$$b_{RA}(v_i) > b_{RN}(v_i)$$

for all $0 < v_i \leq 1$. Intuitively, bidders submit more aggressive bids when they are risk averse than risk neutral, as shown in Exercise 3.4.

(e) *Parametric example.* Evaluate your results in a setting where all bidders are risk averse, and their utility function is $u(x) = x^\alpha$, where $0 < \alpha < 1$.

- In this setting, $u'(x) = \alpha x^{\alpha-1}$, implying that the bid function we found in part (b) becomes

$$b'_{RA}(v_i) = \frac{u(v_i - b(v_i))}{u'(v_i - b(v_i))} \frac{g(v_i)}{G(v_i)}$$

$$= \frac{[v_i - b(v_i)]^\alpha}{\alpha [v_i - b(v_i)]^{\alpha-1}} \frac{g(v_i)}{G(v_i)}$$

$$= \frac{v_i - b(v_i)}{\alpha} \frac{g(v_i)}{G(v_i)}$$

By definition, $G(v_i) = F(v_i)^{N-1}$, implying that

Exercise #3.5: First-Price Auction with Risk-Averse Bidders—General Setting[B]

$$g(v_i) = (N-1)[F(v_i)]^{N-2} f(v_i).$$

To solve the above differential equation, it is convenient to define $G_\alpha \equiv G^{\frac{1}{\alpha}}$. Replacing v_i with x,

$$b'(x)G(x) + \frac{1}{\alpha}b(x)g(x) = \frac{1}{\alpha}xg(x)$$

Let $G(x)^{\frac{1}{\alpha}-1}$ be the integrating factor, and we have

$$b'(x)G(x) \times (G(x)^{\frac{1}{\alpha}-1}) + \frac{1}{\alpha}b(x)g(x) \times (G(x)^{\frac{1}{\alpha}-1}) = \frac{1}{\alpha}xg(x) \times (G(x)^{\frac{1}{\alpha}-1})$$

or

$$b'(x)G(x)^{\frac{1}{\alpha}} + b(x)\underbrace{\frac{1}{\alpha}(G(x)^{\frac{1}{\alpha}-1})g(x)}_{=(G(x)^{\frac{1}{\alpha}})'} = x \times \underbrace{\frac{1}{\alpha}(G(x)^{\frac{1}{\alpha}-1})g(x)}_{=(G(x)^{\frac{1}{\alpha}})'}$$

Applying integration by parts, we obtain $[b(x)G_\alpha(x)]' = x[G_\alpha(x)]'$ or

$$\int \frac{d[b(x)G_\alpha(x)]}{dx} dx = \int x[G_\alpha(x)]' dx$$

which yields

$$b(x)G_\alpha(x) = \int_0^x y[G_\alpha(y)]' dy$$

Rearranging, we obtain the general bidding function under risk aversion.

$$b(x) = \frac{1}{G_\alpha(x)} \int_0^x y[G_\alpha(y)]' dy$$

(f) *Uniformly distributed valuations.* In this parametric example, find the bid function assuming that valuations are distributed according to $F(x) = x$. How does the equilibrium bid change with the number of bidders N and parameter of risk aversion α? Interpret your results.

- Since $F(x) = x$, its density function is $f(x) = 1$ for all x. Therefore, the cumulative distribution function of the highest of $N-1$ values $G_\alpha(x)$ becomes $G_\alpha(x) = x^{\frac{N-1}{\alpha}}$. (Note that $N-1 > \alpha$ since $N \geq 2$ by definition and parameter α satisfies $0 < \alpha < 1$ by assumption.) Plugging this into the above bidding function, we obtain

$$b(x) = \frac{1}{x^{\frac{N-1}{\alpha}}} \int_0^x y[G_\alpha(y)]' \, dy$$

Integrating by parts on the right-hand side[1] yields

$$b(x) = \frac{1}{x^{\frac{N-1}{\alpha}}} \left[x \times x^{\frac{N-1}{\alpha}} - \int_0^x G_\alpha(y) dy \right]$$

$$= x^{-\frac{N-1}{\alpha}} \left[x^{\frac{N-1}{\alpha}+1} - \int_0^x y^{\frac{N-1}{\alpha}} dy \right]$$

$$= x - x^{-\frac{N-1}{\alpha}} \times x^{\frac{N-1}{\alpha}+1} \frac{\alpha}{N-1+\alpha}$$

$$= \frac{N-1}{N-1+\alpha} x$$

which coincides with the bidding function found in Exercise 3.4, part (d).

- As a consequence, comparative statics of the bidding function $b(x)$ also coincide with those in Exercise 3.4. In particular, $b(x)$ is increasing in the number of bidders, N, but decreasing in parameter α, implying that bidders submit less aggressive bids when they become less risk averse.

Exercise #3.6: Efficiency with Risk Aversion[A]

3.6 Consider a situation where bidders with heterogeneous attitudes toward risk compete in a first-price auction. Provide an example of how these bidders can lead to an inefficient allocation of the object.

- Suppose two bidders, A and B, were bidding for an object. Bidder A has a slightly higher valuation than B, but the former is risk loving as opposed to the latter being risk averse. As such, bidder A shades his bid by more than 50%, while B shades his bid by less than 50%. Since bidder A shades his bid by a higher proportion than B does, it is possible that B could outbid A even though bidder A has a higher valuation of the object than B does. This would lead to an inefficient allocation of the object.

---

[1] Recall the integration by parts formula: $\int uv' = uv - \int u'v$. Let $u = x$ and $v' = G_\alpha(x)'$, then $u' = 1$ and $v = G_\alpha(x) = x^{\frac{N-1}{\alpha}}$.

Exercise #3.7: First-Price Auction with Asymmetrically Distributed Valuations[B]

3.7 Previous exercises generally assume that all bidders independently draw their valuation from a *common* distribution, $F(v_i)$. In this exercise, we analyze how our equilibrium results are affected by relaxing this assumption. Consider a first-price auction with two risk-neutral bidders, i and j, independently drawing their valuations for the object from the following cumulative distribution functions: $F_i(v_i) = v_i^\alpha$ and $F_j(v_j) = v_j^\gamma$, respectively, where $\alpha \neq \gamma > 0$. For simplicity, assume that $v_i, v_j \in [0, 1]$.

(a) Find the equilibrium bidding function for bidders i and j.

- *Writing expected utility.* We can write bidder i's expected utility maximization problem as follows:

$$\max_{b_i \geq 0} \Pr(win) \times (v_i - b_i),$$

which denotes the probability of winning the object times bidder i's net payoff from winning, $v_i - b_i$, because he values the object at v_i and pays his bid b_i for it.

- *Finding the probability of winning.* At this point, we write the probability of winning as follows:

$$\Pr(win) = \Pr\{b_i > b_j\}$$
$$= \Pr\{b_i > b_j(v_j)\}$$

and inverting by $b_j^{-1}(\cdot)$ yields

$$\Pr\{b_j^{-1}(b_i) > b_j^{-1}(b_j(v_j))\} = \Pr\{b_j^{-1}(b_i) > v_j\}$$
$$= \left(b_j^{-1}(b_i)\right)^\gamma$$

where the first equality uses $b_j^{-1}(b_j(v_j)) = v_j$, see last term inside the probability expression. The last equality considers that bidder j's valuation is distributed according to $F_j(v_j) = v_j^\gamma$.

Therefore, the above expected utility maximization problem can be rewritten as

$$\max_{b_i \geq 0} \left(b_j^{-1}(b_i)\right)^\gamma \times (v_i - b_i).$$

- *First-order condition.* Differentiating with respect to b_i yields

$$-\left(b_j^{-1}(b_i)\right)^\gamma + \gamma(v_i - b_i)\left(b_j^{-1}(b_i)\right)^{\gamma-1}\frac{\partial b_j^{-1}(b_i)}{\partial b_i} = 0.$$

Because $b_j(v_i) = b_i$, we can write that $b_j^{-1}(b_i) = v_i$. Therefore, $\frac{\partial b_j^{-1}(b_i)}{\partial b_i} = \frac{1}{b'\left(b_j^{-1}(b_i)\right)} = \frac{1}{b'(v_i)}$, implying that the above first-order condition simplifies to

$$-v_i^\gamma + \gamma(v_i - b_i)v_i^{\gamma-1}\frac{1}{b'(v_i)} = 0$$

or

$$\gamma v_i^\gamma = \gamma v_i^{\gamma-1} b_i + v_i^\gamma b'(v_i)$$

The right side is $\frac{\partial[v_i^\gamma b_i(v_i)]}{\partial v_i}$, which helps us rewrite this expression as

$$\gamma v_i^\gamma = \frac{\partial\left[v_i^\gamma b_i(v_i)\right]}{\partial v_i}.$$

Integrating both sides yields

$$\int_0^{v_i} \gamma x^\gamma dx = v_i^\gamma b_i(v_i)$$

and solving for $b_i(v_i)$, we obtain the equilibrium bidding function, as follows:

$$b_i(v_i) = \frac{1}{v_i^\gamma}\int_0^{v_i} \gamma x^\gamma dx.$$

Solving the integral, we can find a more compact expression for this bidding function, that is,

$$\begin{aligned} b_i(v_i) &= \frac{1}{v_i^\gamma}\int_0^{v_i} \gamma x^\gamma dx \\ &= \frac{1}{v_i^\gamma}\left[\frac{\gamma}{1+\gamma}x^{\gamma+1}\right]_0^{v_i} \\ &= \frac{1}{v_i^\gamma}\frac{\gamma}{1+\gamma}v_i^{\gamma+1} \\ &= \frac{\gamma}{1+\gamma}v_i \end{aligned}$$

where, intuitively, the term $0 < \frac{\gamma}{1+\gamma} < 1$ captures the extent of bid shading. Operating similarly for bidder j, we obtain that his equilibrium bidding function is

$$b_j(v_j) = \frac{\alpha}{1+\alpha} v_j.$$

- Finally, comparing $b_i(v_i)$ and $b_j(v_j)$, we claim that bidder i wins the auction if $\frac{\gamma}{1+\gamma} v_i > \frac{\alpha}{1+\alpha} v_j$, which holds if $v_i > \frac{\alpha(1+\gamma)}{\gamma(1+\alpha)} v_j$. Intuitively, this occurs if bidder i has a sufficiently higher valuation than bidder j. If bidders' valuations coincide, $v_i = v_j$, but $\gamma > \alpha$, we have that $F_j(v_j) = v_j^\gamma$ first order stochastically dominates $F_i(v_i) = v_i^\alpha$, and thus bidder j assigns a larger probability weight on high valuations than bidder i does.

(b) *Symmetrically distribution values.* Assume now that $\alpha = \gamma > 0$. How are your above results affected? How are equilibrium bids affected by a marginal increase in α? Interpret.

- When $\alpha = \gamma$, bidder i's bidding function becomes $b_i(v_i) = \frac{\alpha}{1+\alpha} v_i$, and that of bidder j is symmetric, that is, $b_j(v_j) = \frac{\alpha}{1+\alpha} v_j$. In that context, bidder i wins the auction if his valuation is higher than that of bidder j, $v_i > v_j$, and a marginal increase in α induces every bidder to submit (weakly) more aggressive bids because

$$\frac{\partial b_i(v_i)}{\partial \alpha} = \frac{1}{(1+\alpha)^2} v_i \geq 0.$$

Intuitively, an increase in α entails that the common cumulative distribution function assigns a larger probability weight on high valuations. As a consequence, for any given value v_i that bidder i privately observes, he knows that the probability that his rival draws a high valuation is, essentially, increasing in α and thus responds submitting a higher bid b_i.
- For illustration purposes, Fig. 3.4 depicts the cumulative distribution function, and Fig. 3.5 plots the corresponding equilibrium bidding function. If $\alpha = \gamma = 2$, the equilibrium bidding function becomes $b_i(v_i) = \frac{2}{3} v_i$ for every bidder i, which does not coincide with that in the standard first-price auction with two bidders independently drawing their valuation from a common uniform distribution. A similar argument applies if $\alpha = \beta = 3$, where the equilibrium bidding function becomes $b_i(v_i) = \frac{3}{4} v_i$, thus reducing bid shading.
  In the limit where $\alpha \to \infty$, bidders do not shade their bids since

$$\lim_{\alpha \to \infty} b_i(v_i) = \lim_{\alpha \to \infty} \left(1 - \frac{1}{1+\alpha}\right) v_i = v_i$$

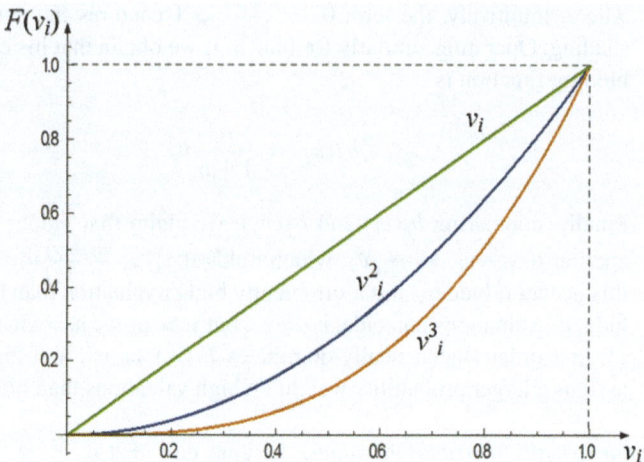

Fig. 3.4 $F(v_i)$ evaluated at $\alpha = 2$ and $\alpha = 3$

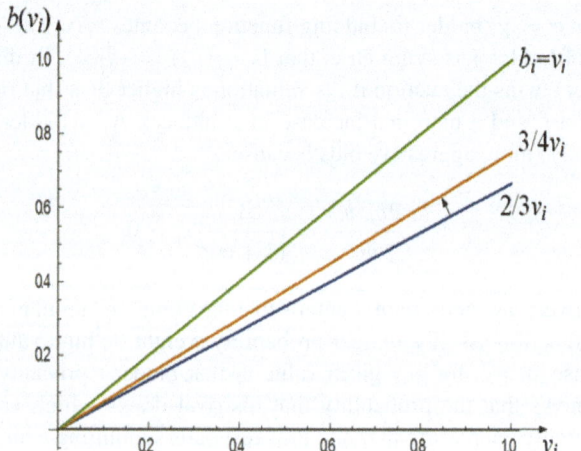

Fig. 3.5 Equilibrium bidding function evaluated at $\alpha = 2$ and at $\alpha = 3$

where the common cumulative distribution function $F_i(v_i)$ assigns all probability weight to $v_i = 1$. In this context, every bidder, knowing for sure that the other bidder has a valuation of $1 with certainty, can only win the object if he does not shade his bid and submits a bid equal to his valuation at $1, thus producing a tie with the other bidder.

(c) *Uniformly distributed valuations.* How are the equilibrium results affected if $\alpha = \gamma = 1$?

- If $\alpha = \gamma = 1$, both bidders' valuations are distributed according to a uniform distribution, that is, $F_i(v_i) = v_i$, for every bidder i. In this context, equilibrium bidding functions simplify to

$$b_i(v_i) = \frac{1}{2}v_i,$$

as in the first-price auction with two bidders independently drawing their valuation from a common uniform distribution, as in Exercise 2.2.

Exercise #3.8: Sequential Version of the First-Price Auction[A]

3.8 Consider an auction with two bidders. Every bidder $i = \{1, 2\}$ privately observes his valuation v_i for the object, drawn from a uniform distribution $U[0, 1]$, which is common knowledge among players. Assume that all bidders are risk neutral. Unlike the standard first-price auction, where players simultaneously and independently submit their bids, let us consider its sequential-move version with the following time structure: bidder 1 submits his bid, b_1, bidder 2 observes b_1 and responds submitting his own bid.
(a) Find the equilibrium bidding strategies for bidder 2 (follower) and bidder 1 (leader).

- Operating by backward induction, we start with bidder 2 and then analyze bidder 1.[2]
- *Bidder 2.* Consider bidder 2, who responds buying the good if his valuation is higher than the bid submitted by bidder 1, that is, $v_2 \geq b_1$. In this case, bidder 2 submits a bid $b_2 = b_1 + \varepsilon$, where $\varepsilon \to 0$. Otherwise, he steps out of the auction (e.g., submitting a bid of zero).
- *Bidder 1.* Anticipating this response from bidder 2, bidder 1 understands that $v_2 \geq b_1$ occurs with probability $1 - b_1$ since player 2's valuation is uniformly distributed in $[0, 1]$. Therefore, bidder 1 submits a bid b_1 expecting to win the object with probability b_1. In this context, bidder 1's expected utility maximization problem is

$$\max_{b_1 \geq 0} EU_1(v_1) = \Pr(win) \times (v_1 - b_1)$$

---

[2] Recall that we apply backward induction to find the subgame perfect equilibrium (or equilibria) of a sequential-move game. When applying backward induction, we start with the last stage of the game, finding optimal actions for the player(s) called to move in this stage. Then, we move to the second-to-last mover who, anticipating equilibrium behavior in the last stage, chooses his optimal action. We can repeat this process by moving one step closer to the first stage of the game, successively finding equilibrium behavior of each stage. For more details on this solution concept, see the Game Theory Appendix at the end of the book, and for more examples, see Munoz-Garcia and Toro-Gonzalez (2020, Chapter 4).

$$= b_1(v_1 - b_1)$$

Differentiating with respect to his bid b_1 yields $v_1 - 2b_1 = 0$, and solving for b_1, we find that his equilibrium bidding function is

$$b_1(v_1) = \frac{1}{2}v_1$$

which indicates a bid shading of 1/2 of his valuation for the object. As a remark, note that this equilibrium bidding function coincides with that in the simultaneous-move version of the first-price auction where two players compete for the object drawing their valuations from a uniform distribution.

(b) Is this auction efficient?

- No. The object could go to bidder 2 even if he has a lower valuation for the object. To see this point, consider valuations $v_1 = 0.7$ and $v_2 = 0.5$. In this setting, bidder 1 submits a bid of $b_1(0.7) = \frac{1}{2}(0.7) = 0.35$, and bidder 2 responds with a slightly higher bid (such as 0.36), letting him win the good and provides him with a positive surplus, $0.5 - 0.36 = 0.14$. Therefore, the object is not necessarily assigned to the bidder with the highest valuation.

Exercise #3.9: First-Price Auctions with Reservation Prices—One Bidder[A]

3.9 This and the following exercise examine the role of reservation prices in first-price auctions, and how the seller can strategically set the reservation price to increase his expected revenue. For simplicity, this exercise considers an artificial setting with only one bidder, and the next exercise generalizes our analysis to $N \geq 2$ bidders. The bidder's valuation, v, is distributed according to a cumulative distribution function $F(v)$, with positive density $f(v) > 0$ in all valuations of its domain $v \in [0, \bar{v}]$. Consider the following two-stage game: in the first stage, the seller sets a reservation price $r \geq 0$ that the bidder must meet for his bid to be considered by the seller, that is, $b_i \geq r$, otherwise his bid is ignored. In the second stage, the bidder observes the reservation price and submits his bid for the object.

(a) *Second stage.* Starting from the second stage, find the optimal bidding function for this bidder, $b(v)$.

- Since a single bidder competes for the object, he does not need to outbid anyone else. In other words, the bidder will only submit a bid equal to the reservation price r,

Exercise #3.9: First-Price Auctions with Reservation Prices—One Bidder[A]

$$b(v) = r,$$

independently on his valuation for the object, v. Bidding more is unnecessary, as no other bidders compete for the object, while bidding less would entail that his bid is not considered by the seller. This result holds regardless of how his valuations are distributed and regardless of his risk preferences.

(b) *First stage.* Anticipating the optimal bidding function $b(v)$ you found in part (a), what is the reservation price r^* that the seller sets in the first stage to maximize his expected revenue from the auction?

- Anticipating bidding function $b(v) = r$, the seller knows that he will receive r for the object, conditional on the bidder participating in the auction, which occurs when $v > r$. This happens with probability $1 - F(r)$, where $F(r) = \Pr\{v \leq r\}$. Therefore, the seller chooses the reservation price r that solves

$$\max_{r \geq 0} \; [1 - F(r)]r$$

where $1 - F(r)$ denotes the probability that the bidder participates in the auction. Differentiating with respect to r yields

$$1 - F(r) = f(r)r.$$

Intuitively, the left side represents the marginal cost of increasing the reservation price, r, since the probability that the bidder participates in the auction goes down. The right side, in contrast, captures the marginal benefit of a higher reservation price, as the seller may earn a higher price for the object. Solving for r, we obtain the optimal reservation price

$$r^* = \frac{1 - F(r)}{f(r)}.$$

- *Second-order conditions.* Note that for this solution to be a maximum, we need that the derivative of our first-order condition, $1 - F(r) = f(r)r$, or $r = \frac{1-F(r)}{f(r)}$, is negative. For that, we need the inverse hazard rate function $\frac{1-F(r)}{f(r)}$ to be decreasing in r or, alternatively, that the hazard rate function $\frac{f(r)}{1-F(r)}$ to be increasing in r. Intuitively, this means that the probability of drawing a specific valuation $v = r$, given that we have drawn valuations above r before, is increasing. For a uniform distribution, for instance, the hazard rate function is $\frac{1}{1-r}$, which is, indeed, increasing in r.

(c) *Uniform distribution.* If valuations are uniformly distributed, $F(v) = v$ in $v \in [0, 1]$, which is the optimal reservation price, r^*?

- In the context of uniformly distributed valuations, the first order condition found in part (b) becomes

$$1 - r = r$$

which, solving for r, yields $r = \frac{1}{2}$. Intuitively, the seller sets a reservation price $r^* = \frac{1}{2}$, anticipating that the bidder submits a bid of exactly $b = \frac{1}{2}$. This is, of course, risky for the seller since he does not know whether the bidder's valuation lies above $\frac{1}{2}$ (in which case the seller earns $\frac{1}{2}$) or below this reservation price (in which case the seller earns zero).

Exercise #3.10: First-Price Auctions with Reservation Prices—Several Bidders[C]

3.10 Consider a first-price auction with $N \geq 2$ bidders. Every bidder i's valuation, v_i, is distributed according to a cumulative distribution function $F(v_i)$, with positive density $f(v_i) > 0$ in all valuations of its domain $v_i \in [0, \overline{v}]$. Consider the following two-stage game: in the first stage, the seller sets a reservation price $r \geq 0$ that every bidder must meet for his bid to be considered by the seller, that is, $b_i \geq r$, otherwise his bid is ignored. In the second stage, every bidder i observes the reservation price and responds independently and simultaneously submitting his bid for the object.

(a) *Second stage.* Starting from the second stage, find the optimal bidding function for bidder i, $b_i(v_i)$.

- For bidder i whose valuation is no less than the reservation price r, his expected utility maximization problem is

$$\max_{b_i \geq 0} EU_i(b_i | b_i \geq r) = \Pr\{win\} \times (v_i - b_i)$$

where the probability of winning the auction is given by the probability that bidder i's bid b_i exceeds the highest bid of all other bidders, that is, $b_i \geq b_j$ for every player $j \neq i$. Considering bidding function $b_i(v_i)$, we can find its inverse $v_i = b_i^{-1}(b_i)$, which helps us represent player i's probability of winning the auction as the probability that his valuation v_i exceeds that of all other bidders, $v_i \geq v_j$ for every player $j \neq i$, or $\Pr\{win\} = [F(v_i)]^{N-1}$. Alternatively, we can express the probability with the inverse bidding function as $b_i^{-1}(b_i) \geq b_j^{-1}(b_j)$ for every player $j \neq i$ as follows:

Exercise #3.10: First-Price Auctions with Reservation Prices—Several Bidders[C]

$$\Pr\{win\} = \left[F\left(b_i^{-1}(b_i)\right)\right]^{N-1}.$$

Therefore, we can represent the above maximization problem as

$$\max_{b_i \geq 0} \left[F\left(b_i^{-1}(b_i)\right)\right]^{N-1} \times (v_i - b_i)$$

Interestingly, player i's objective function does not depend on the reservation price r because, to win the object, his valuation must be above all his rivals' (including those bidders whose valuations fall below the reservation price r). In other words, for bidder i to win the auction, his bid must exceed the reservation price, r, thus making his expected utility maximization problem unaffected by r.

- Differentiating with respect to the bid b_i yields

$$(N-1)\left[F\left(b_i^{-1}(b_i)\right)\right]^{N-2} f\left(b_i^{-1}(b_i)\right) \frac{db_i^{-1}(b_i)}{db_i}(v_i - b_i)$$
$$-\left[F\left(b_i^{-1}(b_i)\right)\right]^{N-1} = 0$$

Consider now that every bidder i uses a symmetric bidding function $b_i(v_i)$, so applying its inverse we obtain

$$b_i(b_i^{-1}(b_i)) = b_i$$

In other words, this result says that the inverse function of a function returns the argument. Differentiating both sides with respect to the bid b_i, and by Chain rule,

$$\frac{db_i\left(b_i^{-1}(b_i)\right)}{db_i^{-1}(b_i)} \frac{db_i^{-1}(b_i)}{db_i} = 1$$

Since the bidder's valuation, v_i, is a one-to-one mapping from the bid b_i, that is, $v_i = b_i^{-1}(b_i)$, we can further rewrite the above expression to yield

$$\frac{db_i^{-1}(b_i)}{db_i} = \frac{1}{\frac{db_i(v_i)}{dv_i}}$$

Substituting this into the first-order condition above, we find

$$(N-1)\left[F(v_i)\right]^{N-2} f(v_i)(v_i - b_i) = b_i'(v_i)\left[F(v_i)\right]^{N-1}$$

After rearranging, we have

$$(N-1)[F(v_i)]^{N-2} f(v_i) b_i(v_i) + [F(v_i)]^{N-1} b'_i(v_i)$$
$$= (N-1)[F(v_i)]^{N-2} f(v_i) v_i$$

Integrating both sides with respect to v_i, and taking the reservation price r as the lower bound of integration (because his bid needs to be above the reservation price to be considered), we obtain

$$\int_r^{v_i} \frac{d\left[[F(x)]^{N-1} b_i(x)\right]}{dx} dx = \int_r^{v_i} (N-1)[F(x)]^{N-2} f(x) x \, dx$$

which simplifies to

$$\left[[F(x)]^{N-1} b_i(x)\right]_r^{v_i} = \left[[F(x)]^{N-1} x\right]_r^{v_i} - \int_r^{v_i} [F(x)]^{N-1} dx$$

where x is the parameter of integration representing the value of bidder i's bid.[3] The above expression can be further simplified to yield

$$[F(v_i)]^{N-1} b_i(v_i) = [F(v_i)]^{N-1} v_i - \int_r^{v_i} [F(x)]^{N-1} dx$$

Dividing both sides by $[F(v_i)]^{N-1}$, the optimal bidding function becomes

$$b_i(v_i) = v_i - \underbrace{\frac{1}{[F(v_i)]^{N-1}} \int_r^{v_i} [F(x)]^{N-1} dx}_{\text{Bid shading}}$$

(b) How does the optimal bidding function $b_i(v_i)$ change in the reservation price r?

- Like in first-price auctions without reservation prices, the second term in bidder i's optimal bidding function represents his bid shading, i.e.,

---

[3] The above expression considers that bidders submitting bids $b_i(v_i) \geq r$ participate in the auction, while those with valuations leading to low bids $b_i(v_i) < r$ do not. In addition, it considers that the participating bidder with the lowest valuation submits a bid $b_i(r) = r$ where his valuation is $v_i = r$. To see this point, note that he must be indifferent between participating and not participating in the auction, so his expected utility must satisfy $F(v_i)^{N-1}(v_i - b_i) = 0$. Substituting $b_i = r$, we obtain $F(v_i)^{N-1}(v_i - r) = 0$, which holds only if $v_i = r$, i.e., if his valuation coincides with the reservation price. Therefore, the participation condition using bids, $b_i(v_i) \geq r$, implies an analogous condition using valuations, which we can write as $v_i \geq r$, explaining that the lower bound of integration is $v_i = r$ in the above expressions.

how much he reduces his bid relative to his true valuation for the object, v_i. Interestingly, the bid shading term decreases in the reservation price r, since the support of the integral becomes narrower, while the integrand is unaffected by r.

- Intuitively, the introduction of a reservation price r limits the number of bidders who participate in the auction (i.e., those with optimal bids satisfying $b_i(v_i) < r$ do not participate) and, in addition, induces participating bidders (those with $b_i(v_i) \geq r$) to submit more aggressive bids than what they would in the absence of reservation prices. Participating bidders submit more aggressive bids because they know that the average valuation (and bids) among participating bidders is larger than that among all bidders in a standard first-price auction with no reservation prices.

  A similar argument applies when the seller increases the reservation price r, as the number of participating bidders decreases, while the bids that participating bidders submit increase. Therefore, the seller faces a trade-off when setting a positive (or, generally, a higher) reservation price. We return to this trade-off below when we find the optimal reservation price that the seller sets in the first stage of the game.

(c) *Example of uniform distribution.* Assume that bidders' valuations are uniformly distributed, that is, $F(v_i) = v_i$ for all $v_i \sim U[0, \overline{v}]$. Evaluate the optimal bidding function found in part (a).

- Since in this context $F(v_i) = v_i$ and $F(x) = x$, the optimal bidding function, $b_i(v_i)$, becomes

$$b_i(v_i) = v_i - \frac{1}{v_i^{N-1}} \int_r^{v_i} x^{N-1} dx$$

and solving the integral, we obtain

$$b_i(v_i) = v_i - \frac{1}{v_i^{N-1}} \left. \frac{x^N}{N} \right|_r^{v_i}$$

$$= v_i - \frac{1}{v_i^{N-1}} \frac{v_i^N - r^N}{N}$$

$$= \frac{(N-1)v_i^N + r^N}{Nv_i^{N-1}}$$

which is increasing in the reservation price, r, as discussed in part (b) of the exercise. When the reservation price is zero, i.e., $r = 0$, the bidding function simplifies to $b_i(v_i) = \frac{N-1}{N} v_i$, which coincides with our results in Exercise 2.3.

(d) *Example of exponential distribution.* Assume that two bidders compete for the object and their valuations follow the exponential distribution, that is, $F(v_i) = 1 - \exp(-\lambda v_i)$, where $v_i \in [0, \infty)$ and $\lambda > 0$. Evaluate the optimal bidding function found in part (a).

- Since $F(v_i) = 1 - \exp(-\lambda v_i)$, the optimal bidding function, $b_i(v_i)$, becomes

$$b_i(v_i) = v_i - \frac{1}{1 - \exp(-\lambda v_i)} \int_r^{v_i} \left[1 - \exp(-\lambda x)\right] dx$$

$$= v_i - \frac{v_i - r + \frac{1}{\lambda} \int_r^{v_i} d\exp(-\lambda x)}{1 - \exp(-\lambda v_i)}$$

$$= v_i - \frac{\lambda(v_i - r) + \exp(-\lambda v_i) - \exp(-\lambda r)}{\lambda \left[1 - \exp(-\lambda v_i)\right]}$$

$$= \frac{-(1 + \lambda v_i)\exp(-\lambda v_i) + \lambda r + \exp(-\lambda r)}{\lambda \left[1 - \exp(-\lambda v_i)\right]}$$

which is increasing in the reservation price r since

$$\frac{\partial b_i(v_i)}{\partial r} = \frac{1 - \exp(-\lambda r)}{1 - \exp(-\lambda v_i)} > 0.$$

(e) *First stage.* Anticipating the optimal bidding function $b_i(v_i)$ you found in part (a), what is the reservation price r^* that the seller sets in the first stage to maximize his expected revenue from the auction?

- We first find the seller's expected revenue from the auction and then differentiate it with respect to the reservation price r, to identify the revenue-maximizing reservation price r^*.
- *Finding the seller's revenue from the auction.* For compactness, let us define $G(x) = (F[x])^{N-1}$ to be the joint cumulative probability density function for $N - 1$ bidders, where valuation x satisfies $x \in [0, \bar{v}]$. Then the above optimal bidding function can be rewritten as

$$b_i(v_i) = v_i - \frac{1}{G(v_i)} \int_r^{v_i} G(x) dx$$

$$= \frac{1}{G(v_i)} \left[G(v_i) v_i - \int_r^{v_i} G(x) dx\right]$$

$$= \frac{1}{G(v_i)} \left[G(r) r + G(v_i) v_i - G(r) r - \int_r^{v_i} G(x) dx\right]$$

$$= \frac{1}{G(v_i)} \left[G(r)r + \int_r^{v_i} xg(x)\,dx \right]$$

From an *ex-ante* point of view (before observing his own valuation for the object), bidder *i*'s expected payment to the seller is given by the probability of winning the auction times the bid he pays for the object upon winning, that is,

$$\pi_i(v_i|v_i \geq r) = \Pr(win) \times b_i(v_i)$$
$$= (F[v_i])^{N-1} \times b_i(v_i)$$
$$= G(v_i) \times b_i(v_i)$$
$$= G(v_i) \times \frac{1}{G(v_i)} \left[G(r)r + \int_r^{v_i} xg(x)\,dx \right]$$
$$= G(r)r + \int_r^{v_i} xg(x)\,dx$$

where the second line indicates that, as discussed in previous parts of the exercise, bidder i wins the auction if his valuation v_i is higher than everyone else's, that is, $v_i \geq v_j$ for every bidder $j \neq i$. The probability of his valuation exceeding that of every other bidder is given by $(F[v_i])^{N-1}$, which we represented more compactly as $G(v_i) = (F[v_i])^{N-1}$; see third line. In the fourth line, we just insert the equilibrium bidding function found above, $b_i(v_i)$, and in the last line we simplify the expression.

- Since the seller cannot observe bidders' values, he finds the expected payment from each bidder i, $E[\pi_i(v_i|v_i \geq r)]$, and then sums up for all N bidders, $\sum_{i=1}^{N} E[\pi_i(v_i|v_i \geq r)]$, which gives us the seller's revenue from the auction (this is, of course, understood from an *ex-ante* perspective since the seller does not observe bidders' valuations). We find the seller's revenue as follows:

$$E[\pi(r)] = \sum_{i=1}^{N} E[\pi_i(v_i|v_i \geq r)]$$
$$= N \int_r^{\bar{v}} \underbrace{\left[rG(r) + \int_r^{v} xg(x)\,dx \right]}_{\text{Expected payment, } \pi_i(v_i|v_i \geq r)} f(z)\,dz$$

Since $rG(r)$ is constant, it is unaffected by the integration, helping us rewrite the seller's revenue as

$$E[\pi(r)] = NrG(r)\int_r^{\bar{v}} f(z)\,dz + N\int_r^{\bar{v}}\int_r^v xg(x)\,dx f(z)\,dz$$

$$= NrG(r)[F(z)]_r^{\bar{v}} + N\int_r^{\bar{v}}\left(\int_v^{\bar{v}} f(z)\,dz\right) vg(v)\,dv$$

$$= NrG(r)[1 - F(r)] + N\int_r^{\bar{v}} (1 - F(v))\,vg(v)\,dv$$

where the first line expands the integral into two parts, and the second line integrates the probability density function into the cumulative distribution function from the reservation price r to the upper bound $\bar{v}$, with the second part exchanging the order of integration that considers the density mass of bidders whose valuation is above v. Finally, the third line simplifies the mathematical expression.

- *Revenue-maximizing reservation price.* We can now differentiate the seller's revenue with respect to the reservation price r,

$$\frac{dE[\pi(r)]}{dr} = N[G(r)(1 - F(r) - rf(r)) + (1 - F(r))rg(r)]$$

$$\qquad - (1 - F(r))rg(r)]$$

$$= N[G(r)(1 - F(r) - rf(r))]$$

$$= NG(r)(1 - F(r))\left[1 - r\frac{f(r)}{1 - F(r)}\right] = 0$$

Intuitively, increasing the reservation price r produces a positive and negative effect on the seller's expected revenue. These effects are rather obscure in the first-order condition we just found, but they are easier to understand in the seller's expected revenue expression

$$E[\pi(r)] = N\int_r^{\bar{v}} \underbrace{\left[rG(r) + \int_r^v xg(x)\,dx\right]}_{\text{Expected payment, } \pi_i(v_i | v_i \geq r)} f(z)\,dz$$

We can more easily describe each effect from a marginal increase in the reservation price, r, as follows:

- The negative effect of increasing the reservation price r arises because the seller is limiting the number of bidders who participate in the auction, which is illustrated in the above expression by the fact that the support of the integral becomes narrower as its lower bound, r, increases.
- The positive effect of increasing the reservation price r, however, emerges from the property that the equilibrium bid $b_i(v_i)$ increases

Exercise #3.10: First-Price Auctions with Reservation Prices—Several Bidders^C

in r, and thus the expected payment, $\pi_i(v_i|v_i \geq r)$, inside the above integral also increases in r.

In short, the seller faces a trade-off when setting a higher reservation price, since that entails reducing the number of participants but being able to induce larger bids from those bidders who do participate in the auction.

- From the above first-order condition, we find that the optimal reservation price r^* solves

$$NG(r)[1 - F(r)]\left[1 - r\frac{f(r)}{1 - F(r)}\right] = 0.$$

Since $G(r) > 0$ (otherwise, $r = 0$, making the reservation price inconsequential) and $[1 - F(r)] > 0$ (otherwise, $r = \overline{v}$, leaving no bidder participating in the auction). As a consequence, we must have that $1 = r\frac{f(r)}{1-F(r)}$, which solving for r yields

$$r^* = \frac{1 - F(r)}{f(r)}.$$

The right-hand side of the above inequality is the inverse hazard rate, $\frac{1-F(r)}{f(r)}$, which measures how sensitive is the distribution of the bidders' valuation $F(\cdot)$ to a change in the reservation price. That is, if the density mass is concentrated in the region above the reservation price r, then $1 - F(r)$ would be large relative to $f(r)$ so that the seller can further increase the reservation price to raise his expected revenue by selling the object to those bidders with higher valuations. We will elaborate on the relationship between the optimal reservation price, stochastic dominance, and the inverse hazard rate in the next part of the exercise.

- *Analogy with monopoly pricing.* As a final remark, note that our result coincides with that in Exercise 3.9 where the seller faces a single bidder. In other words, the seller faces a similar trade-off when setting a positive reservation price as a monopolist pricing a good to consumers. Recall that the monopolist sets a relatively high price, which induces consumers with low valuations not to buy the good but helps the firm extract a larger surplus from those consumers with high valuations for the good. Similarly, the seller sets a positive reservation price r^*, which induces low-value bidders to abstain from participating in the auction but allows the seller to induce larger bids from participating bidders, ultimately extracting a larger surplus from them.

(f) Is the reservation price r^* you found in part (e) positive? Does this lead to an efficient outcome in the first-price auction?

- Reservation price $r^* = \frac{1-F(r)}{f(r)}$ is positive since $f(r) > 0$ for all points in the support of the density function, and $F(r)$ lies between 0 and 1.
- The outcome of the first-price auction with a positive reservation price, however, is *not* efficient since the object may not be sold and thus remain with the seller (whose reservation price is zero). In that context, the seller and at least one buyer could renegotiate since there is room for a mutually beneficial bargain, indicating that the allocation mechanism we used (first-price auction with a positive reservation price) was not allocating the object efficiently.

(f) Consider a seller conducting two auctions for unrelated objects. In auction A, bidders' valuations are drawn from cumulative distribution function $F_A(v_i)$, while in auction B valuations are drawn from cumulative distribution function $F_B(v_i)$, where we assume that both distributions have positive densities in their common support $v_i \in [0, \bar{v}]$. Consider that $F_A(v_i)$ first order stochastically dominates $F_B(v_i)$, that is, $F_A(v_i) \leq F_B(v_i)$ for all v_i. In which auction should the seller set the highest reservation price? Interpret.

- If $F_A(v_i)$ first order stochastically dominates $F_B(v_i)$, $F_A(v_i) \leq F_B(v_i)$, we have that

$$1 - F_A(v_i) \geq 1 - F_B(v_i).$$

Since this holds for all valuations v_i, it must also hold for $v_i = r$, so we can write $1 - F_A(r) \geq 1 - F_B(r)$, implying that

$$\frac{1 - F_A(r)}{f_A(r)} \geq \frac{1 - F_B(r)}{f_B(r)}$$

given that the probability density, $f(\cdot)$, is assumed continuous and of mass zero at any particular point in its support. Therefore, the seller should set a higher reservation price in auction A than in auction B, i.e., $r_A^* \geq r_B^*$.
- Intuitively, cumulative distribution function $F_A(v_i)$ concentrates a higher probability weight on higher values, indicating that it is more likely that bidders assign higher valuation for the object than in cumulative distribution function $F_B(v_i)$. The seller can then set a higher reservation price since it suffers a smaller expected cost from limiting the participation of bidders with low valuations in auction A than in auction B.

(h) *Uniformly distributed valuations.* Evaluate the optimal reservation price r^* you found in part (e) when valuations are uniformly distributed, that is, $F(v_i) = v_i$ for all $v_i \sim U[0, \bar{v}]$.

- Since $F(r) = r$ and $f(r) = 1$, the optimal reservation price r^* solves

$$r = \frac{1 - F(r)}{f(r)} = \frac{1-r}{1}, \text{ or } r = 1 - r$$

which, after solving for r, yields

$$r^* = \frac{1}{2}.$$

In other words, the seller sets a reservation price that leaves half of the bidders not participating in the auction. Only those with valuations in $v_i \in [1/2, 1]$ would submit a positive bid in equilibrium.

(i) *Exponentially distributed valuations.* Evaluate the optimal reservation price r^* you found in part (e) when valuations are exponentially distributed, that is, $F(v_i) = 1 - \exp(-\lambda v_i)$, where $v_i \in [0, \infty)$. Evaluate this price at $\lambda = 2$ and $\lambda = 3$. Interpret your results.

- Since

$$1 - F(v_i) = 1 - \left[1 - \exp(-\lambda v_i)\right] = \exp(-\lambda v_i)$$

and $f(v_i) = \lambda \exp(-\lambda v_i)$, the optimal reservation price r^* solves

$$r = \frac{1 - F(r)}{f(r)} = \frac{\exp(-\lambda v_i)}{\lambda \exp(-\lambda v_i)}$$

which simplifies to

$$r^* = \frac{1}{\lambda}.$$

The seller sets a reservation price that is inversely proportional to the rate of decay of bidders' valuation. In other words, the faster the rate of decay, the more concentrated are bidders' valuations on low values, so that the seller sets a lower reservation price. For instance, when $\lambda = 2$, the seller sets $r^* = \frac{1}{2}$, which decreases to $r^* = \frac{1}{3}$ when the rate of decay increases to $\lambda = 3$.

Exercise #3.11: First-Price Auction with Entry Fees[C]

3.11 Consider a first-price auction with N bidders. Every bidder i's valuation, v_i, is distributed according to a cumulative distribution function $F(v_i)$, with positive support, i.e., $f(v_i) > 0$ for all $v_i \in [0, \overline{v}]$. Consider the following two-stage game: in the first stage, the seller sets an entry fee $E \geq 0$ that every participating bidder must pay, otherwise his bid is ignored, and in the second stage, every bidder i independently and simultaneously submits his bid for the object.

(a) *Second stage.* Starting from the second stage, find the optimal bidding function for bidder i, $b_i(v_i)$. [*Hint*: Assume that there exists a critical bidder whose valuation v_e makes him indifferent between participation and non-participation, given a positive entry fee E.]

- Every bidder i's expected utility maximization problem is

$$\max_{b_i \geq 0} EU_i(b_i) = \Pr\{win\}(v_i - b_i) - E$$

where the entry fee, E, is a constant that bidder i must pay when he participates in the auction, whether he wins the object or not.

- The probability of bidder i winning the object is analogous to the standard first-price auction without entry fees, which is given by

$$\Pr\{win\} = [F(v_i)]^{N-1}$$

when his valuation exceeds that of all other $N-1$ bidders, where $v_i \geq v_j$ for all $j \neq i$ and $j \in \{1, \ldots, N\}$. Note that for a given cumulative distribution function, $F(v_i)$ represents that

$$F(v_i) = F\left(b_i^{-1}(b_i)\right)$$

such that bidder i's expected utility maximization problem becomes

$$\max_{b_i \geq 0} EU_i(b_i) = \left[F\left(b_i^{-1}(b_i)\right)\right]^{N-1}(v_i - b_i) - E$$

- We assume that the bidding function $b_i(v_i)$ is monotonically increasing in v_i, which we will demonstrate later. In addition, let us define a "critical bidder" with valuation v_e and bid $b_e(v_e)$, where v_e solves[4]

$$[F(v_e)]^{N-1}(v_e - b_e) - E = 0$$

which means that his expected utility from participating in the auction (given the entry fee E) is zero.

---

[4]We need this condition to ensure a one-to-one mapping between the entry fee and the critical bidder's valuation. The monotonically increasing bidding function $b_i(v_i)$ ensures that bidders with valuations above v_e obtain a positive utility from participating in the auction (after paying the entry fee E) and thus submit a positive bid for the object.

- Differentiating bidder i's expected utility with respect to his bid b_i yields

$$(N-1)\left[F\left(b_i^{-1}(b_i)\right)\right]^{N-2} f\left(b_i^{-1}(b_i)\right) \frac{db_i^{-1}(b_i)}{db_i} (v_i - b_i)$$
$$- \left[F\left(b_i^{-1}(b_i)\right)\right]^{N-1} = 0$$

Since the inverse of the bidding function yields the bidder's valuation, $b_i^{-1}(b_i) = v_i$, and the derivative of this inverse can be written as $\frac{db_i^{-1}(b_i)}{db_i} = \frac{1}{b_i'\left(b_i^{-1}(b_i)\right)}$, the above expression becomes

$$(N-1)\left[F(v_i)\right]^{N-2} f(v_i)(v_i - b_i) = b_i'(v_i)\left[F(v_i)\right]^{N-1}$$

Further rearranging, we obtain

$$(N-1)\left[F(v_i)\right]^{N-2} f(v_i) b_i(v_i) + \left[F(v_i)\right]^{N-1} b_i'(v_i)$$
$$= (N-1)\left[F(v_i)\right]^{N-2} f(v_i) v_i$$

The left-hand side is $\frac{d\left[[F(v_i)]^{N-1} b_i(v_i)\right]}{dv_i}$. Hence,

$$\frac{d\left[[F(v_i)]^{N-1} b_i(v_i)\right]}{dv_i} = (N-1)\left[F(v_i)\right]^{N-2} f(v_i) v_i$$

- Applying integration by parts to the right-hand side of the above expression with respect to v_i, and taking the indifferent bidder's valuation v_e as the lower bound of integration, yields

$$\int_{v_e}^{v_i} \frac{d\left[[F(x)]^{N-1} b_i(x)\right]}{dx} dx = \int_{v_e}^{v_i} (N-1)\left[F(x)\right]^{N-2} f(x) x \, dx$$

or

$$\left[[F(x)]^{N-1} b_i(x)\right]_{v_e}^{v_i} = \left[[F(x)]^{N-1} x\right]_{v_e}^{v_i} - \int_{v_e}^{v_i} [F(x)]^{N-1} dx$$

We can then reorder the terms in the above expression as follows:

$$[F(v_i)]^{N-1} b_i(v_i) = [F(v_i)]^{N-1} v_i - [F(v_e)]^{N-1} [v_e - b_e(v_e)]$$
$$- \int_{v_e}^{v_i} [F(x)]^{N-1} dx$$

Substituting the indifferent bidder's condition, $[F(v_e)]^{N-1}(v_e - b_e) - E = 0$, into the above expression, we find

$$[F(v_i)]^{N-1}[v_i - b_i(v_i)] = E + \int_{v_e}^{v_i} [F(x)]^{N-1} dx$$

and rearranging, we obtain

$$b_i(v_i) = v_i - \underbrace{\frac{E + \int_{v_e}^{v_i} [F(x)]^{N-1} dx}{[F(v_i)]^{N-1}}}_{\text{bid shading}}$$

Note that when entry fees are absent, $E = 0$, the equilibrium bidding function collapses to the standard expression found in previous exercises.
- Lastly, we show that the equilibrium bidding function, $b_i(v_i)$, is monotonically increasing in the bidder's valuation, v_i, since

$$\frac{db_i(v_i)}{dv_i} = 1$$

$$- \frac{[F(v_i)]^{2N-2} - (N-1)[F(v_i)]^{N-2} \left(E + \int_{v_e}^{v_i} [F(x)]^{N-1} dx\right)}{[F(v_i)]^{2N-2}}$$

$$= \frac{N-1}{[F(v_i)]^N} \left(E + \int_{v_e}^{v_i} [F(x)]^{N-1} dx\right) > 0.$$

(b) How are equilibrium bids affected by an increase in the entry fee E? Do they limit participation in the auction?

- As in other games on first-price auctions, the second term in the equilibrium bid represents bidder i's bid shading, which is increasing in the entry fee E. Intuitively, the entry fee affects all bidders uniformly, inducing those with valuations below v_e not to participate in the auction and reducing the bid of those who participate. In particular, every participating bidder expects a lower expected utility in equilibrium, both if he wins and if he loses the auction, leading him to decrease his bid (larger bid shading) to compensate for his lower expected utility.

(c) *Example of uniform distribution.* Assume that bidders' valuations are uniformly distributed, that is, $F(v_i) = v_i$ for all $v_i \sim U[0, \overline{v}]$. Evaluate the optimal bidding function found in part (a).

- Since in this context $F(v_i) = v_i$ for all $v_i \in [0, \overline{v}]$, the optimal bidding function $b_i(v_i)$ becomes

Exercise #3.11: First-Price Auction with Entry Fees

$$b_i(v_i) = v_i - \frac{E + \int_{v_e}^{v_i} x^{N-1} dx}{v_i^{N-1}}$$

and solving the integral, we obtain bidder i's bidding function, $b_i(v_i)$, as follows:

$$b_i(v_i) = v_i - \frac{NE + [x^N]_{v_e}^{v_i}}{Nv_i^{N-1}}$$

$$= v_i - \underbrace{\frac{NE + v_i^N - v_e^N}{Nv_i^{N-1}}}_{\text{Bid shading}}$$

and the indifferent bidder's bid, $b_e(v_e)$, solves

$$v_e^{N-1}(v_e - b_e(v_e)) = E$$

which can be rearranged as

$$b_e(v_e) = \frac{v_e^N - E}{v_e^{N-1}}$$

(d) *First stage.* Anticipating the optimal bidding function $b_i(v_i)$ you found in part (a), what is the optimal entry fee E^* that the seller sets in the first stage to maximize his expected revenue from the auction? For simplicity, assume that the critical bidder, after paying the entry fee, is indifferent between participating and not participating if he submits a bid of zero, that is, $b_e(v_e) = 0$.

- We first find the seller's expected revenue from the auction and then differentiate it with respect to the entry fee E, to identify the revenue-maximizing entry fee E^*.
- *Finding the seller's revenue from the auction.* For compactness, let us define $G(x) = [F(x)]^{N-1}$ to be the joint cumulative probability density function for $N-1$ bidders, whose valuation x satisfies $x \in [0, \bar{v}]$. Then the above optimal bidding function can be rewritten as

$$b_i(v_i) = v_i - \frac{E + \int_{v_e}^{v_i} G(x) dx}{G(v_i)}$$

$$= \frac{1}{G(v_i)} \left[G(v_i) v_i - E - \int_{v_e}^{v_i} G(x) dx \right]$$

$$= \frac{1}{G(v_i)} \int_{v_e}^{v_i} xg(x)\,dx$$

by the fact that $G(v_e)\,v_e = E$ and the opposite of integration by parts.

From an *ex-ante* point of view (before observing his own valuation for the object), bidder i's expected payment to the seller is given by the probability of winning the auction times the bid he pays for the object upon winning, that is,

$$m(v_i|v_i \geq v_e) = \Pr(win) \times b_i(v_i) + E$$

$$= G(v_i) \times \frac{1}{G(v_i)} \int_{v_e}^{v_i} xg(x)\,dx + E$$

$$= \int_{v_e}^{v_i} xg(x)\,dx + E$$

where the second line indicates that, as discussed in previous parts of the exercise, bidder i wins the auction if his valuation v_i is above everyone else's, that is, $v_i \geq v_j$ for every bidder $j \neq i$; and in addition, he participates in the auction and pays a participation fee E. The probability of his valuation exceeding that of every other bidder is given by $[F(v_i)]^{N-1}$, represented more compactly as $G(v_i) = [F(v_i)]^{N-1}$, and we insert the equilibrium bidding function found above, $b_i(v_i)$.

- Since the seller cannot observe bidders' values, he finds the expected payment from each bidder i, $E[\pi_i(v_i|v_i \geq v_e)]$, and then sums up for all N bidders, $\sum_{i=1}^{N} E[\pi_i(v_i|v_i \geq v_e)]$, which gives us the seller's revenue from the auction (this is, of course, understood from an *ex-ante* perspective since the seller does not observe bidders' valuations). We find the seller's revenue as follows

$$R^1(v_e) = \sum_{i=1}^{N} E[\pi_i(v_i|v_i \geq v_e)]$$

$$= N \int_{v_e}^{\overline{v}} \left[E + \int_{v_e}^{v_i} xg(x)\,dx \right] f(z)\,dz$$

Since the participation fee, E, is a constant, it is unaffected by the integration, helping us to rewrite the seller's revenue as

$$R^1(v_e) = N \times E \int_{v_e}^{\overline{v}} f(z)\,dz + N \int_{v_e}^{\overline{v}} \left(\int_{v_i}^{\overline{v}} f(z)\,dz \right) xg(x)\,dx$$

$$= N \times E[1 - F(v_e)] + N \int_{v_e}^{\overline{v}} [1 - F(v_i)]xg(x)\,dx$$

$$= N v_e G(v_e) [1 - F(v_e)] + N \int_{v_e}^{\bar{v}} [1 - F(v_i)] x g(x) \, dx$$

where the first line expands the integral into two parts. The second line integrates the probability density function into the cumulative distribution function, from the critical type's valuation v_e to the upper bound $\bar{v}$, with the second part exchanging the order of integration that considers bidders whose valuation is above v_e. Finally, the third line stems from the fact that

$$E = [F(v_e)]^{N-1} (v_e - b_e) = G(v_e) v_e$$

where we assume that the indifferent bidder submits a bid of $b_e(v_e) = 0$.

- *Revenue-maximizing reservation price.* We can now differentiate the seller's revenue with respect to the critical valuation v_e,

$$\frac{dR^1(v_e)}{dv_e} = N[G(v_e)(1 - F(v_e) - v_e f(v_e))$$
$$+ (1 - F(v_e)) v_e g(v_e) - (1 - F(v_e)) v_e g(v_e)]$$
$$= NG(v_e)(1 - F(v_e)) \left[1 - v_e \frac{f(v_e)}{1 - F(v_e)}\right]$$

- Assuming interior solutions, we set the above first-order condition equal to zero to obtain

$$v_e = \frac{1 - F(v_e)}{f(v_e)}.$$

The right-hand side of the above inequality is the inverse hazard rate, $\frac{1-F(v_e)}{f(v_e)}$, which measures how sensitive is the distribution of the bidders' valuation $F(\cdot)$ to a change in the critical valuation v_e. That is, if the density mass is concentrated in the region above the critical valuation v_e, then $1 - F(v_e)$ would be large relative to $f(v_e)$ so that the seller can further increase the entry fee E to raise his expected revenue by selling the object to those bidders with higher valuations. Next, we will elaborate on the relationship between the entry fee E and the critical valuation v_e.

- Substituting $v_e = \frac{1-F(v_e)}{f(v_e)}$ into the indifferent bidder's valuation function, the optimal entry fee E^* solves

$$E^* = [F(v_e)]^{N-1} v_e$$
$$= [F(v_e)]^{N-1} \frac{1 - F(v_e)}{f(v_e)}$$

(e) *Uniformly distributed valuations.* Evaluate the optimal bidding function, $b_i^*(v_i)$, and the optimal entry fee E^* you found in parts (c) and (d), respectively, when valuations are uniformly distributed, that is, $F(v_i) = v_i$ for all $v_i \sim U[0, \overline{v}]$. How does the bidding function change with bidder i's valuation, v_i, and the number of bidders, N?

- Evaluating $v_e = \frac{1-F(v_e)}{f(v_e)}$ for a uniformly distributed valuation $v_i \sim U[0,1]$, we obtain

$$v_e = 1 - v_e, \text{ or } v_e^* = \frac{1}{2}$$

so that half of the bidders would participate in the auction in equilibrium.

- Substituting $v_e^* = \frac{1}{2}$ into the expression of E^*, we find

$$E^* = v_e^{N-1}(1 - v_e)$$
$$= \frac{1}{2^N}$$

so that the entry fee E^* decreases in the number of bidders N at a decreasing rate. Note that when N becomes infinitely large, that is, $N \to +\infty$, the profit-maximizing entry fee that the seller sets approaches zero asymptotically, as depicted in Fig. 3.6.

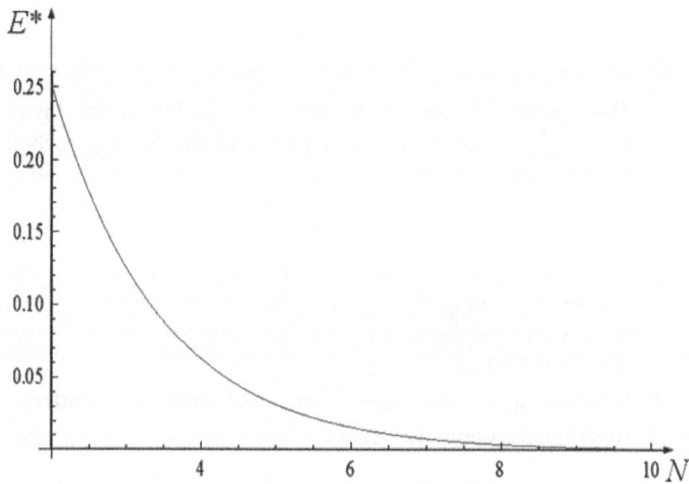

Fig. 3.6 Optimal entry fee E^* as a function of the number of bidders, N

Exercise #3.11: First-Price Auction with Entry Fees^C

For instance, when only $N = 2$ bidders compete for the object, the optimal entry fee becomes $E^* = \frac{1}{4}$ (see vertical intercept at 0.25), while when $N = 3$ bidders compete, the entry fee decreases to $E^* = \frac{1}{8}$. This entry fee quickly approaches zero when the number of bidders further increases.

- Therefore, the optimal bidding function, $b_i^*(v_i)$, becomes

$$b_i^*(v_i) = v_i - \frac{NE^* + v_i^N - (v_e^*)^N}{Nv_i^{N-1}}$$

$$= v_i - \frac{N\frac{1}{2^N} + v_i^N - \frac{1}{2^N}}{Nv_i^{N-1}}$$

$$= \frac{N-1}{N}\left(v_i - \frac{1}{2^N v_i^{N-1}}\right)$$

$$= \frac{N-1}{N}\left[1 - (2v_i)^{-N}\right]v_i.$$

For instance, when only $N = 2$ bidders compete for the object, this optimal bidding function simplifies to

$$b_i^*(v_i) = \frac{v_i}{2} - \frac{1}{8v_i},$$

and when $N = 3$ bidders compete, the optimal bidding function becomes

$$b_i^*(v_i) = \frac{2v_i}{3} - \frac{1}{12v_i^2},$$

while when $N = 10$ bidders compete in the auction, it becomes

$$b_i^*(v_i) = \frac{9v_i}{10} - \frac{9}{10240v_i^9}.$$

Figure 3.7 depicts these bidding functions, where valuations are restricted in $v_i \in \left[\frac{1}{2}, 1\right]$, since the indifferent bidder when valuations are uniformly distributed is $v_e^* = \frac{1}{2}$, as shown above.

Every bidder's bid is increasing in his valuation for the object, v_i, and shifts upward when he competes against a larger number of bidders, N. We next confirm these two points more formally. First, let

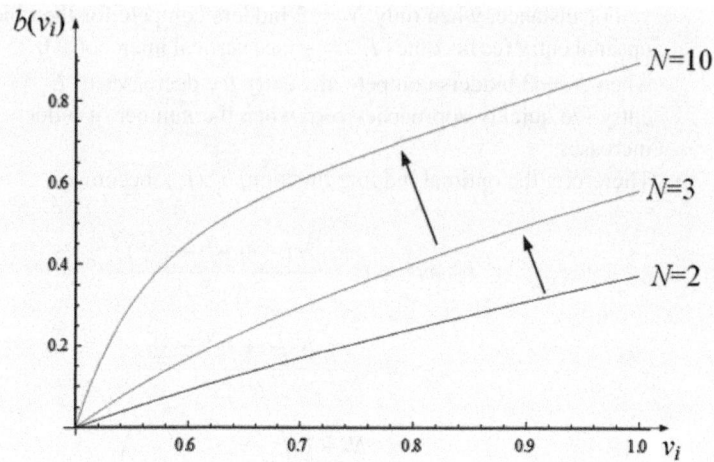

Fig. 3.7 Optimal bidding function shifts up in N

us differentiate the optimal bidding function, $b_i^*(v_i)$, with respect to valuation v_i,

$$\frac{\partial b_i^*(v_i)}{\partial v_i} = \frac{N-1}{N}\left[1+(N-1)(2v_i)^{-N}\right] > 0$$

so that bidder i's bid is increasing in his valuation for the object, v_i. Second, let us differentiate the optimal bidding function, $b_i^*(v_i)$, with respect to the number of bidders, N,

$$\frac{\partial b_i^*(v_i)}{\partial N} = \frac{1}{N^2}\left[1-(2v_i)^{-N}\right]v_i + \frac{N-1}{N}\left[N\log(2v_i)\right]v_i$$

$$= v_i\left[\frac{1-(2v_i)^{-N}}{N^2}+(N-1)\log(2v_i)\right]$$

and a sufficient condition for $\frac{\partial b_i^*(v_i)}{\partial N} \geq 0$ is $v_i \geq \frac{1}{2}$, which entails that $\log(2v_i) \geq 0$. However, we have already shown that bidders who participate in this auction have a private value of $v_i \geq v_e^* = \frac{1}{2}$, such that for those bidders who participate, their equilibrium bids increase when facing competition from more bidders.

Exercise #3.12: First-Price Auction with Discrete Valuations[B]

3.12 Consider a first-price auction with two bidders, each privately observing his valuation for the object, v_i, which is either high (v_H) or low (v_L), where $1 > v_H > v_L > 0$. The probability of bidder i drawing a high valuation, v_H, is $p \in (0, 1)$. If two bidders submit the same bid, assume that the seller randomly assigns the object between the two bidders.

(a) Show that in any bidding strategy (involving pure or mixed strategies), the low-value bidder submits a bid equal to his true value, $b(v_L) = v_L$.

- When submitting a bid $b(v_L) = v_L$, this bidder either loses the auction when facing a high-value bidder (earning a zero payoff) or wins the object with probability 1/2 when facing another low-value bidder, which also yields a zero payoff since

$$\frac{1}{2}(v_L - b(v_L)) = \frac{1}{2}(v_L - v_L) = 0.$$

- Deviations:
  - If this bidder submits a lower bid, $b(v_L) < v_L$, he loses the auction regardless of the rival he faces, thus earning a zero payoff with certainty. Therefore, the low-value bidder does not have strict incentives to shade his valuation.
  - If this bidder submits a higher bid, $b(v_L) > v_L$, he either loses the auction still earning a zero payoff or wins the auction if his bid satisfies $b(v_L) > b(v_H)$, but earning a negative payoff since $v_L - b(v_L) < 0$. Hence, the low-value bidder does not have incentives to submit a bid above his valuation.
- Overall, this bidder does not have incentives to deviate from submitting a bid $b(v_L) = v_L$. Interestingly, our analysis was unaffected by the bidding function that the high-value bidder uses, $b(v_H)$; that is, whether he submits a bid above/below/equal to his valuation or, as we show below, randomizes his bids.

(b) *Discarding dominated strategies.* Show that the high-value bidder's bid, $b(v_H)$, must satisfy $v_L \leq b(v_H) \leq v_H$.

- Bidding strictly above his valuation is dominated by bidding his valuation, so the high-value bidder has no incentives to submit $b(v_H) > v_H$. A similar argument applies to bids below v_L: if the high-value bidder were to submit a bid $b(v_H) < v_L$, he would lose the auction for sure (regardless of the rival he faces), implying that we can find other bidding functions, such as any bid satisfying $b(v_H) > v_L$, which yields a strictly positive expected payoff. Therefore, the high-value bidder's bid must lie between v_L and v_H, as required.

(c) *No pure-strategy bidding profile.* Show that there is no pure-strategy bidding strategy. [*Hint*: Consider all pure-strategy bidding profiles, and show that at least one bidder has a profitable deviation.]

- To understand the high-value bidder's incentives, note that, under complete information, he would submit a bid $b(v_H) = v_L + \varepsilon$, where $\varepsilon \to 0$, when facing a low-value bidder (winning the auction and retaining the highest surplus). However, if the high-value bidder observes that his rival is another high-value bidder, he would submit a bid $b(v_H) = v_H - \varepsilon$, where $\varepsilon \to 0$, to maximize his chances of winning the auction.
- More generally, we next list the different bidding strategy profiles that can arise according to where the bids lie. For each case, we seek to show that we can always find a profitable deviation, so no bidding strategy profile can constitute a pure-strategy equilibrium of the auction:
  - If $b(v_i) < b(v_j) < v_H$, then bidder i loses the auction, but he can increase his bid above that of bidder j, winning the auction as a result, and making a positive margin if his new bid lies below v_H. Then, a profitable deviation exists.
  - If $v_L \leq b(v_j) < b(v_i) < v_H$, then bidder i wins the auction, but he can further decrease his bid and earn a higher surplus. Then, a profitable deviation exists in this case too.
  - If $b(v_j) \leq b(v_i) = v_H$, and by the same argument as above, bidder i wins the auction, but he would have incentives to lower his bid, still winning the auction and retaining a larger surplus.
- We can then conclude that there is no pure-strategy bidding profile where players have no incentives to deviate, so there is no pure-strategy bidding profile that can be sustained as a Bayesian Nash equilibrium of the auction.

(d) *Mixed-strategy bidding profile.* Show that the following mixed-strategy bidding equilibrium can be sustained, where the low-value bidder submits a bid equal to his value, $b(v_L) = v_L$, while the high-value bidder randomizes with cumulative distribution function

$$F(b) = \frac{1-p}{p}\left(\frac{v_H - v_L}{v_H - b} - 1\right)$$

in the interval $[v_L, E[v]]$, where $E[v] \equiv pv_H + (1-p)v_L$ denotes the expected valuation. For simplicity, assume that, if a tie occurs, the seller assigns the object to the individual with the highest valuation.

Exercise #3.12: First-Price Auction with Discrete Valuations[B]

- From part (a), we know that the low-value bidder submits a bid equal to his value, $b(v_L) = v_L$, regardless of the strategy that his rival chooses, so we can focus on the high-value bidder.
- *Lower bound.* The lower bound of the randomizing interval is v_L. Otherwise, the bidder could be submitting an unnecessarily high bid. In other words, he could decrease the lower bound, still win the auction when facing a low-value bidder, and extract a larger surplus.
  - As a remark, note that if the high-value bidder submits a bid $b(v_H) = v_L$ (at the lower bound), there is a tie when facing a low-value bidder, but the object is assigned to him because his valuation is higher (one can assume that, after observing bids, the seller can also observe valuations, thus assigning the object to the bidder with the highest valuation in case of a tie).
- *Upper bound.* When the high-value bidder submits a bid $b(v_H) = v_L$, this bidder wins the auction when facing a low-value bidder (which happens with probability $1 - p$) and does not win the auction when facing another high-value bidder (as his rival bidding at exactly $b(v_H) = v_L$ has a probability that converges to zero). This yields an expected payoff of

$$(1 - p)(v_H - b(v_H))$$

which simplifies to

$$(1 - p)(v_H - v_L).$$

When the high-value bidder submits any other bid $b(v_H) = b > v_L$, he wins the auction when facing a low-value bidder (which occurs with probability $1 - p$) but only wins when facing another high-value bidder if his bid is higher than that of his rival (which occurs with probability $pF(b)$, yielding an expected payoff

$$\underbrace{(1 - p)(v_H - b)}_{\text{if facing a low-value bidder}} + \underbrace{pF(b)(v_H - b)}_{\text{if facing another high-value bidder}}$$

$$= (v_H - b)[1 - p(1 - F(b))].$$

If the high-value bidder is randomizing between v_L and points above v_L, it must be that he is indifferent between the above expected payoffs. At the upper bound, $\overline{b}$, we have that $F(\overline{b}) = 1$, implying that the above expected payoff simplifies to $(v_H - b)$, so the indifference in expected payoff entails

$$(1 - p)(v_H - v_L) = v_H - \overline{b}$$

Rearranging, and solving for $\bar{b}$, yields the upper bound

$$\bar{b} = pv_H + (1-p)v_L \equiv E[v]$$

which can be interpreted as the expected valuation and, therefore, lies above v_L but below v_H, as required from part (b).

- *Cumulative distribution function.* From the above indifference condition, evaluated at any bid in the interval $[v_L, E[v]]$, we obtain that

$$(1-p)(v_H - v_L) = (v_H - b)[1 - p(1 - F(b))].$$

Solving for the cumulative distribution function $F(b)$ yields

$$F(b) = \frac{1-p}{p}\left(\frac{v_H - v_L}{v_H - b} - 1\right)$$

with an associated density of

$$f(b) = F'(b) = \frac{1-p}{p} \frac{v_H - v_L}{(v_H - b)^2}.$$

The cumulative distribution function $F(b)$ is well behaved since (1) it originates at zero at its lower bound, v_L, that is,

$$F(v_L) = \frac{1-p}{p}\left(\frac{v_H - v_L}{v_H - v_L} - 1\right)$$

$$= \frac{1-p}{p}(1-1)$$

$$= 0,$$

(2) it increases in b since $f(b) > 0$, and (3) it is equal to one at its upper bound, $E[v]$, that is,

$$F(E[v]) = \frac{1-p}{p}\left(\frac{v_H - v_L}{v_H - pv_H - (1-p)v_L} - 1\right)$$

$$= \frac{1-p}{p} \frac{p}{1-p}$$

$$= 1.$$

- *No profitable deviations.* From our discussion, the high type is indifferent over all bids in interval $[v_L, E[v]]$. We can now check that he does not have incentives to deviate from this randomization:

Exercise #3.12: First-Price Auction with Discrete Valuations^B

– If he submits a bid $b(v_i) < v_L$, he loses the auction for sure (regardless of the rival he faces), earning a zero payoff. In contrast, with the above randomization, he earns an expected payoff of $(1 - p)(v_H - v_L)$, which is positive since $v_H > v_L$ by assumption.
– If he submits a bid $b(v_i)$ that satisfies $v_H > b(v_i) > E[v]$, he wins the auction but pays more for the object than with the above randomization.

Therefore, the high-value bidder has no incentives to bid below or above the interval $[v_L, E[v]]$. As shown in part (a), the low-value bidder does not have incentives to deviate from submitting a bid equal to his valuation, implying that this strategy profile is a symmetric mixed-strategy equilibrium of the first-price auction with discrete valuations.

(e) How are your equilibrium results affected by a marginal increase in probability p? And by a marginal increase in valuations v_H or v_L? Interpret your results.

- *Higher p.* When the probability of a high valuation, p, increases, the expected value $E[v]$ increases, expanding the support where the high-value bidder randomizes his bid, $[v_L, E[v]]$. In addition, an increase in p shifts $F(b)$ downward since

$$\frac{\partial F(b)}{\partial p} = -\frac{v_H - v_L}{(v_H - b)p^2} < 0,$$

indicating that the high-value bidder assigns less probability weight on low bids when p increases. Technically, for two probabilities p and p', where $p' > p$, $F(b, p')$ first order stochastically dominates $F(b, p)$. Intuitively, this suggests that the high-value bidder becomes more aggressive in his bids as the probability of facing another high-value bidder increases.

- *Higher v_H.* When the high value increases, the expected value $E[v]$ increases, also expanding the support of the randomization and shifting $F(b)$ downward given that

$$\frac{\partial F(b)}{\partial v_H} = -\frac{1-p}{p}\frac{b - v_L}{(v_H - b)^2} < 0.$$

In other words, the bidder randomizes over a larger support and becomes more aggressive in his bids as his valuation (and, thus, the surplus he can retain) increases.

- *Higher v_L.* When the low value increases, the expected value $E[v]$ decreases, shrinking the support of the bid randomization. In addition, a higher v_L produces a downward shift in $F(b)$ since

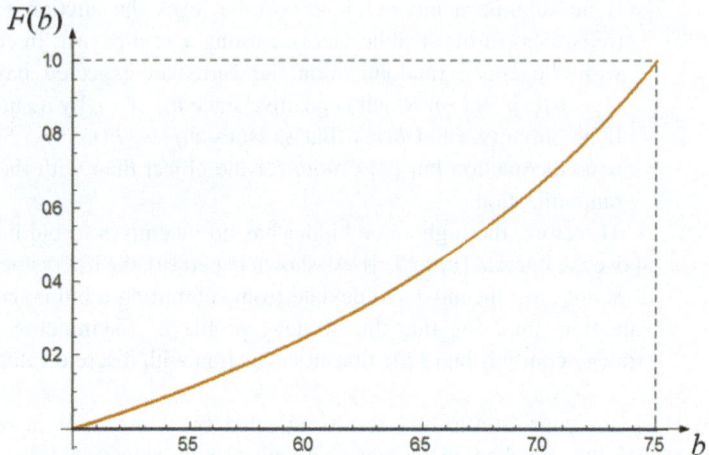

Fig. 3.8 $F(b)$ as a function of b

$$\frac{\partial F(b)}{\partial v_L} = -\frac{1-p}{p}\frac{1}{v_H - b} < 0.$$

Therefore, the high-value bidder bids over a narrower support but becomes more aggressive in his bids as the valuation of the low-value bidder increases.

(f) *Numerical example.* Evaluate your results from part (d) at parameter values $p = 1/2$, $v_H = 10$, and $v_L = 5$.

- The low-value bidder submits a bid equal to his valuation, $b(v_L) = 5$, and the high-value bidder submits a bid in the interval $[5, 7.5]$ since $E[v] = \frac{1}{2}10 + \frac{1}{2}5 = 7.5$ and uses the following cumulative distribution function

$$F(b) = \frac{1-\frac{1}{2}}{\frac{1}{2}}\left(\frac{10-5}{10-b} - 1\right)$$

$$= \frac{b-5}{10-b}.$$

Figure 3.8 depicts $F(b)$ where the horizontal axis considers that bids lie in the interval $[5, 7.5]$. As expected, the cumulative distribution function puts no probability weight at its lower bound, $F(5) = 0$, increases in b, and puts full probability weight at its upper bound, $F(7.5) = 1$.

Exercise #3.13: Collusion in First-Price Auctions, Based on McAfee and McMillan (1992)[B]

3.13 Consider a first-price auction with $N \geq 2$ risk-neutral bidders where every bidder i independently draws his valuation v_i from a uniform distribution, $F(v_i) = v_i$, where $v_i \in [0, 1]$. Assume that all bidders collude, so the bidder with the highest valuation (bidder h) submits a bid equal to zero (or one cent), and all other bidders do not participate (i.e., do not submit their bids).

This type of collusive agreement is, of course, problematic because any bidder $j \neq h$ with a valuation above zero could deviate, submitting a small but positive bid, and win the auction. To prevent defection from this agreement, the literature proposes a two-stage game where, in the first stage, every bidder i submits a bid b_i. The winning bidder in this stage pays his bid to each of the $N - 1$ losing bidders. In the second stage, the winning bidder is the only participant in the auction (with collusion) where he earns the object at a price of zero.

In this exercise, we seek to show that, in the pre-auction stage, every bidder i submitting a bid according to bidding function $b(v_i) = \frac{1}{N+1} v_i$ is a Bayesian Nash equilibrium of the game. The following steps should walk you through the proof where, as in Exercise 2.6, we use the envelope theorem approach.

(a) Write the expected utility of the bidder according to a generic, but symmetric, bidding function $b(z_i)$, where $z_i \neq v_i$.

- If, in the first stage, all bidders submit their bids according to the bidding function $b(\cdot)$, then bidder i, with valuation v_i but bidding according to valuation z_i, faces the following events:
  - He wins when z_i is higher than the valuation of all other $N - 1$ bidders, which happens with probability $F(z_i)^{N-1}$, or z_i^{N-1} since valuations are uniformly distributed. In this event, his payoff is $v_i - (N - 1)b(z_i)$, as he needs to pay his bid $b(z_i)$ to all the other $N - 1$ bidders.
  - If, instead, he loses the first-stage auction, he receives the bid of the winning bidder. This bidder must bid according to z_j, where $z_j > z_i$ since he won the auction. We denote the expected winning bid as

$$E[b(z_j)|z_j > z_i] = \int_{z_i}^{1} b(y)g(y)dy$$

where $g(y) = (N - 1)F(y)^{N-2}$. Since valuations are uniformly distributed, $g(y) = (N - 1)y^{N-2}$, and the expected winning bid becomes $(N - 1) \int_{z_i}^{1} y^{N-2} b(y) dy$.
- Therefore, bidder i's expected utility from this auction is

$$EU(z_i, v_i) = z_i^{N-1}[v_i - (N-1)b(z_i)] + (N-1)\int_{z_i}^{1} y^{N-2}b(y)dy$$

(b) Differentiate bidder i's expected utility with respect to z_i. Then, show that this first-order condition collapses to zero when bidder i bids according to his true valuation, $z_i = v_i$, under the bidding function $b(z_i) = \frac{1}{N+1}z_i$.

- Differentiating bidder i's expected utility with respect to z_i yields

$$(N-1)z_i^{N-2}[v_i - Nb(z_i)] - (N-1)z_i^{N-1}b'(z_i)$$

which we can rearrange more compactly as follows:

$$(N-1)z_i^{N-2}\left[v_i - Nb(z_i) - z_i b'(z_i)\right].$$

- Inserting the bidding function $b(z_i) = \frac{1}{N+1}z_i$ into the above first-order condition, where the derivative of the bidding function is $b'(z_i) = \frac{1}{N+1}$, we obtain

$$(N-1)z_i^{N-2}\left[v_i - N\left(\frac{1}{N+1}z_i\right) - z_i \frac{1}{N+1}\right]$$
$$= (N-1)z_i^{N-2}(v_i - z_i),$$

which is zero when bidder i bids according to his true valuation, $v_i = z_i$. Therefore, every bidder i has incentives to bid according to the bidding function $b(v_i) = \frac{1}{N+1}v_i$ based on his true valuation v_i.

(c) *Comparative statics.* How are equilibrium bids affected by the number of bidders?

- Differentiating the bidding function with respect to N yields

$$\frac{\partial b(v_i)}{\partial N} = -\frac{v_i}{(N+1)^2} < 0.$$

Intuitively, as the number of bidders increases, every bidder i decreases his bid because, upon winning, he would have to pay his bid to a larger number of bidders.
- Figure 3.9 depicts bidding function $b(v_i) = \frac{1}{N+1}v_i$, evaluated at $N = 2$ bidders, which yields $b(v_i) = \frac{1}{3}v_i$; at $N = 3$ bidders, which entails $b(v_i) = \frac{1}{4}v_i$; and at $N = 10$ bidders, which implies $b(v_i) = \frac{1}{11}v_i$. The bidding function rotates clockwise as the number of bidders increases, reflecting that every bidder i becomes more conservative in his bids.

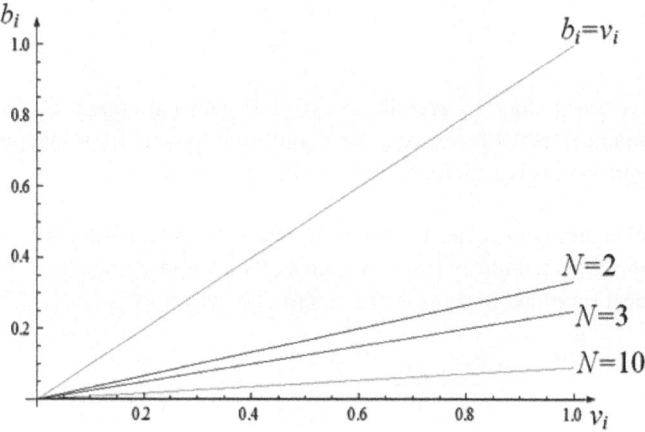

Fig. 3.9 Equilibrium bidding function in the first-price auction with collusion

(d) Find bidder i's equilibrium payoff from participating in this collusive agreement (also known as "bidding ring").

- When using the bidding function $b(v_i) = \frac{1}{N+1} v_i$, bidder i's expected utility becomes

$$EU(b(v_i), v_i) = v_i^{N-1} \left[v_i - (N-1) \overbrace{\frac{1}{N+1} v_i}^{b(v_i)} \right]$$

$$+ (N-1) \int_{v_i}^{1} \overbrace{\left(\frac{1}{N+1} y \right)}^{b(y)} y^{N-2} dy$$

where, relative to the expected utility expression in part (a), we evaluated it at $z_i = v_i$ since every bidder i has incentives to bid according to his valuation, submitting $b(v_I)$ rather than $b(z_i)$ for any $z_i \neq v_i$. This expression simplifies to

$$EU(b(v_i), v_i) = \frac{2v_i^N}{N+1} + \frac{N-1}{N+1} \left[\frac{y^N}{N} \right]_{v_i}^{1}$$

$$= \frac{2v_i^N}{N+1} + \frac{(N-1)(1-v_i^N)}{N(N+1)}$$

$$= \frac{(N+1) v_i^N + N - 1}{N(N+1)}.$$

(e) Show that bidder i's equilibrium payoff from the collusive agreement, as found in part (d), exceeds his equilibrium payoff from independently submitting his bid, as found in Exercise 2.3.

- Comparing the expected utility from collusion found in part (e) against the expected utility from independently submitting bids in a first-price auction found in part (d) of Exercise 2.3, we find that

$$\frac{(N+1) v_i^N + N - 1}{N(N+1)} - \frac{v_i^N}{N} = \frac{N-1}{N(N+1)}$$

which is unambiguously positive, entailing that every bidder i enjoys a higher expected utility from colluding in the first-price auction than independently submitting his bid.
- For an accessible literature review on collusion in different auction formats, see Robinson (1985) and Hendricks and Porter (1989).

Exercise #3.14: First-Price Auctions with Budget Constrained Bidders, Based on Che and Gale (1998)[B]

3.14 Consider again a first-price auction with $N \geq 2$ bidders, but assume now that every bidder privately observes his valuation for the object, v_i, and his budget, w_i. Bidder i's type in this context is, then, a pair (v_i, w_i), where both v_i and w_i are uniformly and independently drawn from the [0, 1] interval. For simplicity, assume that if a bidder wins the auction and the winning price is above his budget, w_i, he cannot afford to pay this price, and the seller imposes a fine on the buyer, $F > 0$, for having to renege.

(a) Show that every bidder i finds it dominated to bid above his budget, $b_i > w_i$.

- Consider a bidder who submits a bid above his own budget, $b_i > w_i$, and wins the auction. In this case, he would have to pay his bid, b_i, which he cannot afford, thus not receiving the object and, in addition, facing a penalty from the seller after reneging (for a total utility of $-F$). Therefore, deviations to any bid satisfying $b_i \leq w_i$ weakly increase bidder i's payoff, implying that bidding according to $b_i > w_i$ is a strictly dominated strategy.

Exercise #3.14: First-Price Auctions with Budget Constrained Bidders, Based...

(b) If bidder i's valuation, v_i, satisfies $\frac{N-1}{N}v_i \leq w_i$, show that bidding according to $b_i(v_i) = \frac{N-1}{N}v_i$ (as in Exercise 2.3) is still a weakly dominant strategy in the first-price auction.

- When bidder i's bid in an unconstrained setting, $b_i(v_i) = \frac{N-1}{N}v_i$, is affordable, i.e., $b_i(v_i) \leq w_i$, the bidder behaves as in Exercise 2.3, where we found that this bidding function was optimal for the unconstrained bidder.

(c) If bidder i's valuation, v_i, satisfies $\frac{N-1}{N}v_i > w_i$, show that submitting a bid equal to his budget, $b_i = w_i$, is a weakly dominant strategy.

- If bidder i's bid in an unconstrained setting, $b_i(v_i) = \frac{N-1}{N}v_i$, is unaffordable, i.e., $b_i(v_i) > w_i$, he cannot submit $b_i(v_i)$. Doing so would make him renege on the bid he submitted, thus not receiving the object and paying a fine F to the seller. Therefore, deviating to $b_i = w_i$ weakly increases bidder i's payoff (never decreases it).
- In addition, deviating from bid $b_i = w_i$ cannot strictly increase bidder i's payoff:
  - If he wins by submitting $b_i = w_i$, he exhausts his budget, yielding a payoff of zero. If he deviates to $b_i < w_i$, he would increase his probability of losing the auction, thus decreasing his expected utility.
  - If he loses by submitting $b_i = w_i$, his payoff is zero. Deviating to a higher bid increases his chances of winning, but he would not be able to pay his bid, not receiving the object and facing a fine from the seller, ultimately producing a negative payoff.

  In summary, if $\frac{N-1}{N}v_i > w_i$, so $v_i > \frac{N}{N-1}w_i$, bidder i submits a bid equal to his budget, $b_i = w_i$, and he cannot strictly increase his payoff by submitting any other bids, making $b_i = w_i$ a weakly dominant strategy in this context.

(d) Combine your results from parts (b) and (c) to describe the equilibrium bidding function in the first-price auction with budget constraints, $b_i(v_i, w_i)$. Depict it as a function of v_i.

- We found that every bidder i submits a bid $\frac{N-1}{N}v_i$ when his budget constraint is not binding (as in part b) and a bid equal to his budget, w_i, otherwise (as in part c). More compactly,

$$b_i(v_i, w_i) = \begin{cases} \frac{N-1}{N}v_i & \text{if } v_i \leq \frac{N}{N-1}w_i \\ w_i & \text{otherwise} \end{cases}$$

For illustration purposes, Fig. 3.10 considers $N = 2$ bidders. In this context, the equilibrium bid becomes $\frac{v_i}{2}$, when $\frac{v_i}{2} \leq w_i$ or $v_i \leq 2w_i$, but

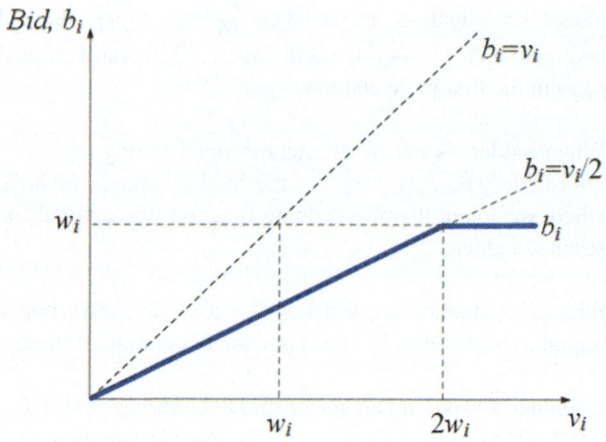

Fig. 3.10 Equilibrium bids in a first-price auction when bidders face budget constraints

it becomes $b_i = w_i$ otherwise (see the flat segment). Graphically, when the budget constraint is not met, the bidder behaves as in a standard first-price auction, but otherwise he submits a bid equal to his budget w_i.

- *Allowing for more bidders.* If more bidders compete in the auction, the diagonal line $\frac{N-1}{N}v_i$ rotates counterclockwise, approaching the 45°-line, reaching it in the limit when $N \to +\infty$. Specifically, the kink of the equilibrium bidding function occurs at $\frac{N-1}{N}v_i = w_i$ or, solving for v_i, at $v_i = \frac{N}{N-1}w_i$. This cutoff for v_i is clearly decreasing in the number of bidders since

$$\frac{\partial \left(\frac{N}{N-1}w_i\right)}{\partial N} = -\frac{w_i}{(N-1)^2} < 0.$$

Figure 3.11 depicts this effect, showing that the kink of the equilibrium bid shifts leftward as more bidders compete for the object and, in addition, the bidding function becomes steeper (and approaching the 45°-line) before the kink showing less bid shading.

For instance, with $N = 3$ bidders, the equilibrium bidding function under no constraints becomes $\frac{2}{3}v_i$, so the kink occurs at $v_i = \frac{3}{2}w_i = 1.5w_i$, as depicted in Fig. 3.11. When the number of bidders increases to $N = 4$, the equilibrium bidding function under no constraints becomes $\frac{3}{4}v_i$, implying that the kink shifts leftward to $v_i = \frac{4}{3}w_i = 1.33w_i$. Similarly, when $N = 10$, the equilibrium bidding function under no constraints becomes $\frac{9}{10}v_i$, entailing that the kink shifts further to the left to $v_i = \frac{10}{9}w_i = 1.11w_i$. In summary, bidders submit more

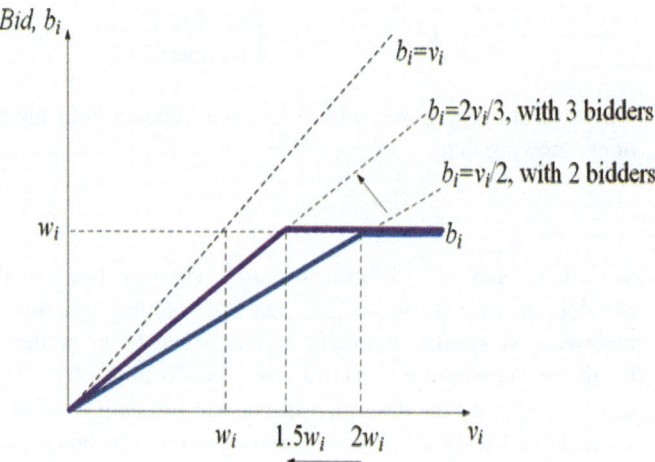

Fig. 3.11 Allowing for more bidders who face budget constraints

Fig. 3.12 Comparing equilibrium bids in the first- and second-price auctions when bidders face budget constraints

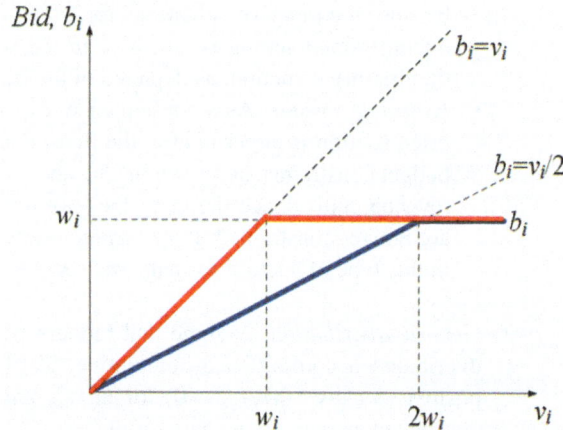

aggressive bids as more bidders compete for the object, but their bids level off at the budget w_i for lower valuations.

(e) Compare the equilibrium bids in the first- and second-price auctions with budget constrained bidders. For simplicity, assume that valuations and budgets are uniformly and independently distributed in both auction formats. Interpret.

- Equilibrium bids in first-price auctions with budget constrained bidders are those described in part (d). In second-price auctions, we know from Exercise 1.12 that the equilibrium bid is

$$b_i^{SPA}(v_i, w_i) = \begin{cases} v_i & \text{if } v_i \leq w_i \\ w_i & \text{otherwise.} \end{cases}$$

For illustration purposes, Fig. 3.12 superimposes both bidding functions, showing that

$$b_i^{SPA}(v_i, w_i) \geq b_i^{FPA}(v_i, w_i)$$

for all v_i and w_i. Specifically, $b_i^{SPA}(v_i, w_i)$ lies strictly above $b_i^{FPA}(v_i, w_i)$ for all $v_i < 2w_i$, but both bidding functions coincide otherwise. A similar argument applies when more bidders compete for the object where $b_i^{SPA}(v_i, w_i)$ lies strictly above $b_i^{FPA}(v_i, w_i)$ for all $v_i < \frac{N}{N-1} w_i$ (as described in part d), but both bidding functions coincide and level off at w_i otherwise. Intuitively, budget constraints are "softer" in the second-price auction than in the first-price auction, as the bidder does not have to pay the bid he submitted, leading him to bid more aggressively than in the first-price auction. This occurs because the range of valuations for which the bidder behaves as in the unconstrained setting is narrower in the second-price auction than in the first-price auction, as depicted in the figure.

- *Expected revenue.* As a consequence, expected revenue in the second-price auction is larger than in the first-price auction when bidders face budget constraints, as shown in Che and Gale (1998), implying that the revenue equivalence theorem does not necessarily hold when bidders are budget constrained. For a detailed description of expected revenue in this type of auction formats, see section 4.2 in Krishna (2009).

(f) *Generic distribution.* Assume that bidders' valuations are independently distributed according to a cumulative distribution function $F(v_i)$ with positive density, $f(v_i) > 0$, for all v_i, but not necessarily uniformly distributed as in previous parts of the exercise. Follow your results in part (d) to characterize the equilibrium bidding function in a first-price auction with budget constraints.

- From Exercise 2.4, we know that the equilibrium bidding function in the first-price auction without budget constraints is

$$b_i(v_i) = v_i - \frac{\int_0^{v_i} F(x)^{N-1} dx}{F(v_i)^{N-1}}$$

where the last term represents bidder i's "bid shading." Following our analysis in part (d), we can say that the bidder affords the unconstrained bid if

$$v_i - \frac{\int_0^{v_i} F(x)^{N-1}dx}{F(v_i)^{N-1}} \leq w_i$$

or, after rearranging, if the bidder's assigns a value to the object exceeding the sum of his budget and his bid shading, that is,

$$v_i \leq w_i + \frac{\int_0^{v_i} F(x)^{N-1}dx}{F(v_i)^{N-1}}.$$

Intuitively, this occurs when his budget is large, when he shades his bid significantly, or both. The bidding function in this context can, then, be summarized as

$$b_i(v_i, w_i) = \begin{cases} v_i - \frac{\int_0^{v_i} F(x)^{N-1}dx}{F(v_i)^{N-1}} & \text{if } v_i \leq w_i + \frac{\int_0^{v_i} F(x)^{N-1}dx}{F(v_i)^{N-1}} \\ w_i & \text{otherwise.} \end{cases}$$

Graphically, when bid shading decreases, approaching zero, the equilibrium bid in the unconstrained case becomes closer to the 45°-line and levels off when v_i approaches $v_i \leq w_i$. When bid shading increases, however, the equilibrium bid rotates clockwise, moving further away from the 45°-line and reaching a height of w_i for a higher valuation, v_i. If bid shading is sufficiently intense, the unconstrained bid never reaches a height of w_i, implying that the equilibrium bid coincides with and without budget constraints.

All-Pay Auctions and Auctions with Asymmetrically Informed Bidders

4

Keywords

All-pay auctions · Quibids.com · R&D race · First-price all-pay auction · Second-price all-pay auction · War of attrition · First-price auction with asymmetrically informed bidders · All-pay auction under complete information · Order of integration · Asymmetrically informed risk-neutral bidders · Asymmetrically informed risk-averse bidders

Introduction

In this chapter, we study all-pay auctions. As in the auction formats we examined in previous chapters, every bidder submits his bid and the bidder submitting the highest bid wins the object. However, unlike in auctions where only the winning bidder must pay for the object (either the highest or second-highest bid), in the all-pay auction *every bidder* must pay the bid that he submitted. As expected, this makes bidders less aggressive in their bids than in other auction formats, which we demonstrate to hold under different settings. Examples of all-pay auctions include contests, political campaigns, awarding of monopoly licenses, and R&D races where players (e.g., firms) cannot recover their participation costs upon losing. More recently, we also find this type of auctions being used by internet sellers such as QuiBids.com, which requires bidders to purchase tokens/points before an auction starts, and then submit their token bids, which cannot be recovered regardless of the outcome of the auction.

Exercise 4.1 considers an all-pay auction under complete information, where every bidder can observe his rivals' valuations. We first show that a pure-strategy bidding equilibrium cannot be sustained, but a mixed-strategy equilibrium can be supported, where players randomize their bidding strategy in an interval. In all remaining exercises, we assume that bidders cannot observe their rivals' valuations. Exercise 4.2 starts by testing if a specific bidding strategy can be sustained as an equilibrium of the all-pay auction. Exercises 4.3 and 4.4 find the bidding function that bidders use in equilibrium, where Exercise 4.3 (4.4) uses the Envelope Theorem (Direct) approach. We compare the bidding function with that in the first- and second-price auction, showing that bid shading is more intense in the all-pay auction. We also analyze how this bidding function changes in the number of bidders, and we find that bidding becomes more convex, in the sense that bidders are more conservative (aggressive) when their valuation is relatively low (high), as the probability that other bidders have a high valuation is relatively high.

Exercise 4.5 evaluates whether all-pay auctions are efficient, showing that the object goes to the individual with the highest value, as in auction formats studied in previous chapters. Exercise 4.6 then finds the seller's expected revenue in the all-pay auction, and we show that it coincides with that when he runs a first- or second-price auction (revenue equivalence).

Exercise 4.7 considers a rule change in the all-pay auction, where the winning bidder must now pay the second-highest bid (rather than his own bid), while losing bidders still pay the bids they submitted. This auction format is known as the "second-price all-pay auction," as opposed to the "first-price all-pay auction" where the winning bidder must pay the highest bid (his own). We find the equilibrium bidding function, and compare it against the first-price all-pay auction, showing that, as expected, bidders become more aggressive in this auction format since they anticipate that, if winning, the bid they must pay is lower.

Exercise 4.8 analyzes a war of attrition, which shares several features with all-pay auctions. Firms in a mature industry compete against each other, but there is enough demand to support one of the two firms making positive profits. If both firms stay in the industry, each of them makes losses, so every firm considers at every period whether to stay or exit the industry. In this context, we find that the equilibrium probability of leaving the industry at any given period to be decreasing in the monopoly profit that every firm earns.

Exercises 4.9 and 4.10 consider a first-price auction with asymmetrically informed bidders, that is, every bidder has a probability of observing his rival's valuation for the object before the auction starts. We show that some bidders may need to randomize, and this randomization expands as his rival is more likely informed. Finally, Exercise 4.10 examines how this equilibrium results are affected when players are risk averse, finding that risk aversion makes bidders less sensitive to the probability that his rivals are informed about their valuation.

Exercise #4.1: All-Pay Auction Under Complete Information, Based on Baye et al. (1996)[B]

4.1 Consider an all-pay auction between $N \geq 2$ bidders who can perfectly observe each other's valuations. The player submitting the highest bid wins the object, but all players must pay the bid they submitted.
   (a) *No pure-strategy equilibrium*. Show that there is no pure-strategy bidding profile in this auction.

- Every bidder i has no incentive to bid above his valuation v_i, as that would lead to a negative payoff both when the bidder wins and loses the auction. Similarly, every bidder i cannot be submitting a zero bid, as then each would have an incentive to marginally increase his bid to $b(v_i) = \varepsilon > 0$, where $\varepsilon \to 0$, and win the auction. We can then restrict our attention to strategy profiles where bids satisfy $0 < b(v_i) \leq v_i$.
- In this context, take a bidder who submitted a bid $b(v_i)$. If he loses the auction, he would have an incentive to deviate from $b(v_i) > 0$ to $b(v_i) = 0$ in order to reduce his payment (recall that he must pay the bid he submitted). If, instead, he wins the auction, two cases can arise:
  - If he won because $b(v_i) > b(v_j)$ for every other bidder $j \neq i$, bidder i has the incentive to lower his bid to decrease his payment.
  - If he won because $b(v_i) = b(v_j)$, that is, there was a tie in the two highest bids, the object is randomly assigned to bidder i, yielding a payoff of $\frac{1}{2}(v_i - b(v_i))$. In this context, bidder i can marginally increase his bid to $b(v_i) + \varepsilon$ and win the auction, earning a payoff $v_i - b(v_i) - \varepsilon$. Since $\varepsilon \to 0$, his payoff increases; that is,

$$v_i - b(v_i) - \varepsilon > \frac{1}{2}(v_i - b(v_i))$$

  simplifies to $\frac{1}{2}(v_i - b(v_i)) > \varepsilon$, which holds for all $\varepsilon \to 0$.
- Overall, we showed that every bidder has an incentive to deviate from his bid $b(v_i)$, which is any generic bid in pure strategies, and which holds both when he wins and when he loses the auction. Therefore, a pure-strategy equilibrium cannot be sustained in the all-pay auction when every bidder observes his rivals' valuations.

(b) *Mixed-strategy equilibrium.* Find a mixed-strategy equilibrium where every bidder i randomizes his bids according to the cumulative distribution function $F(b) = \Pr\{b_i < b\}$, with positive density in all points of the interval $[0, v]$. For simplicity, assume that bidders have the same valuation, $v_i = v_j = v$ for every bidder $j \neq i$.

- If bidders $j \neq i$ randomize their bids according to $F(b)$ and bidder i submits a bid b, he wins the auction with probability

$$\underbrace{\Pr\{b_1 < b\} \times \cdots \times \Pr\{b_N < b\}}_{N-1 \text{ times}} = \underbrace{F(b) \times \cdots \times F(b)}_{N-1 \text{ times}} = F(b)^{N-1}$$

so his expected payoff is

$$\underbrace{F(b)^{N-1}(v - b)}_{\text{if he wins}} + \underbrace{\left[1 - F(b)^{N-1}\right](-b)}_{\text{if he loses}} = F(b)^{N-1}v - b$$

For bidder i to randomize, he must be indifferent between participating and not participating in the auction, earning a zero payoff, that is,

$$F(b)^{N-1}v - b = 0.$$

Solving for the cumulative distribution function $F(b)$, yields

$$F(b) = \left(\frac{b}{v}\right)^{\frac{1}{N-1}}$$

We can check that, at its lower bound, we have that $F(0) = \left(\frac{0}{v}\right)^{\frac{1}{N-1}} = 0$; $F(b)$ increases in b; and at its upper bound, $F(v) = \left(\frac{v}{v}\right)^{\frac{1}{N-1}} = 1$, as required. Differentiating with respect to b, we find its associated density function

$$f(b) = \frac{1}{N-1}\left(\frac{b}{v}\right)^{-\frac{N-2}{N-1}}$$

which is positive for all points in the support $[0, v]$.
- **Remark:** For a characterization of this mixed-strategy equilibrium in a context where every bidder has a different valuation, v_i, see (Baye et al. 1996, page 296).

(c) *Comparative statics.* How is the randomization you found in the mixed-strategy equilibrium of part (b) affected by the number of bidders, N?

- Differentiating $F(b)$ with respect to N, yields

$$\frac{\partial F(b)}{\partial N} = -\frac{1}{(N-1)^2}\ln\left(\frac{b}{v}\right) \times \left(\frac{b}{v}\right)^{\frac{1}{N-1}}$$

which is positive since $\ln\left(\frac{b}{v}\right) \leq 0$ given that $b \leq v$.
- Graphically, this result entails that an increase in N shifts $F(b)$ upward, making it more concave, as depicted in Fig. 4.1 (which assumes a valuation $v = \frac{1}{2}$). Intuitively, this upward shift indicates that every bidder i uses a more aggressive randomization as more bidders compete for the object. In particular, Fig. 4.1 illustrates that, when $N = 2$, $F(b) = \left(\frac{b}{1/2}\right)^{\frac{1}{2-1}} = 2b$; when the number of bidders increases to $N = 4$, we find $F(b) = \left(\frac{b}{1/2}\right)^{\frac{1}{4-1}} = (2b)^{1/3}$; and when $N = 10$, we have $F(b) = \left(\frac{b}{1/2}\right)^{\frac{1}{10-1}} = (2b)^{1/9}$.

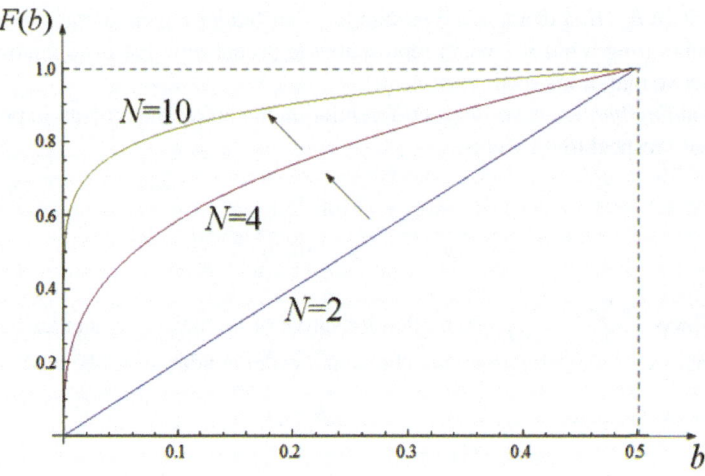

Fig. 4.1 $F(b)$ shifts as N increases

Exercise #4.2: Testing a Bidding Function in a First-Price, All-Pay Auction[A]

4.2 Consider the following all-pay auction with two bidders privately observing their valuation for the object. Valuations are uniformly distributed $v_i \sim U[0, 1]$. The player submitting the highest bid wins the object, but all players must pay the bid they submitted. Find the optimal bidding strategy, taking into account that it is of the form $b_i(v_i) = m \times v_i^2$, where m is a positive constant.

- *Writing bidder i's problem.* We know that bidder i's expected utility from choosing bid b_i is

$$EU_i(b_i|v_i) = \Pr(win) \times v_i + \Pr(lose) \times 0 - b_i(v_i)$$
$$= \Pr(win) \times v_i - b_i(v_i)$$

where $b_i(v_i)$ expresses bidder i's bidding function.

- *Probability of winning.* Since valuations are uniformly distributed in the unit line, we know that $\Pr(win) = b_i^{-1}(v_i)$; the cumulative probability that $v_i > v_j$. This is analogous to the probability of winning in exercises we analyzed in previous chapters where we also assumed that bidders' valuations were uniformly distributed according to $U[0, 1]$.

Thus, bidder i must choose a bid b_i that maximizes his expected utility,

$$\max_{b_i \geq 0} EU_i(b_i|v_i) = b_i^{-1}(b_i) \times v_i - b_i(v_i)$$

where $b_i^{-1}(b_i)$ denotes the probability that bidder i wins the auction when submitting a bid b_i, which represents the probability that his valuation lies above that of his rival, $v_i > v_j$.
- *Finding bidder i's strategy.* Differentiating the above maximization problem with respect to b_i, yields

$$\frac{\partial EU_i(b_i|v_i)}{\partial b_i} = \frac{\partial b_i^{-1}(b_i)}{\partial b_i} v_i - 1 = 0.$$

Since $\frac{\partial b_i^{-1}(b_i)}{\partial b_i} = \frac{1}{b_i'(v_i)}$ (i.e., the derivative of the inverse coincides with the inverse of the derivative), the above first order condition can be rewritten as

$$\frac{v_i}{b_i'(v_i)} = \frac{v_i - b_i'(v_i)}{b_i'(v_i)} - 1 = 0$$

Because $b_i'(v_i) > 0$, this expression holds when its numerator is zero, that is,

$$b_i'(v_i) = v_i$$

Integrating both sides of this expression gives

$$b_i(v_i) = \int_0^{v_i} v_i dv_i$$

which yields

$$b_i(v_i) = \int_0^{v_i} v_i dv_i = \frac{1}{2} v_i^2$$

Therefore, each bidder bids according to $b_i(v_i) = m \times v_i^2$, where $m = 1/2$. More generally, in a setting with N bidders, the equilibrium bidding function in an all-pay auction is $b_i(v_i) = \frac{N-1}{N} v_i^N$, which is increasing and convex in bidder i's valuation for the object v_i, as we formally show in Exercise 4.3.

Exercise #4.3: Finding the Equilibrium Bidding Function in the First-Price All-Pay Auction Using the Envelope Theorem Approach[C]

4.3 Modify the first-price, sealed-bid auction so that the loser also pays his bid (but does not win the object). This modified auction is called the all-pay auction (APA), since all bidders must pay their bids, regardless of whether they win the object or not.
(a) Find the equilibrium bidding function allowing for valuations to be distributed according to a general cumulative distribution function $F(v_i)$ with

an associated density function $f(v_i) > 0$ in all v_i. Assume that N bidders participate in this auction.

- APA with $N \geq 2$ bidders, and independent private valuations drawn from a uniform distribution with support on $[0, 1]$. In this case the utility of bidder i when his valuation is v_i becomes

$$u_i(b_i, b_{-i}, v_i) = \begin{cases} v_i - b_i & \text{if } b_i > \max_{j \neq i} b_j \\ \frac{v_i - b_i}{k} & \text{if } b_i = \max_{j \neq i} b_j \\ -b_i & \text{if } b_i < \max_{j \neq i} b_j \end{cases}$$

where $k \geq 2$ denotes the number of bidders who submitted the highest bid. This payoff function coincides with that of the first-price auction (FPA), except for the case in which bidder i's bid is lower than that of the highest competing bidder, i.e., $b_i < \max_{j \neq i} b_j$, and thus loses the auction, since in the APA bidder i must still pay his bid. Indeed, his payoff in this event is $-b_i$, as opposed to a payoff zero in the FPA.

- The expected payoff of bidder i with a real valuation of v_i but bidding according to a different valuation of $z_i \neq v_i$ is

$$EU_i(v_i, z_i) = \Pr(win) \underbrace{[v_i - b_i(z_i)]}_{\text{Payoff from winning}} + [1 - \Pr(win)] \underbrace{(-b_i(z_i))}_{\text{Payoff from losing}}$$

where, if winning, bidder i obtains a net payoff $v_i - b_i(z_i)$; but, if losing, which occurs with probability $1 - \Pr(win)$, he must pay the bid he submitted. Similar to the FPA, his probability of wining is given by

$$\Pr(win) = \Pr\left(\max_{j \neq i} b_j(v_j) < b_i(z_i)\right)$$

$$= \Pr\left(\max_{j \neq i} v_j < z_i\right)$$

$$= F(z_i)^{N-1}$$

since we assume a symmetric and monotonic bidding function. Hence, the expected utility from participating in the APA can be rewritten as

$$EU_i(v_i, z_i) = F(z_i)^{N-1}[v_i - b_i(z_i)] + \left[1 - F(z_i)^{N-1}\right](-b_i(z_i))$$

or, rearranging, as

$$EU_i(v_i, z_i) = F(z_i)^{N-1} v_i - b_i(z_i)$$

Intuitively, if winning, bidder i enjoys his valuation for the object, v_i, but he must pay the bid he submitted both when winning and losing the auction.

- Since the bidder is supposed to be utility-maximizing, we can find the following first-order condition using the Envelope Theorem:

$$\frac{dEU_i(v_i, z_i)}{dz_i} = (N-1) F(z_i)^{N-2} f(z_i) v_i - \frac{db_i(z_i)}{dz_i} = 0$$

Hence, for $b_i(v_i)$ to maximize bidder i's utility, it should not be optimal for him to pretend to have a valuation z_i different from his real one, v_i. Therefore, $z_i = v_i$ is the optimal solution to the above first-order condition, implying that

$$(N-1) F(v_i)^{N-2} f(v_i) v_i = \frac{d\widehat{b}_i(v_i)}{dv_i}$$

Given that this equality holds for every valuation, v_i, we can apply integrals on both sides of the equality to obtain,

$$b_i(v_i) = \int_0^{v_i} (N-1) F(x)^{N-2} f(x) x\, dx + C$$

where C is the constant of integration, which is equal to 0 given that $\widehat{b}(0) = 0$, i.e., every bidder submits a bid of zero when his valuation is the lowest.

Hence, using the property that $dF(x)^{N-1} = (N-1) F(x)^{N-2} f(x)$, we obtain the optimal bidding function for this APA in the form of

$$b_i^{APA}(v_i) = \int_0^{v_i} x\, dF(x)^{N-1}$$

(b) *Comparison with first-price auction.* Compare the equilibrium bidding function you found in part (a) against that in a first-price auction. In which auction format the bidder submits more aggressive bids?

- From Chap. 2 (see Exercise 2.4), the equilibrium bidding function in the first-price auction (FPA) is

$$b_i^{FPA}(v_i) = v_i - \frac{\int_0^{v_i} F(x)^{N-1} dx}{F(v_i)^{N-1}}$$

$$= \frac{v_i F(v_i)^{N-1} - \int_0^{v_i} F(x)^{N-1} dx}{F(v_i)^{N-1}}$$

$$= \frac{\int_0^{v_i} x \, dF(x)^{N-1}}{F(v_i)^{N-1}}$$

This bidding is, then, higher than that in the all-pay auction, $b_i^{APA}(v_i) = \int_0^{v_i} x \, dF(x)^{N-1}$, since $b_i^{FPA}(v_i) > b_i^{APA}(v_i)$ entails

$$\frac{\int_0^{v_i} x \, dF(x)^{N-1}}{F(v_i)^{N-1}} > \int_0^{v_i} x \, dF(x)^{N-1}$$

given that $F(v_i)^{N-1} < 1$. Intuitively, a bidder in the all-pay auction must pay his bid whether he wins the object or not, leading him to bid more conservatively than in the first-price auction.

(c) *Uniform distribution*. For the rest of the exercise assume that valuations are uniformly distributed in $[0, 1]$. Show that, in the case of two bidders, the equilibrium bidding function you found in part (a) can be represented as a *convex* function of his private valuation v_i for the object.

- When valuations are drawn from a uniform distribution $F(x) = x$ with density $f(x) = 1$ and support $[0, 1]$, we obtain that $F(x)^{N-1} = x^{N-1}$ and its derivative is $dF(x)^{N-1} = (N-1)x^{N-2}$, implying that the optimal bidding function becomes

$$b_i^{APA}(v_i) = \int_0^{v_i} x(N-1)x^{N-2} dx$$

$$= (N-1) \int_0^{v_i} x^{N-1} dx$$

$$= (N-1) \left[\frac{x^N}{N} \right]_0^{v_i}$$

$$= \frac{N-1}{N} v_i^N$$

Therefore, in the case of $N = 2$ bidders, the optimal bidding function for the APA will be $b_i^{APA}(v_i) = \frac{1}{2}v_i^2$, which is increasing and convex in bidder i's valuation, as depicted in Fig. 4.2. As a reference, the figure also includes the 45°-line where $b_i = v_i$.

(d) *Comparative statics*. Still assuming $v_i \sim U[0, 1]$, analyze how the optimal bidding function varies in the number of bidders, N.

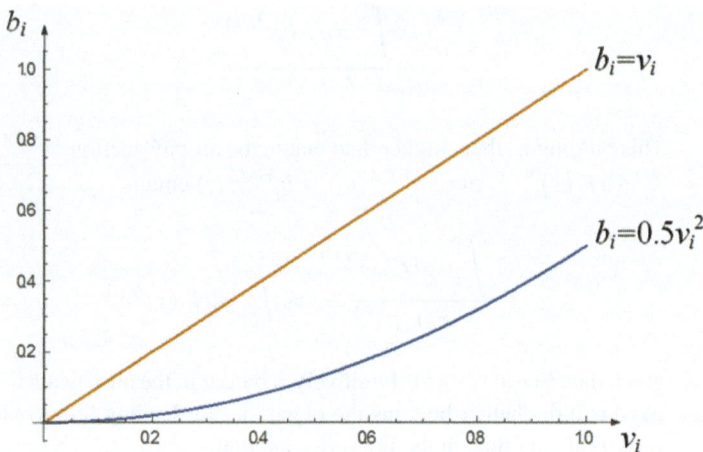

Fig. 4.2 Equilibrium bidding function in the all-pay auction with two bidders

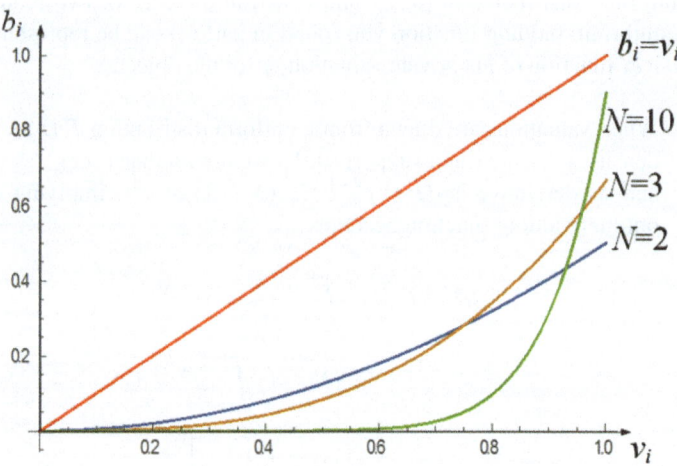

Fig. 4.3 Equilibrium bidding function in the all-pay auction—More bidders

- When increasing the number of bidders, N, we obtain a more convex function. As depicted in Fig. 4.3, the optimal bidding function in the APA with two bidders is $\frac{v_i^2}{2}$, with three bidders becomes $\frac{2}{3}v_i^3$, and with ten bidders this bidding function is $\frac{9}{10}v_i^{10}$. Intuitively, as more bidders compete for the object, bid shading becomes more substantial when bidder i has a relatively low valuation, but induces him to bid more aggressively when his valuation is likely the highest among all other players, i.e., when $v_i \to 1$ (see right-hand side of Fig. 4.3).

Exercise #4.3: Finding the Equilibrium Bidding Function in the First-Price All-...

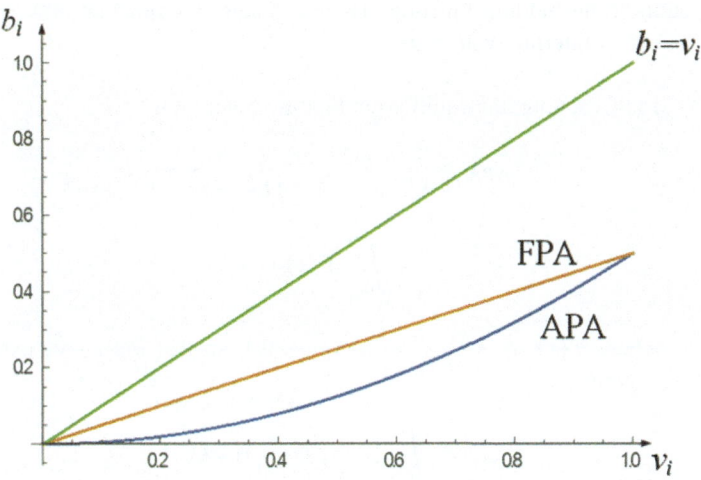

Fig. 4.4 Equilibrium bidding functions in first- and all-pay auction

(e) *Comparison.* Still assuming $v_i \sim U[0, 1]$, how do players' bids compare to those in the first-price auction? What is the intuition behind the difference in bids?

- The equilibrium bidding function in the FPA, $b^{FPA}(v_i) = \frac{N-1}{N} v_i$, lies above that in the APA, $b^{APA}(v_i) = \frac{N-1}{N} v_i^N$, because the difference

$$b^{FPA}(v_i) - b^{APA}(v_i) = \frac{(N-1)\left(1 - v_i^{N-1}\right) v_i}{N}$$

is positive, since $N \geq 2$ and $0 \leq v_i^{N-1} \leq 1$ (graphically, the 45°-line lies above any convex function such as v^{N-1}). Hence, every bidder i in the APA bids less aggressively than in the FPA since he has to pay the bid he submits.
- For illustration purposes, Fig. 4.4 depicts the equilibrium bidding function in the first- and all-pay auctions, evaluated at $N = 2$, so that $b^{FPA}(v_i) = \frac{1}{2} v_i$ and $b^{APA}(v_i) = \frac{1}{2} v_i^2$. Note that both functions coincide at $v_i = 0$ and at $v_i = 1$, but $b^{FPA}(v_i) > b^{APA}(v_i)$ for all other valuations. A similar argument applies if we evaluate both expressions at $N = 3$ or any other number of bidders.

(f) *All-pay auction with exponentially distributed valuations.* Assume that valuations are distributed according to an exponential distribution $F(v_i) = 1 - \exp(-\lambda v_i)$, where $\lambda > 0$, and there are $N = 2$ bidders. Find the equilibrium bidding function. Then, evaluate this function at $\lambda = 1$ and at $\lambda = 2$. Interpret your results.

- In this setting, the equilibrium bidding function is

$$b_i^{APA}(v_i) = \int_0^{v_i} (2-1) F(x)^{2-2} f(x) x\, dx$$

$$= \int_0^{v_i} f(x) x\, dx$$

where $f(x) \equiv \frac{\partial F(v_i)}{\partial v_i} = \lambda \exp(-\lambda v_i)$, so that the above expression becomes

$$b_i^{APA}(v_i) = \int_0^{v_i} \left[\lambda \exp(-\lambda x)\right] x\, dx$$

$$= -\int_0^{v_i} x\, d\left[\exp(-\lambda x)\right]$$

$$= -\left[x \exp(-\lambda x)\right]_0^{v_i} + \int_0^{v_i} \exp(-\lambda x)\, dx$$

$$= -v_i \exp(-\lambda v_i) - \frac{1}{\lambda}\left[\exp(-\lambda x)\right]_0^{v_i}$$

$$= -v_i \exp(-\lambda v_i) - \frac{1}{\lambda}\exp(-\lambda v_i) + \frac{1}{\lambda}$$

$$= \frac{1 - (1 + \lambda v_i)\exp(-\lambda v_i)}{\lambda}$$

We do not include the integration constant since $b_i(0) = 0$.

- Evaluating this bidding function at $\lambda = 1$, we obtain

$$b_i^{APA}(v_i) = 1 - (1 + v_i)\exp(-v_i)$$

while evaluating it at $\lambda = 2$ yields

$$b_i^{APA}(v_i) = \frac{1 - (1 + 2v_i)\exp(-2v_i)}{2}$$

Figure 4.5 depicts these bidding functions, showing that, as the exponential distribution assigns a larger probability weight to low valuations (higher λ), every bidder i bids less aggressively. (Recall that, as described in Exercise 1.7, parameter λ represents the rate parameter, describing how quickly the decay of the exponential function is, and that the expected value, $E[x] = \frac{1}{\lambda}$, and its variance, $Var[x] = \frac{1}{\lambda^2}$, are both decreasing in λ, implying that bidders' values are more concentrated at

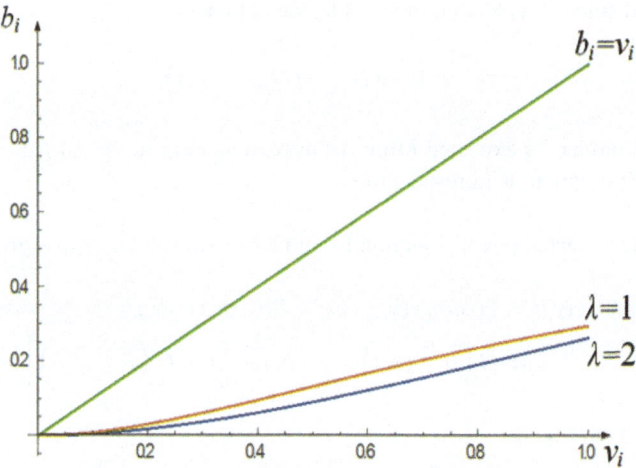

Fig. 4.5 Equilibrium bidding in the all-pay auction when valuations are exponentially distributed

the lower bound.) Graphically, an increase in λ shifts the equilibrium bidding function downward.

Exercise #4.4: Finding the Equilibrium Bidding Function in the First-Price All-Pay Auction Using the Direct Approach[B]

4.4 Consider the all-pay auction we studied in Exercise 4.3. In this exercise, rather than using the envelope approach, we seek to directly differentiate with respect to bidder i's bid, b_i, and rearrange, to obtain his equilibrium bidding function. We know that bidder i wins the auction if his bid, b_i, exceeds that of the highest competing bidder, $\beta(Y_1)$, where Y_1 represents the highest valuation among the $N-1$ remaining bidders (the second order statistic). Answer the following questions to find the equilibrium bidding function:

(a) Find bidder i's probability of winning as a function of b_i.

- The probability of winning is given by

$$\Pr(win) = \Pr(\beta(Y_1) \leq b_i)$$

applying the inverse bidding function on both sides of the inequality yields

$$\Pr(win) = \Pr\left(Y_1 \leq \beta^{-1}(b_i)\right)$$

and since Y_1 is a random variable, we obtain

$$\Pr(win) = G\left(\beta^{-1}(b_i)\right)$$

(b) Write bidder i's expected utility from participating in the all-pay auction, and differentiate with respect to b_i.

- The expected payoff function for bidder i with valuation v_i is given by

$$EU_i(v_i) = \Pr(win)(v_i - b_i) + \Pr(lose)(-b_i)$$
$$= G\left(\beta^{-1}(b_i)\right)(v_i - b_i) + \left[1 - G\left(\beta^{-1}(b_i)\right)\right](-b_i)$$

or, rearranging,

$$EU_i(v_i) = G\left(\beta^{-1}(b_i)\right)v_i - b_i$$

We can now take first-order conditions with respect to the optimal bid of this player, b_i, and obtain

$$\frac{dEU_i(v_i)}{db_i} = \frac{g\left(\beta^{-1}(b_i)\right)}{\beta'\left(\beta^{-1}(b_i)\right)}v_i - 1 = 0$$

Multiplying both sides by $\beta'\left(\beta^{-1}(b_i)\right)$, and rearranging,

$$g\left(\beta^{-1}(b_i)\right)v_i - \beta'\left(\beta^{-1}(b_i)\right) = 0$$

(c) In equilibrium, we have that $\beta(v_i) = b_i$. We can then invert this bidding function to obtain $v_i = \beta^{-1}(b_i)$. Use this property in the first-order condition you found in part (b), and rearrange, to obtain the equilibrium bidding function in an all-pay auction.

- Using $v_i = \beta^{-1}(b_i)$, we can rewrite the above expression as

$$g(v_i)v_i - \beta'(v_i) = 0$$

Since this equality holds for any value of v, we can integrate both sides of the equality to obtain

$$\beta(v_i) = \int_0^{v_i} g(y)\,y\,dy + C$$

where C is the integration constant and is zero at $\beta(0) = 0$. Thus, the optimal bidding function in an all-pay auction becomes

$$\beta(v_i) = \int_0^{v_i} g(y)\, y\, dy$$

(d) Show that the equilibrium bidding function found in part (c), using the Direct approach, coincides with that found in Exercise 4.3, using the Envelope Theorem approach.

- First, note that $G(x) = F(x)^{N-1}$ and, hence, its derivative is $g(x) = (N-1)F(x)^{N-2} f(x)$. Substituting into the expression of the optimal bidding function that we found using the Direct approach yields

$$\beta(v_i) = \int_0^{v_i} x(N-1)F(x)^{N-2} f(x)\, dx$$

$$= \int_0^{v_i} x\, dF(x)^{N-1}$$

$$= b^{APA}(v_i)$$

since $(N-1)F(x)^{N-2} f(x) = dF(x)^{N-1}$. Hence,

$$\beta(v_i) = b^{APA}(v_i)$$

and both methods produce equivalent bidding functions.

Exercise #4.5: Efficiency in All-Pay Auctions[A]

4.5 Consider the all-pay auction in Exercise 4.3. Argue that the object, in equilibrium, is assigned to the bidder with the highest valuation.

- In this auction format, every bidder i uses a bidding function

$$\beta(v_i) = \int_0^{v_i} g(y)\, y\, dy$$

where $g(x) = (N-1)F(x)^{N-2} f(x)$. This bidding function is monotonically increasing in the bidder's valuation, v_i (i.e., it expands the interval of integration without affecting the integrand). Then, the individual submitting the highest bid, h, satisfies $\beta_h(v_h) > \beta_j(v_j)$ for every bidder $j \neq h$, implying that bidder h assigns the highest value to the object, $v_h > v_j$. In summary, the bidder who wins the auction, h, also has the highest value for the object, entailing that the all-pay auction is efficient.
- Recall from Exercise 1.5 (analyzing second-price auctions) and 2.6 (studying first-price auctions) that the notion of efficiency in auctions means that the seller cannot reassign the object to another bidder and increase his expected

revenue. Similarly, allowing for trade between bidders once the auction is over would not improve the payoff of at least one bidder without reducing the payoff of any other bidders.

Exercise #4.6: Finding the Expected Revenue in the First-Price All-Pay Auction[B]

4.6 Consider the first-price all-pay auction studied in Exercise 4.3.
   (a) Find the seller's expected revenue.

- The seller's expected revenue in an all-pay auction is the expected bid from bidder i, $\int_0^1 b^{APA}(x) f(x) dx$, times N bidders since all of them must pay the bid they submit. That is,

$$R^{APA} = N \int_0^1 b^{APA}(x) f(x) dx$$

Inserting the equilibrium bid we found in Exercises 4.3 and 4.4, $b^{APA}(v_i) = \int_0^{v_i} x dF(x)^{N-1}$, yields

$$R^{APA} = N \int_0^1 \underbrace{\left[\int_0^{v_i} x dF(x)^{N-1} \right]}_{b^{APA}(v_i)} f(v_i) dv_i$$

$$= N \int_0^1 \left(\int_0^{v_i} x \underbrace{(N-1) F(x)^{N-2} f(x) dx}_{dF(x)^{N-1}} \right) f(v_i) dv_i$$

$$= N(N-1) \int_0^1 \left(\int_0^{v_i} x F(x)^{N-2} f(x) dx \right) f(v_i) dv_i$$

since $dF(x)^{N-1} = (N-1) F(x)^{N-2} f(x) dx$. We can now exchange the order of integration, so we integrate the probability density function into the cumulative distribution function from the lower bound v_i to the upper bound 1, as follows:

$$R^{APA} = N(N-1) \int_0^1 \left(\int_{v_i}^1 f(x) dx \right) v_i F(v_i)^{N-2} f(v_i) dv_i$$

We can now consider that the winning bid, which is distributed on the unit line [0, 1], must be above all the other $N-1$ bids, to obtain

Exercise #4.6: Finding the Expected Revenue in the First-Price All-Pay Auction[B]

$$R^{APA} = N(N-1) \int_0^1 [F(x)]_{v_i}^1 v_i F(v_i)^{N-2} f(v_i) dv_i$$

Rearranging, finally yields

$$R^{APA} = N(N-1) \int_0^1 v_i [1 - F(v_i)] F(v_i)^{N-2} f(v_i) dv_i$$

This expected revenue coincides, as expected, with that in the first- and second-price auctions. We return to this coincidence of expected revenues across different auction formats in Chap. 6, where we introduce the Revenue Equivalence principle.

(b) *Uniformly distributed values.* Evaluate the seller's expected revenue found in part (a) assuming that bidders' valuations are all drawn from a uniform distribution $U[0, 1]$.

- When valuations are uniformly distributed, $F(v_i) = v_i$ and $f(v_i) = 1$, so the expected revenue found in part (a) simplifies to

$$\begin{aligned}
R^{APA} &= N(N-1) \int_0^1 v_i (1 - v_i) v_i^{N-2} dv_i \\
&= N(N-1) \left[\left(\frac{1}{N} - \frac{v_i}{N+1} \right) v_i^N \right]_0^1 \\
&= N(N-1) \left(\frac{1}{N} - \frac{1}{N+1} \right) \\
&= N(N-1) \frac{1}{N(N+1)} \\
&= \frac{N-1}{N+1}
\end{aligned}$$

which, as expected, coincides with the expected revenue in first- and second-price auctions.

(c) *Exponentially distributed valuations.* Evaluate the seller's expected revenue found in part (a) assuming two bidders whose valuations are drawn from an exponential distribution, $F(v_i) = 1 - \exp(-\lambda v_i)$, where $v_i \in [0, \infty)$.

- Since $1 - F(v_i) = 1 - [1 - \exp(-\lambda v_i)] = \exp(-\lambda v_i)$ and $f(v_i) = \lambda \exp(-\lambda v_i)$, the seller's expected revenue becomes

$$R^{APA} = 2 \int_0^1 v_i [1 - F(v_i)] [F(v_i)]^{2-2} f(v_i) dv_i$$

$$= 2\lambda \int_0^1 v_i \exp(-2\lambda v_i)\, dv_i$$

which coincides with the expected revenue in the second-price auction (see part (e) of Exercise 1.7) and in the first-price auction (see part (d) of Exercise 3.1). Rearranging this expression following the same algebra steps as in Exercise 1.7, part (e), we obtain

$$R^{APA} = \frac{1}{2\lambda}$$

which decreases in λ at a decreasing rate.

- Intuitively, as described in Exercise 4.3(f), the expected value of the exponential distribution, $E[x] = \frac{1}{\lambda}$, and its variance, $Var[x] = \frac{1}{\lambda^2}$, are both decreasing in parameter λ, implying that bidders' values are more concentrated at the lower bound of the distribution. As a consequence, the seller earns a lower expected revenue when parameter λ increases since expected valuations are lower and, in addition, more concentrated around zero.

Exercise #4.7: Finding Equilibrium Bids in the Second-Price All-Pay Auction[B]

4.7 Consider again the all-pay auction studied in Exercise 4.3, but assume now that only two bidders compete for the object. The bidder submitting the highest bid still wins the object but pays the second-highest bid (as in a second-price auction).

(a) Find the equilibrium bids in this second-price all-pay auction. [*Hint*: Use the Envelope Theorem approach, considering a bidder with valuation v_i who bids according to valuation z.]

- If a bidder with valuation v_i bids according to valuation z, he submits a bid $b_i(z)$. Given that there are only two bidders in this auction, he wins with probability $F(z)$. His expected utility maximization problem from submitting a bid $b_i(z)$ is then

$$\max_{z \geq 0} \underbrace{F(z) \times \left(v_i - \frac{1}{F(z)} \int_0^z b_i(x) f(x) dx \right)}_{\text{if winning}} - \underbrace{[1 - F(z)] \times b_i(z)}_{\text{if losing}}$$

where $\frac{1}{F(z)} \int_0^z b_i(x) f(x)\, dx$ denotes the expected bid that bidder i pays if he wins, which occurs when his bid exceeds that of his rival, $b_i(z) >$

Exercise #4.7: Finding Equilibrium Bids in the Second-Price All-Pay Auction

$b_j(v_j)$, or $z > v_j$, which happens with probability $F(z)$. Otherwise, he loses and pays his bid $b_i(z)$ with probability $1 - F(z)$.

- Differentiating with respect to z, and evaluating the first-order condition at $z = v_i$, yields

$$v_i f(v_i) - b_i(v_i) f(v_i) + b_i(v_i) f(v_i) - [1 - F(v_i)] b_i'(z) = 0$$

which simplifies to

$$v_i f(v_i) = [1 - F(v_i)] b_i'(z)$$

Solving for $b_i'(z)$, we find that

$$b_i'(z) = v_i \frac{f(v_i)}{1 - F(v_i)}$$

Integrating both sides, we obtain the equilibrium bidding function in the second-price all-pay auction

$$b_i(v_i) = \int_0^{v_i} x \frac{f(x)}{1 - F(x)} dx$$

(b) Compare equilibrium bids in the second-price all-pay auction from part (a) and in the first-price all-pay auction from Exercise 4.3. Interpret your results.

- The equilibrium bidding function the first-price all-pay auction found in Exercise 4.3 was

$$\beta(v_i) = \int_0^{v_i} x (N-1) F(x)^{N-2} f(x) dx$$

which in the context of $N = 2$ bidders simplifies to $\beta(v_i) = \int_0^{v_i} x f(x) dx$. Comparing it with the equilibrium bidding function in the second-price all-pay auction found in part (a), we obtain that

$$\int_0^{v_i} x \frac{f(x)}{1 - F(x)} dx > \int_0^{v_i} x f(x) dx$$

since $F(x)$ is a probability, that is, $F(x) \in [0, 1]$, implying that $\frac{1}{1-F(x)} > 1$. Therefore, bidding is more aggressive in the second-price all-pay auction than in the first-price all-pay auction.

(c) *Uniformly distributed valuations.* Assume now that valuations are uniformly distributed, $F(v_i) = v_i$, where $v_i \in [0, 1]$. Evaluate the equilibrium

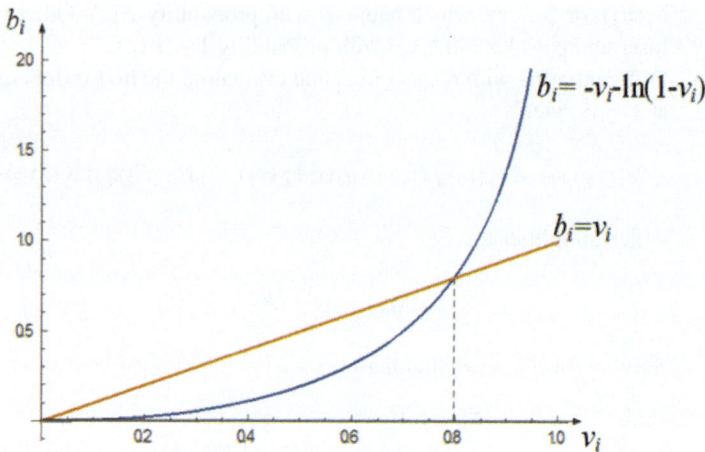

Fig. 4.6 Equilibrium bidding function in the second-price all-pay auction

bidding function in the first-price all-pay auction and compare it against that in the second-price all-pay auction.

- The equilibrium bidding function in the first-price all-pay auction in this context is

$$\beta(v_i) = \int_0^{v_i} x\, dx = \frac{1}{2} v_i^2$$

while that in the second-price all-pay auction is

$$b_i(v_i) = \int_0^{v_i} x \frac{1}{1-x} dx$$
$$= \int_0^{v_i} \left(-1 + \frac{1}{1-x}\right) dx$$
$$= -v_i - \ln(1-v_i)$$

Figure 4.6 depicts this bidding function, along with the 45°-line where $b_i = v_i$, showing that bidder i shades his bid when his valuation is relatively low ($v_i < 0.8$), but otherwise he submits a bid above his valuation.

For comparison purposes, Fig. 4.7 includes the equilibrium bidding function in the first-price all-pay auction and that in the second-price all-pay auction. As described in part (b), bidding behavior is more aggressive in the second-price version than in the first-price version of the all-pay auction.

Exercise #4.7: Finding Equilibrium Bids in the Second-Price All-Pay Auction[B]

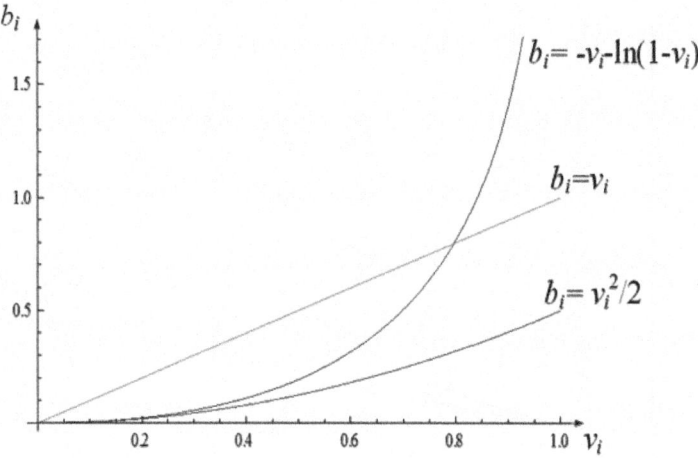

Fig. 4.7 Equilibrium bidding function in the first-price all-pay auction and in the second-price all-pay auction—Uniform distribution

(d) *Exponentially distributed valuations.* Assume now that valuations follow exponential distribution, $F(v_i) = 1 - \exp(-\lambda v_i)$, where $\lambda > 0$. Evaluate the equilibrium bidding function in this context, and compare it to the first-price all-pay auction.

- Since $1 - F(v_i) = 1 - [1 - \exp(-\lambda v_i)] = \exp(-\lambda v_i)$ and $f(v_i) = \lambda \exp(-\lambda v_i)$, the equilibrium bid in the second-price all-pay auction becomes

$$b_i(v_i) = \int_0^{v_i} x \frac{f(x)}{1 - F(x)} dx$$

$$= \int_0^{v_i} x \frac{\lambda \exp(-\lambda v_i)}{\exp(-\lambda v_i)} dx$$

$$= \lambda \int_0^{v_i} x \, dx$$

$$= \frac{\lambda v_i^2}{2}$$

Comparing to the equilibrium bid to that in the first-price all-pay auction, $\beta(v_i) = \frac{1 - (1 + \lambda v_i) \exp(-\lambda v_i)}{\lambda}$, as shown in Exercise 4.3(f), we see that bidders bid more aggressively under the second-bid all-pay auction for all valuations $v_i > 0$.

- Figure 4.8 depicts the equilibrium bidding function in the first-price all-pay auction and that in the second-price all-pay auction. As discussed in part (b), bidding behavior is more aggressive in the second-price version than in the first-price version of the all-pay auction. For illustration

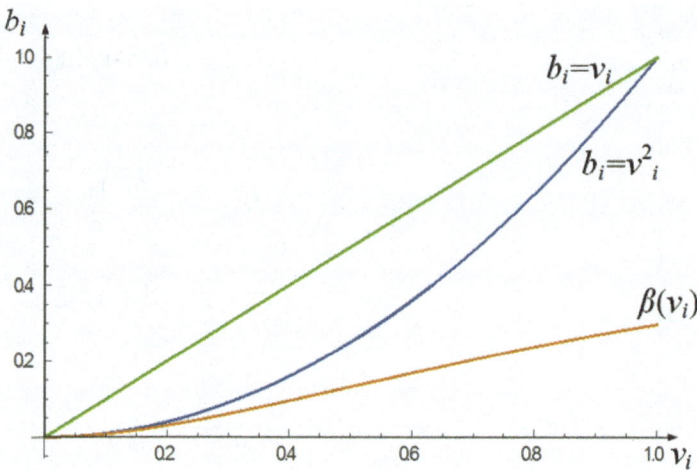

Fig. 4.8 Equilibrium bidding function in the first-price all-pay auction and in the second-price all-pay auction—Exponential distribution

purposes, we assume $\lambda = 2$, yielding bidding function $\beta(v_i) = \frac{1-(1+2v_i)\exp(-2v_i)}{2}$ in the first-price all-pay auction, and $b_i(v_i) = v_i^2$ in its second-price version.

Exercise #4.8: War of Attrition[A]

4.8 The all-pay auction is often used to study settings exhibiting "war of attrition" characteristics, where firms in an industry incur costs in each period whether they remain as the monopolist in a future period or not (similar to bidders in the all-pay auction, who must pay their bids regardless of whether they win the object or not). In this exercise, we study equilibrium behavior in a war of attrition game between two symmetric firms. For a more detailed analysis, see Fudenberg and Tirole (1986) and Bulow and Klemperer (1999), among others.

Assume that every firm $i = \{1, 2\}$ incurs a cost of $c > 0$ from staying in the industry for one more period, which we can normalize to one, for simplicity. Intuitively, this market is mature and cannot sustain two firms making positive profits. When only firm i remains, it earns monopoly profit $\pi^m > c$, and both firms' discount factor is $\delta \in (0, 1)$.

(a) If both firms remain in the industry until period $t-1$, and only firm i remains in period t, which is firm i's discounted net profit (including π^m and its losses in all previous periods)?

- At period t, firm i becomes the monopolist, so it earns a discounted profit of $\delta^t \pi^m$. However, its losses are

Exercise #4.8: War of Attrition[A]

$$-1 - \delta - \delta^2 - \cdots - \delta^{t-1} = -\frac{1-\delta^t}{1-\delta}$$

so its discounted net profit is

$$\delta^t \pi^m - \frac{1-\delta^t}{1-\delta}$$

(b) Assume that firms use a symmetric strategy. Show that a pure-strategy equilibrium cannot be sustained.

- If both firms leave the industry with certainty in every period t (i.e., using pure strategies), then one of the two firms has an incentive to deviate in period $t = 1$, staying in the industry instead and becoming a monopolist, earning π^m.
- Similarly, if both firms stay in the market with certainty in every period t, one of the two firms has incentive to deviate, leaving the industry and avoiding the waiting cost (normalized to one) in every subsequent period. Note that this holds because, according to this pure strategy profile, its rival stays in the market with certainty in all periods.

(c) Assume that firms use a symmetric strategy. Find a mixed-strategy equilibrium where every firm leaves the industry with probability p.

- At any period t, every firm i must be indifferent between leaving the industry, suffering a loss $-\frac{1-\delta^t}{1-\delta}$, or staying in the market, which yields an expected payoff

$$\underbrace{p\left(\delta^t \pi^m - \frac{1-\delta^t}{1-\delta}\right)}_{\text{Firm } j \text{ leaving in period } t} + \underbrace{(1-p)\left(-\frac{1-\delta^{t+1}}{1-\delta}\right)}_{\text{Firm } j \text{ staying in period } t}$$

where the first term denotes the net profit that firm i earns when becoming a monopolist in period t, which occurs when firm j leaves the industry; while the second term represents that firm j stays in the industry entailing that firm i will suffer a certain loss in the next period.

Therefore, firm i is indifferent between leaving and staying in the market when

$$\underbrace{-\frac{1-\delta^t}{1-\delta}}_{\text{Firm } i \text{ leaving in period } t} = \underbrace{p\left(\delta^t \pi^m - \frac{1-\delta^t}{1-\delta}\right) + (1-p)\left(-\frac{1-\delta^{t+1}}{1-\delta}\right)}_{\text{Firm } i \text{ staying in period } t}$$

Rearranging, and noting that $\frac{1-\delta^{t+1}}{1-\delta} = \frac{1-\delta^t}{1-\delta} + \delta^t$, we obtain

$$-\frac{1-\delta^t}{1-\delta} = p\delta^t \pi^m - p\left(\frac{1-\delta^t}{1-\delta}\right) - (1-p)\left(\frac{1-\delta^t}{1-\delta}\right) - (1-p)\delta^t$$

which simplifies to

$$p\delta^t \pi^m = (1-p)\delta^t$$

or more compactly as

$$p\pi^m = 1 - p$$

Intuitively, the left side represents the marginal benefit of staying in the industry for one more period (the expected monopoly profit, which is only realized if firm j leaves), while the right side captures the marginal cost of staying in the market for one more period.

- Solving for p, we find that every firm i randomizes between leaving and staying with probability

$$p^* = \frac{1}{1+\pi^m}$$

(d) Check that the probability you found in part (c), p^*, lies between zero and one. Is p^* increasing or decreasing in monopoly profit π^m?

- Probability $p^* = \frac{1}{1+\pi^m}$ is positive and lower than one since $1 < 1+\pi^m$ simplifies to $0 < \pi^m$, which holds by assumption.
- In addition, probability p^* decreases in π^m since

$$\frac{\partial p^*}{\partial \pi^m} = -\frac{1}{(1+\pi^m)^2} < 0$$

Intuitively, as the monopoly profit increases, every firm i is willing to stay in the industry for more periods, implying that the probability of leaving, p^*, decreases. For illustration purposes, Fig. 4.9 depicts probability p^* as a function of π^m.

(e) Is the probability of a firm leaving the industry at a given period t affected by the cumulative losses in previous periods? Interpret your results in terms of sunk costs.

- Probability p^* is unaffected by the cumulative losses in previous periods, as captured by $-\frac{1-\delta^t}{1-\delta}$. These losses are sunk, and thus should not affect the firm's decision on whether to stay for one more period or to leave. Otherwise, the firm would be falling prey of the sunk-cost fallacy.

Exercise #4.8: War of Attrition

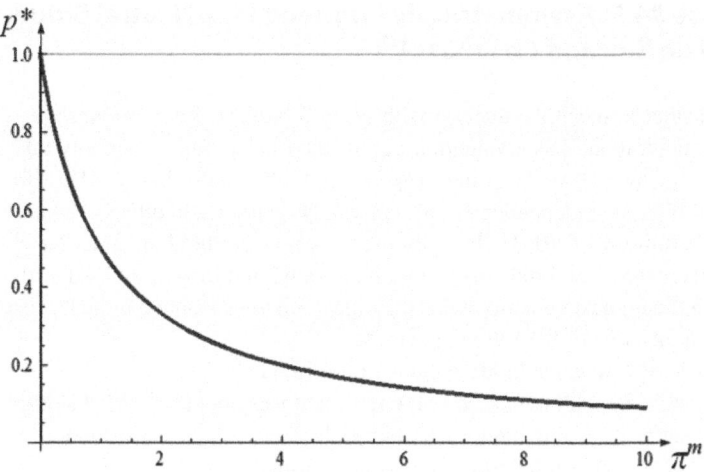

Fig. 4.9 Probability p^* as a function of π^m

(f) Is the probability of a firm leaving the industry at a given period t affected by the cost of staying for one more period (waiting cost, c)? Interpret.

- Probability p^* is unaffected by the waiting cost c. In our above analysis, we normalized this cost to one, for simplicity, but the same exact probability p^* arises if we do not normalize this cost, leaving it as c. Indeed, the indifference condition in this setting becomes

$$-\frac{1-\delta^t}{1-\delta}c = p\left(\delta^t \pi^m - \frac{1-\delta^t}{1-\delta}c\right) + (1-p)\left(-\frac{1-\delta^{t+1}}{1-\delta}c\right)$$

Rearranging, and noting that $\frac{1-\delta^{t+1}}{1-\delta}c = \frac{1-\delta^t}{1-\delta}c + \delta^t c$, we obtain

$$-\frac{1-\delta^t}{1-\delta}c = p\delta^t \pi^m - p\left(\frac{1-\delta^t}{1-\delta}c\right) - (1-p)\left(\frac{1-\delta^t}{1-\delta}c\right) - (1-p)\delta^t c$$

which simplifies to $p\delta^t \pi^m = (1-p)\delta^t$, yielding probability $p^* = \frac{1}{1+\pi^m}$, which is unaffected by the waiting cost c.

- This result suggests that increasing or decreasing the waiting cost does not change the probability with which every firm stays in the industry at any given period. Only the profits from monopolizing the market affect this probability.

Exercise #4.9: Asymmetrically Informed Risk-Neutral Bidders, Based on Kim and Che (2014)[B]

4.9 Consider a first-price auction with $N = 2$ bidders. Every bidder $i = \{1, 2\}$ is risk neutral and has a valuation v_K for the object, which can only take on two values: low ($v_L = 1$) or high ($v_H = 2$), both equally likely. We also assume that bidders independently and privately learn each other's valuation with probability $\alpha \in (0, 1)$. Bidders use a symmetric bidding function $b(v_K, \alpha)$. In the case of tied bids between a high-value and low-value bidder, the bidder with the highest valuation wins the object. Given the above model, every bidder i can face one of the following cases:
  (i) he is low-value bidder who is uninformed.
  (ii) he is low-value bidder who knows he faces another low-value bidder.
  (iii) he is low-value bidder who knows he faces a high-value bidder.
  (iv) he is high-value bidder who is uninformed.
  (v) he is high-value bidder who knows he faces a low-value bidder.
  (vi) he is high-value bidder who knows he faces another high-value bidder.
  Answer the following questions:
(a) Find every bidder i's bidding function, $b(v_K, \alpha)$. [*Hint*: The high-value bidder may need to randomize in some cases.]

- *Low-value bidders.*
  – The three types of low-value bidders (uninformed, informed that he faces another low-value bidder, or informed that he faces a high-value bidder) clearly submit a bid equal to their valuation, $v_L = \$1$.
  – Submitting a bid above $1 would imply bidding above his own valuation for the object, which is a strictly dominated strategy for the low-value bidder.
  – On the other hand, bidding below $1 is never a best response for low-value bidders, since in equilibrium no other bidder is bidding below one. Hence, submitting a bid below $1 would guarantee an expected utility of zero.
- *High-value bidders.*
  – *Informed and facing a low-value bidder.* The high-value bidder who is informed that his opponent is a low-value bidder bids $1, since he knows that in equilibrium his low-value opponent will never bid above $1.
  – *Uninformed.* Let us now analyze which are the equilibrium bidding strategies for the uninformed high-value bidder, who randomizes over an interval $[1, b']$, according to a distribution function $F(b)$, and where the upper bound b' satisfies $1 \leq b' \leq 2$, so the bidder never submits a bid above the high valuation $v_H = 2$.
  One necessary condition for an equilibrium is that the uninformed high-value bidder is indifferent among all the elements of the support

of the distribution $F(b)$, i.e., among all the elements in $[1, b']$. Hence, there must be a constant K such that

$$K = \underbrace{\frac{1}{2}(2-b)}_{\text{low-value bidder}} + \frac{1}{2}[\underbrace{(1-\alpha)F(b)(2-b)}_{\text{uninformed}} + \underbrace{\alpha 0}_{\text{informed}}]$$
$$\underbrace{}_{\text{high-value bidder}}$$

since there is $1/2$ chance of facing a low-value bidder, earning a net payoff of $(2-b)$; a chance $1/2 \times (1-\alpha)$ of facing another uninformed high-value bidder to win the object with probability $F(b)$; and a $1/2 \times \alpha$ probability of facing an informed high-value opponent to whom this bidder will lose the object.

Rearranging and solving for $F(b)$, we obtain

$$F(b) = \frac{b + 2K - 2}{(1-\alpha)(2-b)} \tag{4.1}$$

We conjecture that there is no mass point at the lower bound of the randomization, $b = 1$, so that $F(1) = 0$ since $b \in [1, b']$. We can then evaluate expression (4.1) at $b = 1$, that is,

$$0 = \frac{1 + 2K - 2}{(1-\alpha)(2-1)}$$

which simplifies to $0 = 2K - 1$. Solving for K, we obtain $K = \frac{1}{2}$. Additionally, the upper bound of the support of $F(b)$, b', must satisfy $F(b') = 1$. Using this condition and the value of $K = 1/2$ found above, we obtain that

$$1 = \frac{b' + 2\frac{1}{2} - 2}{(1-\alpha)(2-b')}$$

Solving for b', we find

$$b' = \frac{3 - 2\alpha}{2 - \alpha}$$

- *Informed and facing another high-value bidder.* Let us now analyze the equilibrium bidding strategy for the informed high-value bidder who faces an opponent with high valuation. This informed bidder, in equilibrium, randomizes over an interval $[b', b^*]$ according to a distribution $G(b)$. A necessary condition for an equilibrium is that this informed high-value bidder facing another high-value bidder is indifferent among all the elements in the support of distribution $G(b)$.

Hence, there must exist a constant M such that

$$M = \underbrace{(1-\alpha)(2-b)}_{\text{uninformed high-value bidder}} + \underbrace{\alpha G(b)(2-b)}_{\text{informed high-value bidder}}$$

since there is $(1-\alpha)$ chance of facing an uninformed high-value bidder, earning net payoff of $(2-b)$; and a chance α of facing another informed high-value bidder, thus winning the object with probability $G(b)$.

Rearranging and solving for $G(b)$, we have

$$G(b) = \frac{M - (1-\alpha)(2-b)}{\alpha(2-b)} \quad (4.2)$$

We conjecture that there is no mass point at the lower bound of the randomization $b = b'$, so that $G(b') = 0$. We can then evaluate expression (4.2) at $b = b'$, that is,

$$0 = \frac{M - (1-\alpha)(2-b')}{\alpha(2-b')}$$

which simplifies to

$$0 = M - (1-\alpha)(2-b')$$

Solving for M, we obtain

$$M = (1-\alpha)(2-b')$$

Substituting $b' = \frac{3-2\alpha}{2-\alpha}$ into the above expression, and simplifying

$$M = (1-\alpha)\left(2 - \overbrace{\frac{3-2\alpha}{2-\alpha}}^{M}\right)$$

$$= \frac{1-\alpha}{2-\alpha}$$

Additionally, the upper bound of the support of $G(b)$, b^*, must satisfy $G(b^*) = 1$. Using this condition and the value of M found above, we find

$$1 = \frac{\overbrace{\left(\dfrac{1-\alpha}{2-\alpha}\right)}^{M} - (1-\alpha)(2-b^*)}{\alpha(2-b)}$$

Simplifying, we obtain

$$\alpha(2-\alpha)(2-b^*) = (1-\alpha)\left[1 - (2-\alpha)(2-b^*)\right]$$

Solving for b^*, we have

$$b^* = \frac{3-\alpha}{2-\alpha}$$

- *Sufficiency.* To check that the above necessary conditions for an equilibrium bidding function are also sufficient, we still need to check the following points:
  - First, we check that an informed high-value bidder does not want to submit any bid below b'. To see this, note that by construction an uninformed high-value bidder is indifferent among all bids in the range $[0, b']$. But an uninformed high-value bidder, unlike a high-value bidder who knows that he faces another high-value bidder, puts positive probability on bidding against a low-value bidder, against whom he always gains from lowering his bid. An informed high-value bidder facing another high-value bidder lacks this benefit and hence strictly loses from lowering his bid below b'.
  - Furthermore, an uninformed high-value bidder has no incentive to increase his bid above b'. To see this, suppose he did. Then he would get a payoff which is strictly decreasing in b.
  - A similar analysis for b^* shows sufficiency for the above distribution functions.
- *Summary.* In a first-price auction with $N = 2$ asymmetrically informed risk-neutral bidders, the equilibrium bidding function prescribes that
  - All low-value bidders, either informed or uninformed, submit a bid equal to their valuation ($v_L = 1$) using pure strategies.
  - The high-value bidder who is informed that his opponent is a low-value bidder bids $v_L = 1$ using pure strategies.
  - The high-value bidder who is uninformed randomizes over the interval $[1, b']$, according to a cumulative distribution function, from expression (4.1), of

$$F(b) = \frac{b + 2\overbrace{\frac{1}{2}}^{K} - 2}{(1-\alpha)(2-b)}$$

$$= \frac{b-1}{(1-\alpha)(2-b)}$$

where $b' = \frac{3-2\alpha}{2-\alpha}$.

- The high-value bidder who is informed that his competitor is also a high-value bidder randomizes over the interval $[b', b^*]$, with a cumulative distribution function, from expression (4.2), of

$$G(b) = \frac{\overbrace{\left(\frac{1-\alpha}{2-\alpha}\right)}^{M} - (1-\alpha)(2-b^*)}{\alpha(2-b)}$$

$$= \frac{(1-\alpha)[1-(2-\alpha)(2-b)]}{(2-\alpha)(2-b)}$$

where $b^* = \frac{3-\alpha}{2-\alpha}$.

(b) *Comparative statics.* How is bidding behavior affected by a marginal increase in probability α? Interpret.

- The bidding of all low-value bidders and that of the high-value bidder who is informed of facing a low-value bidder are both constant and equal to \$1, thus being unaffected by an increase in probability α.
- The uninformed high-value bidder and the informed high-value bidder facing another high-value bidder are, however, affected by this probability, as we next describe. Differentiating the upper bound of the cumulative distribution of the uninformed high-value bidder, b', yields

$$\frac{\partial b'}{\partial \alpha} = \frac{-2(2-\alpha) + 3 - 2\alpha}{(2-\alpha)^2}$$

$$= -\frac{1}{(2-\alpha)^2} < 0$$

Intuitively, conditional on bidder i being uninformed, an increase in the probability that his rival (bidder j) is informed, bidder i shrinks the support of his randomization $[1, b']$, thus submitting lower bids on average.

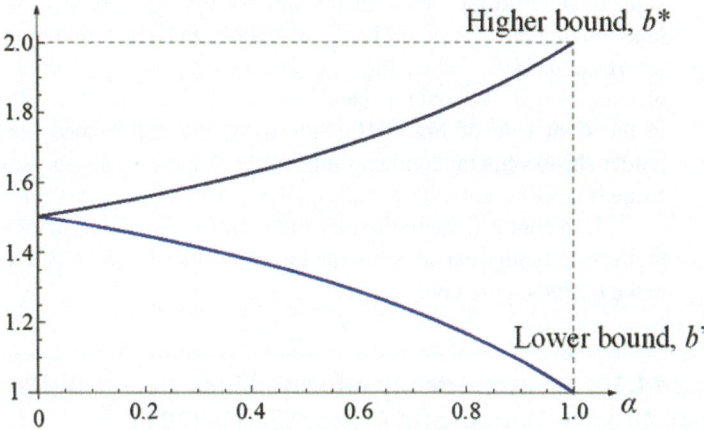

Fig. 4.10 Bounds b' and b^* as a function of α

However, differentiating the upper bound of the cumulative distribution of the uninformed high-value bidder, b^*, yields

$$\frac{\partial b^*}{\partial \alpha} = \frac{-(2-\alpha)+3-\alpha}{(2-\alpha)^2}$$

$$= \frac{1}{(2-\alpha)^2} > 0$$

Therefore, the informed bidder facing another high-value bidder expands the interval over which he randomizes, $[b', b^*]$, since b' decreases in α while b^* increases in α. Figure 4.10 depicts these upper and lower bounds, as a function of probability α, confirming that the support of $G(b)$ expands in α.

(c) *Extreme cases.* Evaluate equilibrium bidding behavior when bidders are symmetrically uninformed about each other's valuations, $\alpha \to 0$, and when they are symmetrically and perfectly informed, $\alpha \to 1$. Interpret your results.

- *Uninformed bidders.* When symmetrically uninformed, $\alpha = 0$, the bidding behavior of all low-value bidders and that of the high-value bidder who is informed of facing a low-value bidder is unaffected, as shown in part (a). For the other two bidders, however, we obtain that $b' = b^* = \frac{3}{2}$, as depicted in the vertical intercept of Fig. 4.10. Intuitively, the uninformed high-value bidder expands his randomizing interval $[1, b']$ to $[1, \frac{3}{2}]$ to make his bids more competitive. In contrast, the high-

value bidder informed about facing another high-value bidder shrinks his randomizing interval $[b', b^*]$ to $\frac{3}{2}$ (a degenerated, pure strategy).
- *Informed bidders.* When bidders observe each other's values for the object, $\alpha = 1$, we obtain that $b' = 1$ while $b^* = 2$, as illustrated in the right side of Fig. 4.10. Intuitively, the uninformed high-value bidder shrinks his randomizing interval $[1, b']$ to 1 (a degenerated, pure strategy), so he submits a bid equal to that of the low-value bidder, $v_L = 1$. In contrast, the high-value bidder informed about facing another high-value bidder expands his randomizing interval $[b', b^*]$ to $[1, 2]$ to make his bids more competitive.

Exercise #4.10: Asymmetrically Informed Risk-Averse Bidders, Based on Orozco-Aleman and Munoz-Garcia (2011)[B]

4.10 Consider the setting in Exercise 4.9, but assume now that bidders are risk averse. For simplicity, we consider that bidders exhibit constant relative risk aversion (CRRA). That is, their utility function is of the form $u(x) = x^\omega$, where $x \geq 0$ is the player's income, and $0 < \omega < 1$, so that the coefficient of relative risk aversion is

$$r_R = -x \frac{u''(x)}{u'(x)} = -x \frac{(\omega-1)\omega x^{\omega-2}}{\omega x^{\omega-1}} = 1 - \omega$$

In other words, a bidder's utility function becomes more concave, implying greater relative risk aversion, as ω decreases (or, alternatively, as $1 - \omega$ increases). Answer the following questions:
(a) Find every bidder i's bidding function, $b(v_K, \alpha, \omega)$.

- *Low-value bidders.*
    - The three types of low-value bidders (uninformed, informed that he faces another low-value bidder, or informed that he faces a high-value bidder) clearly submit a bid equal to their valuations, $v_L = \$1$.
    - Submitting a bid above $1 would imply bidding above his own valuation for the object, which is a strictly dominated strategy for the low-value bidder.
    - On the other hand, bidding below $1 is never a best response for low-value bidders, since in equilibrium no other bidder is bidding below $1. Hence, submitting a bid below one would guarantee an expected utility of zero.
- *High-value bidders.*
    - *Informed and facing a low-value bidder.* The high-value bidder who is informed that his opponent is a low-value bidder bids $1, since he knows that in equilibrium his low-value opponent will never bid above $1.

- *Uninformed.* Let us now analyze which are the equilibrium bidding strategies for the uninformed high-value bidder, who randomizes over an interval $[1, b']$, according to a distribution function $F(b)$. One necessary condition for an equilibrium is that the uninformed high-value bidder is indifferent among all the elements of the support of the distribution $F(b)$, i.e., among all the elements in $[1, b']$. Hence, there must be a constant K such that

$$ K = \underbrace{\frac{1}{2}(2-b)^{\omega}}_{\text{low-value bidder}} + \frac{1}{2}[\underbrace{(1-\alpha)F(b)(2-b)^{\omega}}_{\text{uninformed}} + \underbrace{\alpha 0}_{\text{informed}}] $$

$$\underbrace{\hspace{6cm}}_{\text{high-value bidder}}$$

since there is a 1/2 chance of facing a low-value bidder, earning a utility of $(2-b)^{\omega}$; a probability $1/2 \times (1-\alpha)$ of facing an uninformed high-value bidder to win the object with probability $F(b)$; and a $1/2 \times \alpha$ chance of facing an informed high-value opponent to whom this bidder will lose the object, as we see below.

Solving for $F(b)$, we obtain

$$ F(b) = \frac{2K - (2-b)^{\omega}}{(1-\alpha)(2-b)^{\omega}} \qquad (4.3) $$

We conjecture that there is no mass point at the lower bound of the randomization, $b = 1$, so that $F(1) = 0$. Evaluating expression (4.3) at $b = 1$, yields

$$ 0 = \frac{2K - (2-1)^{\omega}}{(1-\alpha)(2-1)^{\omega}} $$

which simplifies to $0 = 2K - 1$. Solving for K, we obtain $K = \frac{1}{2}$. Additionally, the upper bound of the support of $F(b)$, b', must satisfy $F(b') = 1$, so expression (4.3) becomes

$$ 1 = \frac{2K - (2-b')^{\omega}}{(1-\alpha)(2-b')^{\omega}} $$

and solving for b', yields

$$ b' = 2 - (2-\alpha)^{-\frac{1}{\omega}} $$

- *Informed and facing another high-value bidder.* Let us now analyze the equilibrium bidding strategy for the informed high-value bidder who faces an opponent with high valuation. This informed bidder, in equilibrium, randomizes over an interval $[b', b^*]$ according to a distribution $G(b)$. A necessary condition for an equilibrium is that this informed high-value bidder facing another high-value bidder is indifferent among all the elements in the support of distribution $G(b)$. Hence, there must exist a constant M such that

$$M = \underbrace{(1-\alpha)(2-b)^{\omega}}_{\text{uninformed high-value bidder}} + \underbrace{\alpha G(b)(2-b)^{\omega}}_{\text{informed high-value bidder}}$$

since there is a $(1-\alpha)$ chance of facing an uninformed high-value bidder, earning a utility of $(2-b)^{\omega}$; and α chance of facing another informed high-value bidder to win the object with probability $G(b)$.

Solving for $G(b)$, we find

$$G(b) = \frac{M - (1-\alpha)(2-b)^{\omega}}{\alpha(2-b)^{\omega}} \qquad (4.4)$$

We conjecture that there is no mass point at the lower bound of the support, $b = b'$, so that $G(b') = 0$. Evaluating expression (4.4) at $b = b'$, yields

$$0 = \frac{M - (1-\alpha)\left(2-b'\right)^{\omega}}{\alpha(2-b')^{\omega}}$$

which simplifies to $0 = M - (1-\alpha)\left(2-b'\right)^{\omega}$. Solving for M, we obtain

$$M = (1-\alpha)\left(2-b'\right)^{\omega}$$

Substituting $b' = 2 - (2-\alpha)^{-\frac{1}{\omega}}$ into the above expression, we have

$$M = (1-\alpha)\left(2 - \underbrace{\left(2 - (2-\alpha)^{-\frac{1}{\omega}}\right)}_{b'}\right)^{\omega}$$

$$= \frac{1-\alpha}{2-\alpha}$$

Additionally, the upper bound of the support of $G(b)$, b^*, must satisfy $G(b^*) = 1$, enabling us to solve for b^* and obtain

$$1 = \frac{\left(\frac{1-\alpha}{2-\alpha}\right)^{\overbrace{}^{M}} - (1-\alpha)(2-b^*)^{\omega}}{\alpha(2-b^*)^{\omega}}$$

which is rearranged to yield

$$(2-\alpha)(2-b^*)^{\omega} = 1-\alpha$$

Simplifying, we find

$$b^* = 2 - \left(\frac{1-\alpha}{2-\alpha}\right)^{\frac{1}{\omega}}$$

- *Sufficiency.* To check that the above necessary conditions for an equilibrium bidding function are also sufficient, we still need to check the following points:
  - First, we check that an informed high-value bidder does not want to submit any bid below b'. To see this, note that by construction an uninformed high-value bidder is indifferent among all bids in the range $[1, b']$. This uninformed high-value bidder, unlike a high-value bidder who knows that he faces another high-value bidder, puts positive probability on bidding against a low-value bidder, against whom he always gains from lowering his bid. An informed high-value bidder facing another high-value bidder lacks this benefit and hence strictly loses from lowering his bid below b'.
  - Furthermore, an uninformed high-value bidder has no incentive to increase his bid above b'. To see this, suppose he did. Then he would get a payoff which is strictly decreasing in b.
  - A similar analysis for b^* shows sufficiency for the above distribution functions.
- *Summary.* In a first-price auction with $N = 2$ asymmetrically informed risk-averse bidders, the equilibrium bidding function prescribes that
  - All low-value bidders, either informed or uninformed, submit a bid equal to their valuation ($v_L = 1$) using pure strategies.
  - The high-value bidder who is informed that his opponent is a low-value bidder bids $v_L = 1$ using pure strategies.
  - The high-value bidder who is uninformed randomizes over the interval $[1, b']$ according to a cumulative distribution function

$$F(b) = \frac{1 - (2-b)^{\omega}}{(1-\alpha)(2-b)^{\omega}}$$

where $b' = 2 - (2-\alpha)^{-\frac{1}{\omega}}$.

- The high-value bidder who is informed that his competitor is also a high-value bidder randomizes over the interval $[b', b^*]$, with a cumulative distribution function

$$G(b) = \frac{\overbrace{\left(\frac{1-\alpha}{2-\alpha}\right)}^{M} - (1-\alpha)(2-b)^{\omega}}{\alpha(2-b)^{\omega}}$$

$$= \frac{(1-\alpha)\left[1 - (2-\alpha)(2-b)^{\omega}\right]}{\alpha(2-\alpha)(2-b)^{\omega}}$$

where $b^* = 2 - \left(\frac{1-\alpha}{2-\alpha}\right)^{\frac{1}{\omega}}$.

(b) *Comparative statics.* We now seek to understand how equilibrium bidding behavior is affected by a marginal increase in parameter ω. Evaluate the bidding function you found in part (a) at $\omega = 1$ and at $\omega = 1/2$. Compare and interpret your results.

- The bidding of all low-value bidders and that of the high-value bidder who is informed of facing a low-value bidder are both constant and equal to $v_L = \$1$, thus being unaffected by an increase in the parameter ω. The uninformed high-value bidder and the informed high-value bidder facing another high-value bidder are, however, affected by risk aversion, as we next describe.
- *Uninformed high-value bidder.* Figure 4.11 depicts the upper bound of the cumulative distribution of the uninformed high-value bidder, b', at $\omega = 0.9, 0.5,$ and $\omega = 0.1$. Intuitively, as the bidder becomes more risk averse (lower ω), the upper bound of his randomization shifts up, thus indicating more aggressive bids and, simultaneously, a "flattening" effect occurs, so the bidder becomes more insensitive to a marginal increase in the probability of facing another high-value bidder, α. Evaluating the upper bound of the cumulative distribution of the uninformed high-value bidder, b', at $\omega = 1$, yields

$$b' = 2 - (2-\alpha)^{-\frac{1}{1}} = \frac{3 - 2\alpha}{2 - \alpha}$$

which coincides with what we found in Exercise 4.9 for risk-neutral bidders.

Similarly, evaluating this upper bound at $\omega = 1/2$, we obtain

$$b' = 2 - (2-\alpha)^{-2} = \frac{7 - 8\alpha + 2\alpha^2}{(2-\alpha)^2}$$

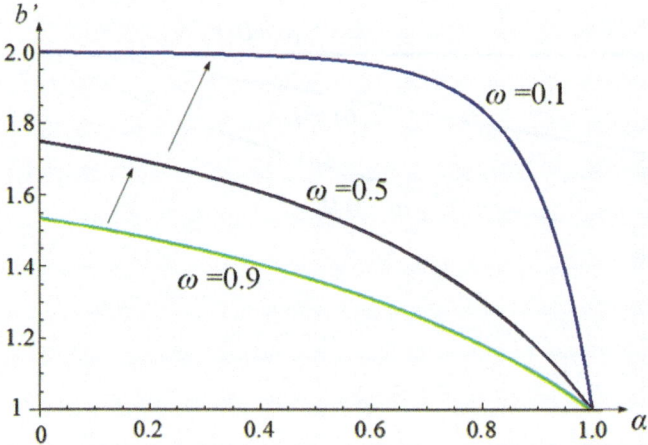

Fig. 4.11 Upper bound b', for different values of ω

thus being higher than that at $\omega = 1$ because

$$(2-\alpha)^{-2} < (2-\alpha)^{-1}$$

simplifies to $2-\alpha > 1$, which holds true since $\alpha < 1$. Intuitively, when the bidder is uninformed about his opponent's valuation and becomes more risk averse, he is so concerned about not losing the auction that he widens his randomization in interval $[1, b']$.

- *Informed high-value bidder facing another high-value bidder.* Figure 4.12 illustrates the upper bound of the cumulative distribution of the high-value bidder, b^*, evaluated at $\omega = 0.9, 0.5$, and $\omega = 0.1$. As with the upper bound b' in Fig. 4.11, we find that risk aversion shifts b^* upward, indicating that the bidder submits more aggressive bids; and that b^* is flatter, suggesting that the randomization becomes more insensitive to the probability of dealing with another high-value bidder, α. Evaluating the upper bound of the cumulative distribution of this high-value bidder, b^*, at $\omega = 1$, yields

$$b^* = 2 - \left(\frac{1-\alpha}{2-\alpha}\right)^{\frac{1}{1}} = \frac{3-\alpha}{2-\alpha}$$

which coincides with what we found in Exercise 4.9 for risk-neutral bidders. Evaluating this upper bound at $\omega = 1/2$, we obtain

$$b^* = 2 - \left(\frac{1-\alpha}{2-\alpha}\right)^2 = \frac{7 - 6\alpha + \alpha^2}{(2-\alpha)^2}$$

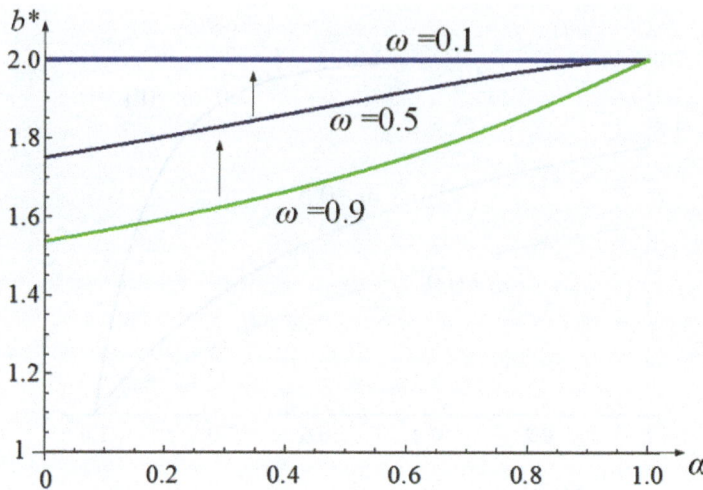

Fig. 4.12 Lower bound b^*, for different values of ω

thus being lower than that at $\omega = 1$ because

$$\left(\frac{1-\alpha}{2-\alpha}\right)^2 < \frac{1-\alpha}{2-\alpha}$$

simplifies to $1 - \alpha < 2 - \alpha$, which holds true.

(c) *Extreme risk aversion.* Find equilibrium bidding behavior when bidders are extremely risk averse, $\omega \to 0$. Interpret your results.

- First, note that when bidders become extremely risk averse, both $b' \to 2$ and $b^* \to 2$. This implies that:
  - The informed high-value bidder concentrates all his randomizations at his own private valuation $v_i = 2$, i.e., his distribution function $G(b)$ becomes degenerated at $v_i = 2$ as he becomes highly risk averse.
  - On the contrary, the uninformed high-value bidder randomizes on the full interval $[1, 2]$ when he is extremely risk averse. Therefore, the upper support of this randomization coincides with his own valuation for the object, $v_i = 2$, regardless of the probability that his opponent is informed or not. This strong preference for winning the auction, at the cost of reducing the expected payoff if winning, is a standard result when dealing with risk aversion (see Exercises 3.4 and 3.5 for more examples about the effect of risk aversion in the equilibrium bidding in first-price auctions). In addition, player i's

bidding strategy is now unaffected by changes in the uncertainty surrounding the auction (i.e., the probability that the other bidder know his valuation, α).
- The above results hold for all values of probability α. Therefore, a seller running an auction in which bidders are extremely risk averse would neither modify equilibrium bids nor his expected revenue by distributing information about bidders' valuations (by increasing probability α).

Third-Price Auctions, kth-Price Auctions, and Lotteries

5

Keywords

Third-price auction · Kth-price auction · Lotteries · Political campaigns · Beta distribution · Revenue equivalence principle

Introduction

This chapter generalizes previous auction formats by allowing the winning bidder to pay the kth highest bid, while all losing bidders pay zero. This implies that in the first-price auction, we have that $k = 1$, as the winning bidder pays the highest bid; and in the second-price auction, $k = 2$, as he pays the second-highest bid. A similar argument applies to the third-price auction, where $k = 3$, as the winning bidder pays the third-highest bid, and, more generally, to any other auction format where $k > 3$. While kth-price auctions where $k \geq 3$ are uncommon in real life, they provide us a general structure to analyze specific auction formats as special cases of this model, helping us understand how changes in k affect equilibrium bidding behavior.

Exercise 5.1 considers the third-price auction in a stylized setting, where bidders draw only three possible valuations for the object, showing that submitting a bid equal to your valuation is not an equilibrium bidding strategy; as opposed to the second-price auction where this bid could be supported in equilibrium.

Exercise 5.2 finds the equilibrium bidding function in a third-price auction, allowing for valuations to be distributed according to a generic distribution. We show that bidders submit a bid above their own valuation, and, of course, they bid more aggressively than in the first- and second-price auction. For illustration purposes, Exercise 5.3 considers that valuations are drawn from a uniform or exponential distribution, evaluating the equilibrium bidding function found in Exercise 5.2 and running comparative statics.

Exercise 5.4 extends our analysis in Exercise 5.2 to a more general setting where the winning bidder pays the kth-highest bid showing that, as expected, every player bids more aggressively as the price he pays upon winning is lower (higher k). Exercise 5.5 evaluates the efficiency of this auction format, showing that the object goes to the bidder with the highest valuation and that efficiency is unaffected by the price he pays upon winning, as captured by k.

Finally, Exercises 5.6 and 5.7 consider lottery auctions, where the probability of winning is a ratio of how bidder i's bid compares relative to all bids, and all bidders must pay their bids (as in the all-pay auction). This auction format is often used to study firms' R&D investment or advertising in political campaigns. The probability of winning ratio implies that even if a bidder submits the highest bid, he does not necessarily win the auction, so that lottery auctions are not efficient (as shown in Exercise 5.8). In Exercise 5.6 we assume two bidders, while in Exercise 5.7 we allow for $N \geq 2$ bidders, finding the equilibrium bidding function, and how this function varies in the number of bidders.

Exercise #5.1: Third-Price Auction, A Numerical Example[A]

5.1 Consider a third-price auction, where the winner is the bidder who submits the highest bid, but he/she only pays the third highest bid. Assume that you compete against two other bidders, whose valuations you are unable to observe, and that your valuation for the object is $10. Valuations are discrete, being high, medium, or low with probabilities p, q, and $1 - p - q$, respectively.

(a) Show that bidding *above* your valuation (with a bid of, for instance, $15) can be a best response to the other bidders' bid, while submitting a bid that coincides with your valuation ($10) might not be a best response to your opponents' bids.

- To show that bidding your valuation may not be optimal, we separate this analysis into three parts, where you have the lowest, middle, or highest valuation for the object among the three bidders in the auction.
    - *Bidder i has the lowest valuation.* In this situation, you have no incentive to bid above your valuation. Were you to bid high enough to win the auction, you would pay the valuation of what would have been the second highest bid (now the third highest bid). Since their valuation exceeds yours, winning the auction would provide you with a negative payoff. Thus, there is no incentive to deviate from bidding your valuation when it is the lowest among the three.
    - *Bidder i has the middle valuation.* In this situation, you have incentive to bid above your valuation. Since the winner of the object pays the valuation of the third highest bidder, the bidder with the middle valuation (you in this case) receives a positive payoff by outbidding the bidder with the highest valuation. For instance, if the lowest bidder had a valuation of $5, you had a valuation of $10, and the other bidder

had a valuation of $15, bidding $20 (above your valuation) would win the auction, but you would only pay $5, leading to a payoff of $10 − $5 = $5.
- *Bidder i has the highest valuation.* By a similar logic to the situation where bidder i had the middle valuation, when you have the highest valuation, you have the chance to be outbid by a middle valuation bidder when you bid your valuation, so that you have an incentive to bid above your valuation as well.

(b) Show that bidding *above* your valuation (with a bid of, for instance, $15) can be a best response to the other bidders' bid.

- As shown in part (a), bidder i does not have incentives to deviate from submitting a bid equal his valuation when his has the lowest valuation. Otherwise, bidder i has incentives to submit a bid above his own valuation.

Exercise #5.2: Finding the Equilibrium Bidding Function in a Third-Price Auction[C]

5.2 Consider an auction with $N \geq 3$ bidders, each of them privately observing his valuation of the good, $v_i \in [0, 1]$. Each bidder independently and simultaneously submits his own bid, b_i, and the bidder submitting the highest bid is selected as the winner of the auction and receives the good. In this case, however, let us assume that the winning bidder pays the *third* highest bid, thus explaining why this auction format is referred to as the "third-price auction."
(a) Assume a cumulative distribution function $F(v_i)$, with positive density for all valuations, that is, $f(v_i) > 0$ for all $v_i \in [0, 1]$. Find bidder i's equilibrium bidding function $b_i(v_i)$ in this auction.

- As in other auction formats, assume that there is a symmetric and increasing bidding function $b_i(v_i)$. In addition, consider that the payoff from the bidder with the lowest valuation, $v_i = 0$, is zero. We then have that, for every valuation v_i, the payment of bidder i to the seller is

$$m(v_i) = \int_0^{v_i} yg(y)dy.$$

This is a direct result from the Revenue Equivalence Theorem, which Chap. 6 presents in more detail; see, for instance, Exercises 6.5 and 6.6. We now seek to write bidder i's payment, $m(v_i)$, using an alternative expression, so that we can ultimately set it equal to the right side of the above equality. In particular, the seller's expected revenue is

$$\Pr(win) \times E[b_i(X_2)|X_1 < v_i]$$

where X_1 (X_2) is the first-order (second-order) statistic. That is, X_1 (X_2) is the highest (second highest) valuation among all $N-1$ remaining bidders. We can then use these statistics to describe that:
(i) Bidder i wins the auction if his valuation is higher than that of all other $N-1$ bidders, that is, $X_1 < v_i$, which happens with probability $F_1^{N-1}(v_i)$; and
(ii) His expected bid is

$$E[b_i(X_2)|X_1 < v_i] = \int_0^{v_i} b_i(y) f_2^{N-1}(y|X_1 < v_i) dy$$

where density $f_2^{N-1}(y|X_1 < v_i)$ is conditional on bidder i's valuation being larger than that of all $N-1$ remaining bidders, $X_1 < v_i$. Specifically, according to the conditional probability density of the kth-order statistic (see Exercise 2.10), we can rewrite this conditional density as

$$f_2^{N-1}(y|X_1 < v_i)$$
$$= \underbrace{\frac{1}{F_1^{N-1}(v_i)}}_{\Pr\{X_1 < v_i\}} (N-1)(N-2)[F(v_i) - F(y)] F^{N-3}(y) f(y)$$

where $(N-1)[F(v_i) - F(y)]$ is the probability that $X_1 < v_i$ but $X_1 > y$, and $f_2^{N-1}(y)$ is the density of the highest valuation of the remaining $N-2$ bidders.

- Using points (i) and (ii), we can write the seller's expected revenue in the third-price auction, $\Pr(win) \times E[b_i(X_2)|X_1 < v_i]$, as follows:

$$F_1^{N-1}(v_i) \int_0^{v_i} b_i(y) \overbrace{\frac{1}{F_1^{N-1}(v_i)} (N-1)(N-2)[F(v_i) - F(y)] F^{N-3}(y) f(y)}^{f_2^{N-1}(y|X_1 < v_i)} dy$$

$$= (N-1)(N-2) \int_0^{v_i} b_i(y) [F(v_i) - F(y)] F^{N-3}(y) f(y) dy$$

since $\frac{1}{F_1^{N-1}(v_i)}$ can be taken out of the integral. We are now ready to use this expression in the left side of $m(v_i)$ to obtain

$$(N-1)(N-2) \int_0^{v_i} b_i(y) [F(v_i) - F(y)] F^{N-3}(y) f(y) dy = \int_0^{v_i} y g(y) dy$$

Exercise #5.2: Finding the Equilibrium Bidding Function in a Third-Price Auction^C

Differentiating with respect to v_i yields

$$(N-1)(N-2) f(v_i) \int_0^{v_i} b_i(y) F^{N-3}(y) f(y) dy = v_i g(v_i)$$

where $g(v_i)$ is the density of the cumulative distribution function $G(v_i) = F(v_i)^{N-1}$, thus implying that $g(v_i) = (N-1) F^{N-2}(v_i) f(v_i)$. Using this term on the right side, we obtain

$$(N-1)(N-2) f(v_i)$$
$$\times \int_0^{v_i} b_i(y) F^{N-3}(y) f(y) dy = (N-1) v_i F^{N-2}(v_i) f(v_i)$$

Cancelling out $(N-1) f(v_i)$ on both sides and rearranging yields

$$(N-2) \int_0^{v_i} b_i(y) F^{N-3}(y) f(y) dy = v_i F^{N-2}(v_i).$$

At this point, recall that our goal is to solve for the bidding function $b_i(y)$. To eliminate the integral operator from the left-hand side, we differentiate both sides with respect to valuation v_i, which yields

$$(N-2) b_i(v_i) F^{N-3}(v_i) f(v_i) = F^{N-2}(v_i) + (N-2) v_i F^{N-3}(v_i) f(v_i)$$

Finally, solving for the bidding function $b_i(v_i)$, we find that

$$b_i(v_i) = \frac{F^{N-2}(v_i)}{(N-2) F^{N-3}(v_i) f(v_i)} + \frac{(N-2) v_i F^{N-3}(v_i) f(v_i)}{(N-2) F^{N-3}(v_i) f(v_i)}$$

which simplifies to

$$b_i(v_i) = v_i + \frac{1}{N-2} \frac{F(v_i)}{f(v_i)}.$$

(b) Show that, in equilibrium, every bidder i submits a bid above his valuation. Interpret.

- The equilibrium bidding function that we found in part (a) satisfies $b_i(v_i) > v_i$ since

$$v_i + \frac{1}{N-2} \frac{F(v_i)}{f(v_i)} > v_i$$

simplifies to

$$\frac{1}{N-2}\frac{F(v_i)}{f(v_i)} > 0$$

which holds since $N \geq 3$ by assumption, $F(v_i) \in [0, 1]$, and $f(v_i) > 0$ for all v_i.

- Intuitively, every bidder anticipates that, upon winning, he only pays the third-highest bid, thus reducing his expected price and providing him with incentives to submit aggressive bids above his own valuation for the object.
- Finally, note that when the number of bidders increases enough, $N \to +\infty$, the term that captures the overbidding strength, $\frac{1}{N-2}\frac{F(v_i)}{f(v_i)}$, approaches zero, implying that every bidder submits a bid that approaches his own valuation, $b_i(v_i) \to v_i$.

(c) *Comparison with other auction formats.* Compare the equilibrium bidding functions in the first-, second-, and third-price auctions. Interpret.

- Our result in part (b), $b_i(v_i) > v_i$, differs from that in the first-price auction, where we showed that bid shading occurs ($b_i(v_i) < v_i$, as shown in Exercise 2.4) since

$$b_i^{1st}(v_i) = v_i - \frac{\int_0^{v_i} F(x)^{N-1}dx}{F(v_i)^{N-1}} < v_i < v_i + \underbrace{\frac{1}{N-2}\frac{F(v_i)}{f(v_i)}}_{\text{Overbidding}} = b_i^{3rd}(v_i).$$

- Similarly, our result in part (b), $b_i(v_i) > v_i$, could not be supported in the second-price auction, where we found that every bidder submits a bid equal to his valuation ($b_i(v_i) = v_i$, as shown in Exercise 1.2), so that

$$b_i^{2nd}(v_i) = v_i < v_i + \underbrace{\frac{1}{N-2}\frac{F(v_i)}{f(v_i)}}_{\text{Overbidding}} = b_i^{3rd}(v_i).$$

As described in part (b), bidders submit a bid above their own valuation in the third-price auction because their expected price upon winning is lower than in the first- and second-price auctions.

Exercise #5.3: Equilibrium Bidding Function in a Third-Price Auction with Uniformly or Exponentially Distributed Values[B]

5.3 From Exercise 5.2, we know that the equilibrium bidding function in the third-price auction is

Exercise #5.3: Equilibrium Bidding Function in a Third-Price Auction with...

$$b_i(v_i) = v_i + \frac{1}{N-2}\frac{F(v_i)}{f(v_i)}.$$

In this exercise, we evaluate this bidding function in two commonly used contexts: the uniform and exponential distributions.

(a) *Uniformly distributed values.* Consider that valuations are distributed according to a uniform distribution, so that $F(v_i) = v_i$. Evaluate the equilibrium bidding function using this distribution function.

- Inserting $F(v_i) = v_i$ and $f(v_i) = 1$ into the equilibrium bidding function, we obtain

$$\begin{aligned} b_i(v_i) &= v_i + \frac{1}{N-2}\frac{F(v_i)}{f(v_i)} \\ &= v_i + \frac{1}{N-2}v_i \\ &= \frac{N-1}{N-2}v_i \end{aligned}$$

which is increasing in bidder i's valuation, v_i.

(b) Still considering uniformly distributed valuations, how does $b_i(v_i)$ change in the number of bidders N? Interpret.

- The equilibrium bidding function found above, $b_i(v_i) = \frac{N-1}{N-2}v_i$, is decreasing in the number of bidders since

$$\frac{\partial b_i(v_i)}{\partial N} = -\frac{1}{(N-2)^2} < 0.$$

Hence, unlike in first-price auctions (whereby an increase in the number of bidders raises bids, making them more aggressive), in a third-price auction equilibrium bids *decrease* in N. Intuitively, as more bidders participate in the auction, the third highest bid, b_j, which is the price that the winning bidder i must pay for the object, is more likely to be above bidder i's own valuation, $b_j > v_i$. To account for this increased risk of overpaying for the object, every bidder i reduces his bid when more bidders participate in the auction.

- Figure 5.1 depicts this equilibrium bidding function evaluated at three different values of N. For illustration purposes, the figure also includes the 45°-line, where $b_i = v_i$, showing that $b_i(v_i) > v_i$ for all values of N. Specifically, for $N = 3$ bidders, $b_i(v_i) = \frac{3-1}{3-2}v_i = 2v_i$; while when $N = 10$ bidders, $b_i(v_i) = \frac{10-1}{10-2}v_i = \frac{9}{8}v_i$, thus approaching the 45°-line.

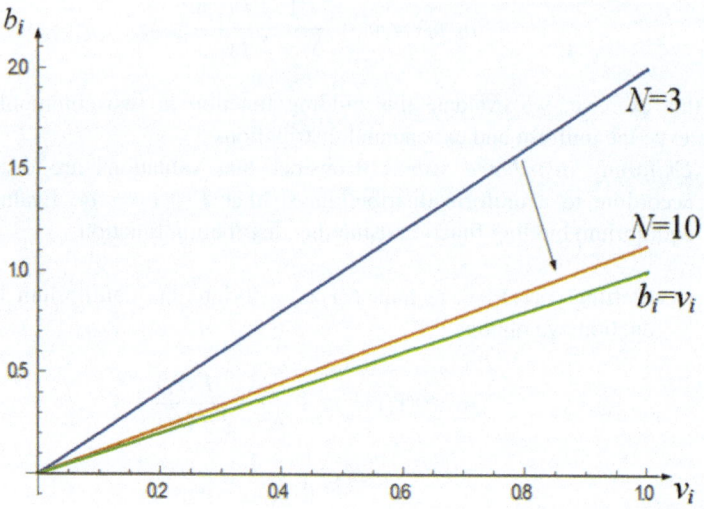

Fig. 5.1 Equilibrium bidding function in the third-price auction—more bidders

- In this case, we confirm our results from Exercise 5.2, where we showed that the equilibrium bidding function satisfies $b_i(v_i) > v_i$. Indeed, the difference

$$b(v_i) - v_i = \frac{N-1}{N-2}v_i - v_i$$

$$= \frac{v_i}{N-2}$$

is positive since $N \geq 3$ by definition.

(c) Still considering uniformly distributed values, compare the equilibrium bidding functions in the first-, second-, and third-price auctions. Interpret.

- Recall that, in a first-price auction with uniformly distributed valuations, the equilibrium bidding function is $b_i^{1st}(v_i) = \frac{N-1}{N}v_i$ (see Exercise 2.5, part a); whereas, in a second-price auction it is $b_i^{2nd}(v_i) = v_i$.[1] Hence,

$$b_i^{3rd}(v_i) > b_i^{2nd}(v_i) \geq b_i^{1st}(v_i)$$

since

---

[1] In the second-price auction, this equilibrium bidding function $b_i(v_i) = v_i$ holds both when valuations are uniformly distributed and otherwise.

Exercise #5.3: Equilibrium Bidding Function in a Third-Price Auction with...

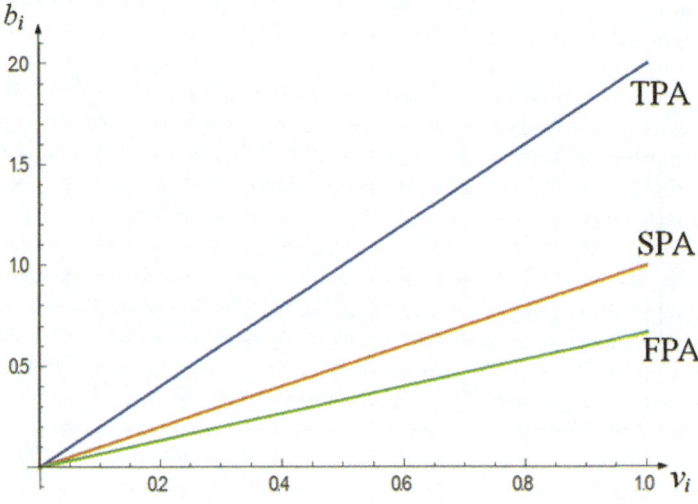

Fig. 5.2 Equilibrium bidding function in three auction formats

$$\frac{N-1}{N-2} v_i > v_i > \frac{N-1}{N} v_i$$

for all values of N. Moving from a first- to a second-price auction, bids increase as every bidder anticipates that, if winning, he will only pay the second-highest bid. A similar argument applies when we move from the second- to the third-price auction, as now every bidder knows that, upon winning, he will only need to pay the third highest bid.

- Figure 5.2 depicts equilibrium bidding functions in the first-price auction (FPA), second-price auction (SPA), and third-price auction (TPA) where, for simplicity, we assume $N = 3$ bidders. In this setting, $b_i^{1st}(v_i) = \frac{3-1}{3} v_i = \frac{2}{3} v_i$ in the first-price auction, $b_i^{2nd}(v_i) = v_i$ in the second-price auction, and $b_i^{3rd}(v_i) = \frac{3-1}{3-2} v_i = 2 v_i$ in the third-price auction.

(d) *Exponentially distributed values.* Consider now that individual valuations are drawn from an exponential distribution, $F(v_i) = 1 - \exp(-\lambda v_i)$ where $v_i \in [0, +\infty)$, and $N = 3$ bidders. Find the equilibrium bidding function $b_i(v_i)$ in this context. Evaluate your results at different values of λ, and interpret your results.

- Since $F(v_i) = 1 - \exp(-\lambda v_i)$, we obtain that $f(v_i) = \lambda \exp(-\lambda v_i)$. Plugging these results into the equilibrium bidding function, along with $N = 3$, yields

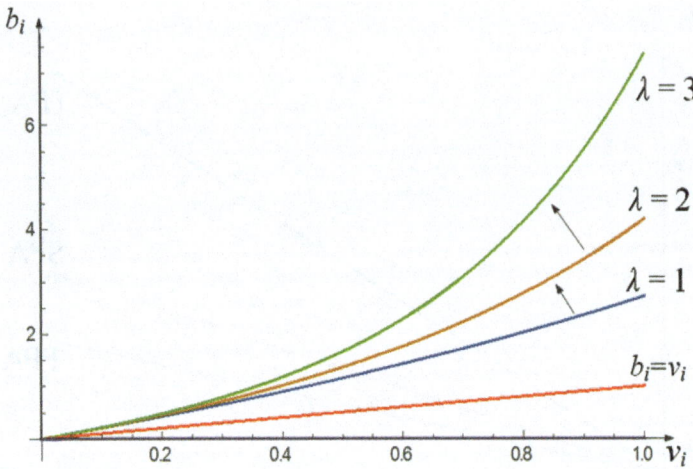

Fig. 5.3 Equilibrium bidding function in the third-price auction—exponentially distributed values

$$b_i(v_i) = v_i + \frac{1}{N-2}\frac{F(v_i)}{f(v_i)}$$

$$= v_i + \frac{1}{3-2}\frac{1-\exp(-\lambda v_i)}{\lambda \exp(-\lambda v_i)}$$

$$= v_i + \frac{\exp(\lambda v_i)-1}{\lambda}.$$

- Evaluating this bidding function at $\lambda = 1$, we obtain $b_i(v_i) = v_i + \exp(v_i) - 1$; at $\lambda = 2$, yields $b_i(v_i) = v_i + \frac{\exp(2v_i)-1}{2}$; and, similarly, at $\lambda = 3$, entails a bidding function of $b_i(v_i) = v_i + \frac{\exp(3v_i)-1}{3}$. Figure 5.3 plots these bidding functions and, as a reference, the line where bids satisfy $b_i = v_i$. The figure illustrates that, as the exponential distribution puts a larger probability weight on small valuations (higher λ), every bidder submits more aggressive bids (i.e., bidding function $b_i(v_i)$ shifts upward in λ), enlarging the overbidding that we identified in Exercise 5.2.

(e) *Other distribution forms.* Consider the following distribution function,

$$F(v_i) = (1+\alpha)v_i - \alpha v_i^2$$

where $v_i \in [0, 1]$, and parameter α satisfies $\alpha \in [-1, 1]$. When $\alpha = 0$, this function collapses to the uniform distribution, $F(v_i) = v_i$; when $\alpha > 0$, it becomes concave, thus putting more probability weight on low valuations;

and when $\alpha < 0$, it is convex, assigning more probability weight on high valuations.

Find the equilibrium bid in this setting, and explain its change in α. For simplicity, you may assume $N = 3$ bidders.

- Since $F(v_i) = (1 + \alpha) v_i - \alpha v_i^2$ and $f(v_i) = F'(v_i) = 1 + \alpha - 2\alpha v_i$, the equilibrium bidding function becomes

$$b_i(v_i) = v_i + \frac{1}{3-2} \frac{F(v_i)}{f(v_i)}$$

$$= v_i + \frac{(1+\alpha) v_i - \alpha v_i^2}{1 + \alpha - 2\alpha v_i}.$$

- Differentiating $b_i(v_i)$ with respect to α, we obtain

$$\frac{\partial b_i(v_i)}{\partial \alpha} = v_i \frac{(1 - v_i)(1 + \alpha - 2\alpha v_i) - (1 - 2v_i)(1 + \alpha - \alpha v_i)}{(1 + \alpha - 2\alpha v_i)^2}$$

$$= \left(\frac{v_i}{1 + \alpha - 2\alpha v_i} \right)^2 > 0.$$

Therefore, as α increases, the distribution function becomes more concave, and bidders put a larger probability weight on low valuations, submitting more aggressive bids.

- Figure 5.4 depicts this bidding function, evaluated at three different values of α:
  - When $\alpha = 0$, the equilibrium bid simplifies to $b_i(v_i) = 2v_i$, thus coinciding with our results when valuations are uniformly distributed (part c) given that $\frac{N-1}{N-2} v_i = \frac{3-1}{3-2} v_i = 2v_i$.
  - When $\alpha = -1/2$, the equilibrium bid is $b_i(v_i) = \frac{v_i(2+3v_i)}{1+2v_i}$ that falls below $b_i(v_i) = 2v_i$, so that bidders submit less aggressive bids because they assign a larger probability weight on high valuations.
  - When $\alpha = 1/2$, the equilibrium bid becomes $b_i(v_i) = \frac{3v_i(2-v_i)}{3-2v_i}$ that lies above $b_i(v_i) = 2v_i$, so that bidders submit more aggressive bids because they assign a larger probability weight on low valuations.

Exercise #5.4: kth-Price Auction[C]

5.4 Consider an auction with $N \geq 2$ risk-neutral bidders and the following rules. Every bidder i privately observes his valuation for the object, v_i, drawn from a uniform distribution $F(v_i) = v_i$, where $v_i \in [0, 1]$. The bidder submitting the highest bid wins the auction and pays the kth-highest price, where k satisfies $N \geq k \geq 1$. This auction format is known as the kth-price auction and embodies

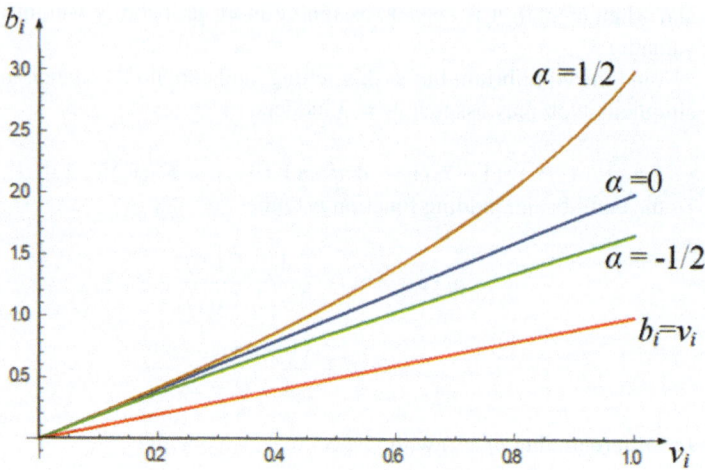

Fig. 5.4 Equilibrium bidding function—other cumulative distributions

other auction formats as special cases. For instance, when $k = 1$, this auction becomes a first-price auction; when $k = 2$, it coincides with a second-price auction, and similarly for higher values of k.

(a) Consider that, in equilibrium, every bidder i uses a symmetric bidding function $b(v_i) = \alpha_k v_i$, which is linear in his valuation, v_i, and α_k captures whether the bidder shades (if $0 \leq \alpha_k < 1$) or inflates (if $\alpha_k > 1$) his bid. Find the value of α_k in this equilibrium bidding function. [*Hint*: First, find the expected value of the $(k-1)$-order statistic of N random variables drawn from $U[0, 1]$. Then, apply the revenue equivalence principle.]

- *Expected value of an order statistic.* The density function of the r-order statistic (in this setting with i.i.d. valuations drawn from a uniform distribution), as found in Exercise 2.10, is

$$f^{[r]}(v_i) = N\binom{N-1}{r-1}v_i^{r-1}(1-v_i)^{N-r}$$

$$= \frac{N!}{(r-1)!(N-r)!}v_i^{r-1}(1-v_i)^{N-r}$$

for all $v_i \in [0, 1]$, which coincides with the Beta distribution, that is, $v_i^{[r]} \sim Beta(r, N-r+1)$. Using the expected value of a random variable distributed according to the Beta distribution, we obtain that the expected value of the r-order statistic in our setting is

$$E\left[v^{[r]}\right] = \frac{r}{N+1}.$$

Exercise #5.4: kth-Price Auction^C

Setting $r = N - k + 1$, we find an expected value of

$$E\left[v^{[N-k+1]}\right] = \frac{N-k+1}{N+1}.$$

- *Revenue equivalence principle.* By the revenue equivalence principle, we know that the expected revenue of the kth-price auction must coincide with that of any other auction format analyzed above, such as the second-price auction where all bidders drawn their valuations from a uniform distribution.
  - Inserting $k = 2$ in the above expected value, we can find the expected revenue in a second-price auction when bidders draw their valuations from a uniform distribution, as follows:

$$R^2 = E\left[v^{[N-2+1]}\right] = E[v^{[N-1]}] = \frac{N-2+1}{N+1} = \frac{N-1}{N+1}.$$

  - We now set the expected revenue in both the second- and kth-price auctions equal to each other, as follows:

$$E\left[v^{[N-1]}\right] = E\left[b\left(v^{[N-k+1]}\right)\right]$$

Since $R^2 = E\left[v^{[N-1]}\right] = \frac{N-1}{N+1}$ and the bidding function satisfies $b(v_i) = \alpha_k v_i$, this expression becomes

$$\frac{N-1}{N+1} = \alpha_k E\left[v^{[N-k+1]}\right].$$

We can now insert $E[v^{[N-k+1]}] = \frac{N-k+1}{N+1}$ that we found above, to obtain

$$\frac{N-1}{N+1} = \alpha_k \frac{N-k+1}{N+1}$$

Solving for α_k yields

$$\alpha_k = \frac{N-1}{N-k+1}$$

which entails that the equilibrium bidding function in the kth-price auction is

$$b(v_i, k) = \frac{N-1}{N-k+1} v_i$$

(b) Show that every bidder submits a bid above his own valuation, v_i, for all $k \geq 3$.

- Bidder i's equilibrium bid is higher than his own valuation if and only if

$$\frac{N-1}{N-k+1}v_i > v_i$$

which simplifies to

$$N - 1 > N - k + 1$$

which further collapses to $k > 2$, i.e., for the third-price auction and beyond. This makes sense, as bidder i pays, upon winning, the third-highest bid rather than his own.

(c) Evaluate the equilibrium bidding function that you found in part (a) at $k = 1$, then at $k = 2$, and finally at $k = 3$. Confirm that your results coincide with those in the previous exercises.

- When $k = 1$, the equilibrium bidding function simplifies to

$$b(v_i, 1) = \frac{N-1}{N-1+1}v_i$$
$$= \frac{N-1}{N}v_i$$

which coincides with the bidding function in a first-price auction with N bidders drawing their valuations from a uniform distribution, as found in Exercise 2.3.

- When $k = 2$, the equilibrium bidding function simplifies to

$$b(v_i, 2) = \frac{N-1}{N-2+1}v_i$$
$$= v_i$$

implying that every bidder i submits a bid equal to his valuation, v_i, as shown for the second-price auction in Exercise 1.2.

- When $k = 3$, the equilibrium bidding function simplifies to

$$b(v_i, 3) = \frac{N-1}{N-3+1}v_i$$
$$= \frac{N-1}{N-2}v_i$$

as in the third-price auction analyzed in Exercise 5.3.

- Figure 5.5 plots the equilibrium bidding function considering $N = 4$ bidders and evaluated at four different values of k. The figure indicates

Exercise #5.4: kth-Price Auction^C

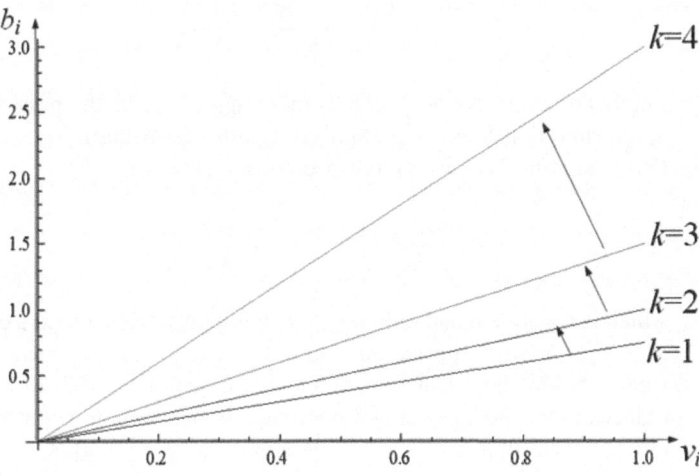

Fig. 5.5 Equilibrium bidding function in the kth-price auction

that the bidding function shifts upward as the price that the bidder pays upon winning is lower (e.g., first-, second-, or third-highest bid).

As a remark, note that when $k = N - 1$, the above bidding function becomes

$$b(v_i, N-1) = \frac{N-1}{N-(N-1)+1} v_i$$

$$= \frac{N-1}{2} v_i.$$

For instance, in the context with $N = 4$ bidders considered in Fig. 5.4, this bidding function simplifies to $\frac{3}{2} v_i$, which coincides with the bidding function $b(v_i, 3) = \frac{N-1}{N-3+1} v_i = \frac{3}{2} v_i$ in this figure since $k = N - 1$ in this setting entails $k = 4 - 1 = 3$. More generally, we can claim that

$$b(v_i, N-1) = \frac{N-1}{2} v_i \leq \frac{N-1}{N-2} v_i = b(v_i, 3)$$

if and only if $N \leq 4$.

(d) *Comparative statics.* How is the equilibrium bidding function that you found in part (a) affected by an increase in k? And by an increase in N? Interpret.

- Differentiating $b(v_i, k)$ with respect to k yields

$$\frac{\partial b(v_i, k)}{\partial k} = \frac{N-1}{(N-k+1)^2} v_i > 0$$

Intuitively, every bidder becomes more aggressive as the price he pays upon winning is lower (e.g., third- or fourth-highest bid).

- Differentiating $b(v_i, k)$ with respect to N yields

$$\frac{\partial b(v_i, k)}{\partial N} = -\frac{k-2}{(N-k+1)^2} v_i$$

which is negative if and only if $k \geq 3$. Intuitively, when bidders compete in a first- or second-price auction ($k = 1$ and $k = 2$), their bids lie weakly below their valuation, and an increase in the number of bidders induces every bidder to submit more aggressive bids. In contrast, when they compete in a third-price auction ($k = 3$) or beyond ($k > 3$), they reduce their bids as more bidders compete for the object. To understand this result, note that, as more bidders compete, it is more likely that the price that bidder i must pay for the object if winning (the kth-highest bid) is close to his own valuation, leading every bidder i to become more conservative in his bids.

Exercise #5.5: Efficiency in kth-Price Auctions[A]

5.5 Consider the kth-price auction in Exercise 5.4. Argue that the object, in equilibrium, is assigned to the bidder with the highest valuation.

- In this auction format, every bidder i uses a bidding function

$$b(v_i, k) = \frac{N-1}{N-k+1} v_i,$$

which is monotonically increasing in the bidder's valuation, v_i, regardless of the number of bidders, N, and the kth-price that the winner must pay.
- Therefore, the individual submitting the highest bid, h, satisfies $b(v_h, k) > b(v_j, k)$ for every bidder $j \neq h$, implying that bidder h assigns the highest value to the object, $v_h > v_j$. In summary, the bidder who wins the auction, h, also has the highest value for the object, entailing that the all-pay auction is efficient.

Exercise #5.6: Lottery Auction, An Introduction[A]

5.6 Consider a situation where an all-pay auction takes place for an item, with its allocation determined by a lottery. Each bidder is able to observe the bids of every other bidder as they make their own. The probability that bidder i wins the auction is

$$p = \frac{b_i}{b_i + B_{-i}}$$

where B_{-i} denotes the total bids made by all other bidders. Suppose that bidder i has a valuation of $v_i = 9$ for this item, and he knows that bids totaling $B_{-i} = \$4$ have already been submitted. Find the optimal bidding strategy for bidder i, b_i, taking into consideration that he must pay his bid regardless of whether he wins the auction.

- Since bidder i can observe all other bids, he simply maximizes his own expected payoff:

$$\max_{b_i \geq 0} EU_i(b_i|9) = \left(\frac{b_i}{b_i + 4}\right)(9) + \left(1 - \frac{b_i}{b_i + 4}\right)(0) - b_i$$

$$= \frac{9b_i}{b_i + 4} - b_i$$

Differentiating this expected utility with respect to b_i yields

$$\frac{9(b_i + 4) - 9b_i}{(b_i + 4)^2} - 1 = 0$$

Rearranging, we obtain

$$9(b_i + 4) - 9b_i = (b_i + 4)^2$$

that further simplifies to

$$(b_i + 4)^2 = 36.$$

Taking the square root on both sides and then subtracting 4, we find bidder i's equilibrium bid $b_i = \sqrt{36} - 4 = 2$. (*Note:* $b_i = -\sqrt{36} - 4 = -10$ is also a solution to this maximization problem, but we assume that only positive bids can be submitted.)

Exercise #5.7: Lottery Auction, A More General Approach[B]

5.7 Consider a lottery auction with $N \geq 2$ bidders competing for an object. For simplicity, let us assume that all bidders assign the same value to the object, v. For a given bidding profile $b = (b_1, \ldots, b_N)$, the probability that bidder i wins the auction is

$$\Pr\{win\} = \frac{b_i}{\sum_{j=1}^{N} b_j},$$

Intuitively, his probability of winning is a function of how his bid compares to the aggregate bids submitted by all players. All bidders must pay their bids, thus making this auction format similar to the all-pay auctions studied in Chap. 4. Therefore, bidder i's utility is

$$EU_i[b_i|v] = \frac{b_i}{\sum_{j=1}^{N} b_j} v - b_i.$$

Lottery auctions are then analogous to contests, where players invest resources (rather than submitting bids), and are often used to model promotions within a firm, where every worker invests time and effort into being selected for a promotion; political campaigns, where candidates invest money and resources to capture a larger share of votes; and research and development races, where firms invest resources into discovering a new product, such as a drug.

(a) Considering a symmetric bidding strategy, find bidder i's equilibrium bidding function.

- We can first rewrite the above utility as follows:

$$EU_i[b_i|v] = \frac{b_i}{b_i + B_{-i}} v - b_i$$

where, for compactness, $B_{-i} \equiv \sum_{j \neq i} b_j$ denotes the sum of all bids from bidder i's rivals.

- Differentiating with respect to b_i, we find

$$\frac{B_{-i}}{(b_i + B_{-i})^2} v - 1 = 0$$

In a symmetric bidding strategy, $b^* = b_1^* = b_2^* = \cdots = b_N^*$, so that $B_{-i}^* = (N-1)b^*$. Therefore, the above first-order condition becomes

$$\frac{(N-1)b^*}{[b^* + (N-1)b^*]^2} v = 1$$

Since $b^* + (N-1)b^* = Nb^*$, we can rearrange this expression as follows:

$$\frac{N-1}{N^2 b^*} v = 1$$

Solving for b^* yields an equilibrium bidding function of

Exercise #5.7: Lottery Auction, A More General Approach[B]

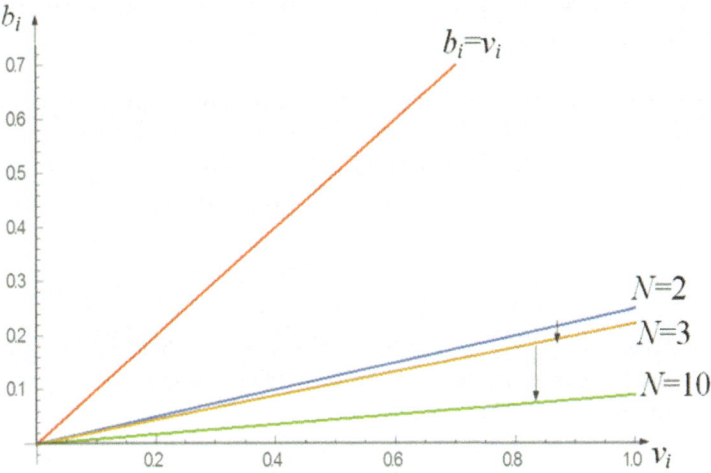

Fig. 5.6 Equilibrium bidding function in a lottery auction

$$b^*(v) = \frac{N-1}{N^2} v.$$

(b) *Comparative statics.* How does bidder i's equilibrium bid changes with v and N?

- Differentiating $b^*(v)$ with respect to v, we find that

$$\frac{\partial b^*(v)}{\partial v} = \frac{N-1}{N^2} \geq 0,$$

implying that bidder i increases his bid when his valuation of the object (and that of all other bidders, since valuations are symmetric in our setting) increases.

- Differentiating $b^*(v)$ with respect to N yields

$$\frac{\partial b^*(v)}{\partial N} = -\frac{N-2}{N^3} v \leq 0$$

Since $N \geq 2$ by assumption, the more bidders competing in the auction, the less that every bidder i bids in equilibrium.

- Figure 5.6 evaluates the equilibrium bidding function $b^*(v)$ at $N = 2$, which yields $b^*(v) = \frac{1}{4}v$; at $N = 3$, entailing $b^*(v) = \frac{2}{9}v$; and at $N = 10$, obtaining $b^*(v) = \frac{9}{100}v$; and shows that this bidding function rotates clockwise as the number of bidders increases.

(c) Verify that the equilibrium bidding function maximizes the expected utility of player i, that is, check second-order conditions.

- The second-order condition of the utility maximization program is

$$\frac{\partial^2 E U_i[b_i|v]}{\partial b_i^2} = \frac{-2B_{-i}}{(b_i + B_{-i})^3} v$$

$$= \frac{-2B_{-i}}{B^3} v$$

where $B \equiv \sum_{j=1}^{N} b_j$ represents the sum of all bidders' bids (including bidder i). Inserting the equilibrium bidding function that we found in part (a), $b^*(v) = \frac{N-1}{N^2} v$, where

$$B^* = N \frac{N-1}{N^2} v = \frac{N-1}{N} v \text{ and}$$

$$B_{-i}^* = (N-1) \frac{N-1}{N^2} v = \left(\frac{N-1}{N}\right)^2 v$$

yields

$$\frac{\partial^2 E U_i[b_i|v]}{\partial b_i^2} = \frac{-2 \left(\frac{N-1}{N}\right)^2 v}{\left(\frac{N-1}{N} v\right)^3} v = -\frac{2N}{(N-1)v}$$

which is negative, as required for concavity in the expected utility function.

(d) Find the bids that bidders would choose if they could coordinate their bidding decisions.

- If bidders choose the bidding profile $(b_1, \ldots, b_N)$ that maximizes their joint expected utility, they would solve

$$\max_{b_1, \ldots, b_N \geq 0} \sum_{i=1}^{N} \left(\frac{b_i}{b_i + B_{-i}} v - b_i\right)$$

$$= \frac{B}{b_i + B_{-i}} v - B$$

$$= v - (b_i + B_{-i}).$$

Differentiating with respect to b_i, we obtain -1, indicating that we face a corner solution. In other words, bidders maximize their joint expected payoff when bidding $b_i^* = b_j^* = 0$ for every bidder $i \neq j$.

(e) Compare the equilibrium bid that you found in part (a) and the bidding profile that maximizes all bidders' joint expected utility that you found in part (d).

- When every bidder simultaneously and independently chooses his bid, he submits a bid of $b^*(v) = \frac{N-1}{N^2} v$, ignoring the negative externality that his bid imposes on the other player; namely, it reduces bidder j's probability of winning the auction. In contrast, when all bidders coordinate their bids, they internalize this externality, reducing their bid to zero.

Exercise #5.8: Efficiency in Lottery Auctions[A]

5.8 Consider the lottery auction in Exercise 5.7. Argue that the object, in equilibrium, is not necessarily assigned to the bidder with the highest valuation.

- In this auction format, every bidder i uses an equilibrium bidding function of

$$b^*(v) = \frac{N-1}{N^2} v,$$

which is monotonically increasing in valuation, v. This property, however, is not sufficient for the winner to be the bidder with the highest valuation. Indeed, the probability of winning in this auction is $\Pr\{win\} = \frac{b_i}{\sum_{j=1}^{N} b_j}$, implying that even if a bidder submits the highest bid, he may lose the auction with probability

$$\Pr\{lose\} = 1 - \Pr\{win\}$$
$$= \frac{B - b_i}{B} > 0$$

where $B \equiv \sum_{j=1}^{N} b_j$. Therefore, the lottery auction does not guarantee that the object is assigned to the individual with the highest valuation, implying that this auction format is not efficient.

The Revenue Equivalence Principle 6

Keywords

Revenue equivalence theorem · Risk-averse bidders · Risk-averse seller · Risk-neutral seller · Second-order stochastic dominance · Misreporting · Revelation principle · Myerson's characterization theorem · Linear utility functions · Expected transfer · Asymmetric probability distributions

Introduction

In this chapter, we evaluate the seller's expected revenue in different auction formats and show under which contexts this revenue is equivalent. First, Exercise 6.1 considers the first-price, second-price, third-price, and all-pay auctions from the previous chapters, evaluating the expected revenue that each of them generate, and showing that the expected revenue coincides. For simplicity, this exercise assumes that valuations are uniformly distributed, but Exercise 6.2 allows for valuations to be drawn from a generic distribution and demonstrates that revenue equivalence between the first- and second-price auction still holds.

Exercise 6.3 shows that these two auction formats do not yield the same expected revenue if bidders are risk-averse—in particular, the first-price auction generates a larger expected revenue than the second-price auction. Similarly, Exercise 6.4 demonstrates that when the seller is risk-averse, he may prefer first-price auction even though both auction formats generate the same expected revenue because the first-price auction exhibits a smaller variance (i.e., a "flatter" distribution of bids).

Exercise 6.5 and 6.6 provide two proofs (one shorter and another longer and a bit more technical) to the so-called Revenue Equivalence Principle, which examines under which conditions the two auction formats yield the same expected revenue for the seller. Informally, this requires: (1) a set of risk-neutral bidders; (2) that both auction formats assign the object to the same bidder; and (3) that the expected utility of the bidder with the lowest valuation must coincide across both auction formats. These three conditions apply to several auctions and, hence, the Revenue Equivalence Principle is a useful tool in the literature.

Exercise #6.1: Revenue Comparison in Four Auction Formats[A]

6.1 Consider a sealed-bid auction with $N \geq 2$ bidders. Every bidder i privately observes his valuation v_i for the object, drawn from uniform distribution $U[0, 1]$, which is common knowledge among players. Assume that all bidders are risk-neutral.
   (a) Find equilibrium bidding functions in the: (1) first-price auction, (2) second-price auction, (3) third-price auction, and (4) all-pay auction. Identify which bidder wins in each auction format.

   - *First-price auction.* From Exercise 2.3, we know that the equilibrium bidding function in this auction with N bidders and uniformly distributed valuations is

   $$b_i(v_i) = \frac{N-1}{N} v_i \text{ for every bidder } i$$

   For instance, in the case of $N = 3$ bidders, this bidding function becomes $b_i(v_i) = \frac{2}{3} v_i$. If bidder j is the individual with the highest valuation for the object, he submits the highest bid (since bids are increasing in valuations) and pays a share $\frac{N-1}{N}$ of his true valuation (this is what we normally denote as "bid shading").
   - *Second-price auction.* From Exercise 1.2, we know that bidding according to his valuation is a weakly dominant strategy in the second-price auction, and also the Bayesian Nash equilibrium of the game, that is, $b(v_i) = v_i$ for every bidder i. Therefore, if bidder i has the highest valuation for the object, that is $v_i > \max_{j \neq i} v_j$, he submits the highest bid, wins the auction, and pays a price that coincides with the bid of the second-highest bidder, $b(v_k) = v_k$ where $v_k = \max_{j \neq i} v_j$ denotes the highest valuation among all bidder i's rivals.
   - *Third-price auction.* From Exercise 5.3, the bidding function is $b_i(v_i) = \frac{N-1}{N-2} v_i$ for every bidder i. The winning bidder, however, pays

   $$\frac{N-2}{N} \times \frac{N-1}{N-2} v_i = \frac{N-1}{N} v_i$$

   to the seller since the third-highest-valued bidder submits a bid that is a fraction $\frac{N-2}{N}$ of this winning bidder.
   - *All-price auction.* From previous exercises (see Exercises 4.3 and 4.4), we know that the equilibrium bidding function in the all-pay auction with uniformly distributed valuations is

Exercise #6.1: Revenue Comparison in Four Auction Formats[A]

$$b_i(v_i) = \frac{N-1}{N} v_i^N \text{ for every bidder } i$$

Since this equilibrium bid is increasing in player i's valuation, v_i, the bidder with the highest bid submits the highest valuation, and wins the object. In the case of $N = 3$ bidders, for instance, the equilibrium bid simplifies to $b_i(v_i) = \frac{2}{3} v_i^3$.

(b) Evaluate the seller's *ex-post* revenue in each auction format (which is a function of realizations of players' valuations). Does the revenue equivalence theorem hold?

- *First-price auction.* From part (a), we know that $b_i(v_i) = \frac{N-1}{N} v_i$, implying that the expected revenue of the seller in this auction format is

$$\frac{N-1}{N} \max\{v_1, v_2, \ldots, v_N\}$$

since the winning bid (and thus the price that the seller receives) is $\frac{N-1}{N}$ of the highest bidder's valuation.

- *Second-price auction.* Recall from part (a) that the individual i with the highest valuation submits the highest bid, $b(v_i) = v_i$ and wins the object, but he only pays a price equal to the bid of the individual with the second-highest valuation, $b(v_j) = v_j$, implying that the seller's expected revenue is v_j.

- *Third-price auction.* From part (a), the winning bidder pays $\frac{N-1}{N} v_i$ to the seller.

- *All-price auction.* From part (a), we know every bidder i submits a bid $b_i(v_i) = \frac{N-1}{N} v_i^N$. In this auction format, the seller keeps all submitted bids (even if the player loses the auction!), implying that his revenue becomes

$$\sum_{i=1}^{N} b_i(v_i) = \sum_{i=1}^{N} \frac{N-1}{N} v_i^N$$

$$= \frac{N-1}{N} \sum_{i=1}^{N} v_i^N$$

For instance, in the case of $N = 3$ bidders, this revenue simplifies to $\frac{2}{3}(v_1^3 + v_2^3 + v_3^3)$.

- *Revenue comparison.* Depending on the specific realization of players' valuations, the actual revenue that the seller receives can be different across auction formats. However, this is not contradictory with the revenue equivalence theorem. This theorem only tells us that, in expectation (that is, before valuations are drawn from the uniform distribution), revenues should coincide across these auction formats. The next part of the exercise tests whether this theorem applies in these three auctions.

(c) Find the seller's expected payments in each auction format (which is not a function of realization of players' valuations but, instead, of the expected value of these valuations), and his expected revenue (summing over all expected payments). Does the revenue equivalence theorem hold?

- *First-price auction.* The *ex-post* profit of the first-price auction that the seller earns from winning bidder is

$$m^1(v_i) = \underbrace{v_i^{N-1}}_{\text{probability of winning}} \times \underbrace{\frac{N-1}{N} v_i}_{\text{value of the bid}}$$

since $\Pr\{b_i > b_j\} = v_i$ for every rival $j \neq i$, entailing that bidder i wins with probability v_i^{N-1}. Simplifying, we obtain an *ex-post* profit (also referred to as the bidder's "payment") of

$$m^1(v_i) = \frac{N-1}{N} v_i^N$$

Therefore, the expected payment from bidder i is

$$E[m^1(v_i)] = \int_0^1 \underbrace{\frac{N-1}{N} v_i^N}_{m^1(v_i)} dv_i$$

$$= \frac{N-1}{N} \int_0^1 v_i^N dv_i$$

$$= \frac{N-1}{N} \left[\frac{1}{N+1} v^{N+1} \right]_0^1$$

$$= \frac{N-1}{N} \left(\frac{1}{N+1} - 0 \right)$$

$$= \frac{N-1}{N(N+1)}$$

Finally, the seller's expected revenue (or *ex-ante* profit from the auction) is

$$R^1 = \sum_{i=1}^{N} E[m^1(v_i)]$$

$$= \sum_{i=1}^{N} \frac{N-1}{N(N+1)}$$

$$= N \frac{N-1}{N(N+1)}$$

$$= \frac{N-1}{N+1}$$

- *Second-price auction.* The *ex-post* profit of the second-price auction from the winning bidder (i.e., the winner's payment) is

$$m^2(v_i) = \underbrace{v_i^{N-1}}_{\text{probability of winning}} \times \underbrace{v_j}_{\text{value of the second-highest bid}}$$

$$= v_i^{N-1} \frac{N-1}{N} v_i$$

$$= \frac{N-1}{N} v_i^N$$

where the second line originates from the fact that uniform distribution sets v_j to be $\frac{N-1}{N}$ below v_i. Therefore, the winner's payment coincides with that in the first-price auction.

Applying the same calculations as above, we can show that the seller's expected revenue (also known as the *ex-ante* profit) from running a second-price auction is also

$$R^2 = \frac{N-1}{N+1}$$

- *Third-price auction.* Since the payment in this auction format satisfies

$$m^3(v_i) = v_i^{N-1} \times \frac{N-1}{N} v_i$$

$$= m^1(v_i)$$

the seller's expected revenue must also coincide with that in the first- and second-price auctions.

- *All-pay auction.* In the all-pay auction, every bidder i pays the bid he submitted, implying that the payment from every bidder i coincides with his bid,

$$m^A(v_i) = b_i(v_i) = \frac{N-1}{N} v_i^N$$

Unlike the first- and second-price auctions, this payment occurs with certainty (probability one), as all bidders must pay their bids. The expected payment from every bidder i is, then,

$$\begin{aligned}
E[m^A(v_i)] &= \int_0^1 \left[\frac{N-1}{N} v_i^N\right] dv_i \\
&= \frac{N-1}{N} \int_0^1 v_i^N dv_i \\
&= \frac{N-1}{N} \left[\frac{v_i^{N+1}}{N+1}\right]_0^1 \\
&= \frac{N-1}{N} \left(\frac{1}{N+1} - 0\right) \\
&= \frac{N-1}{N(N+1)}
\end{aligned}$$

As a consequence, the seller's expected revenue in the all-pay auction becomes

$$\begin{aligned}
R^A &= \sum_{i=1}^{N} E[m^A(v_i)] \\
&= N \frac{N-1}{N(N+1)} \\
&= \frac{N-1}{N+1}
\end{aligned}$$

which coincides with that in the first-price and second-price auctions, verifying that the Revenue Equivalence Theorem holds across all auction formats. As a practice, confirm that requirements (1)–(3) of the Revenue Equivalence Theorem, listed in the Introduction section of this chapter, hold in these three auction formats.

Exercise #6.2: Revenue Comparison Between First- and Second-Price Auctions[B]

6.2 Consider again the first- and second-price auctions with $N \geq 2$ bidders competing for an object. Assume that every bidder i independently draws his valuation

Exercise #6.2: Revenue Comparison Between First- and Second-Price Auctions

v_i from a cumulative distribution function $F(v_i)$ with an associated density function $f(v_i) > 0$ for all $v_i \in [0, 1]$.

(a) Show that the seller earns the same expected revenue in both auction formats. [*Hint:* Use the expected revenue in a first-price auction found in Exercise 3.1, but rearrange it to make it a function of bidder i's equilibrium bid in this auction format. Then use integration by parts to obtain an expression that coincides with that of the expected revenue in the second-price auction as found in Exercise 1.7.]

- The expected revenue in the first-price auction, as shown in Exercise 3.1, is

$$E[\pi] = \int_0^1 b(z) \left[N F(z)^{N-1} f(z) \right] dz$$

Intuitively, the seller earns bid $b(z)$ from the bidder with the highest valuation. Since valuations are independently distributed, the probability that all valuations are below a given valuation z is $F(z)^N$ with an associated density of $N F(z)^{N-1} f(z)$; as written in the bracket of the above expression.

This expected revenue can be rewritten as

$$R^1 = N \int_0^1 \left[b(z) F(z)^{N-1} \right] f(z) \, dz$$

Using integration by parts, we obtain

$$R^1 = N \int_0^1 \left[b(z) F(z)^{N-1} \right] dF(z)$$

$$= N \, b(z) F(z)^N \Big|_0^1 - N \int_0^1 F(z) \frac{\partial \left[b(z) F(z)^{N-1} \right]}{\partial z} dz$$

$$= N \left[b(1) F(1)^N - b(0) F(0)^N \right] - N(N-1)$$

$$\times \int_0^1 b(z) F(z)^{N-1} f(z) \, dz.$$

Since $F(1) = 1$, $F(0) = 0$, and $b(0) = 0$ by definition, from part (b) of Exercise 2.6 we know that $b(1) = (N-1) \int_0^1 z F(z)^{N-2} f(z) \, dz$, yielding

$$R^1 = N(N-1) \int_0^1 z F(z)^{N-2} f(z) \, dz - N(N-1)$$

$$\times \int_0^1 zF(z)^{N-1} f(z)\,dz$$

$$= N(N-1) \int_0^1 zF(z)^{N-2}[1-F(z)]f(z)\,dz$$

which coincides with the expected revenue in the second-price auction in Exercise 1.7.

(b) Show that the expected revenue in the first- and second-price auctions is increasing in the number of bidders competing for the object, N.

- We can rewrite the expected revenue from part (a) as

$$R^1 = N(N-1) \int_0^1 z[F(z)^{N-2} - F(z)^{N-1}]f(z)\,dz$$

and since $\frac{\partial[NF(z)^{N-1} - (N-1)F(z)^N]}{\partial z} = N(N-1)[F(z)^{N-2} - F(z)^{N-1}]f(z)$ we obtain

$$R^1 = \int_0^1 z \frac{\partial[NF(z)^{N-1} - (N-1)F(z)^N]}{\partial z}\,dz$$

which can be rewritten as

$$R^1 = z\Big[NF(z)^{N-1} - (N-1)F(z)^N\Big]\Big|_0^1$$
$$- \int_0^1 \Big[NF(z)^{N-1} - (N-1)F(z)^N\Big]dz$$
$$= 1 - \int_0^1 \Big[NF(z)^{N-1} - (N-1)F(z)^N\Big]dz$$

The first term is a constant, so we can focus on the second term to check if it is increasing in N. Define $A(N) \equiv -\int_0^1 B(N)dz$, where

$$B(N) \equiv NF(z)^{N-1} - (N-1)F(z)^N$$

so we can measure the difference $B(N+1) - B(N)$ as follows:

$$B(N+1) - B(N) = \Big[(N+1)F(z)^N - NF(z)^{N+1}\Big]$$
$$- \Big[NF(z)^{N-1} - (N-1)F(z)^N\Big]$$

which simplifies to

$$B(N+1) - B(N) = -NF(z)^{N+1} + 2NF(z)^N - NF(z)^{N-1}$$
$$= -NF(z)^{N-1}\left[1 - 2F(z) + F(z)^2\right]$$
$$= -NF(z)^{N-1}[1 - F(z)]^2 \leq 0$$

Since the difference $B(N+1) - B(N)$ is negative, the difference $A(N+1) - A(N)$ must be positive. Therefore, $A(N)$ is increasing in N and as a result, the seller's expected revenue increases in the number of bidders.

Exercise #6.3: The Revenue Equivalence Principle with Risk Averse Bidders[A]

6.3 Consider again the setting in Exercise 6.1 with $N \geq 2$ bidders, each drawing his valuation from a uniform distribution. Assume now, however, that bidders are risk-averse, with utility function $u(w) = w^\alpha$, where $w > 0$ denotes income and $0 < \alpha < 1$.

(a) *Second-price auction.* Find the seller's expected revenue in the second-price auction.

- As described in Exercise 1.4, in second-price auctions every bidder i submits an equilibrium bid $b_i(v_i) = v_i$ that coincides with his valuation of the object, and this result is unaffected by bidders' risk preferences. Therefore, the *ex-post* revenue in the second-price auction is still

$$R^2 = \frac{N-1}{N}v_i^N$$

as shown in Exercise 6.1. We can then conclude that expected revenue satisfies $R_{RN}^2 = R_{RA}^2$, where subscript RN (RA) denotes risk neutrality (aversion).

(b) *First-price auction.* Show that the seller's expected revenue in the first-price auction is larger with risk-averse than with risk-neutral preferences. [*Hint:* You do not need to find the expected revenue from the auction.]

- From Exercise 3.4, we know that in the first-price auction with risk-averse bidders, every bidder i's equilibrium bid is $b_i^{RA}(v_i) = \frac{N-1}{N-1+\alpha}v_i$, which is higher than his equilibrium bid under risk-neutral preferences, $b_i^{RN}(v_i) = \frac{N-1}{N}v_i$ when $\alpha = 1$, since

$$\frac{N-1}{N-1+\alpha} v_i > \frac{N-1}{N} v_i$$

simplifies to $1 > \alpha$ that holds. Therefore, expected revenue satisfies $R_{RA}^1 > R_{RN}^1$.

(c) *Revenue comparison.* Show that when bidders are risk-averse, the first-price auction generates a larger expected revenue than the second-price auction.

- From parts (a) and (b), we found that expected revenue satisfies

$$R_{RA}^2 = R_{RN}^2 = R_{RN}^1 < R_{RA}^1$$

implying that, when bidders are risk-averse, the first-price auction generates a larger expected revenue than the second-price auction.

Exercise #6.4: The Revenue Equivalence Principle with Risk Averse Sellers[B]

6.4 A seller considers to either run a first- or second-price auction to sell an object among $N \geq 2$ risk-neutral bidders, each of them drawing his valuation from a uniform distribution $F(v_i) = v_i$, where $v_i \in [0, 1]$. In Exercise 6.1, we showed that a risk-neutral seller earns the same expected revenue from both auctions, $R^1 = R^2 = \frac{N-1}{N+1}$. In this exercise, we seek to show that if the seller is risk-averse, he prefers to sell the object using a first- than a second-price auction.

(a) *Second-price auction.* Find the variance of the payment in the second-price auction, where the seller anticipates that every bidder i submits a bid equal to his valuation, $b(v_i) = v_i$.

- The payment in the second-price auction is the second-highest bid, which given that $b(v_i) = v_i$, coincides with the second-highest valuation (i.e., the second-order statistic). From Exercise 2.9, the variance of the second-order statistic, $v^{[2]}$, is

$$Var\left(v^{[2]}\right) = \frac{2(N-1)}{(N+1)^2(N+2)}$$

(b) *First-price auction.* Find the variance of submitted bids in the first-price auction, where the seller anticipates that every bidder i submits a bid, $b(v_i) = \frac{N-1}{N} v_i$.

- The payment in the first-price auction coincides with the winning bid, that is, $b(v_i^{[1]}) = \frac{N-1}{N} v_i^{[1]}$, where $v_i^{[1]}$ denotes the first-order statistic. From Exercise 2.8, the variance of this payment is

$$Var\left(b\left(v_i^{[1]}\right)\right) = \left(\frac{N-1}{N}\right)^2 Var\left(v_i^{[1]}\right)$$

$$= \left(\frac{N-1}{N}\right)^2 \frac{N}{(N+1)^2(N+2)}$$

$$= \frac{(N-1)^2}{N(N+1)^2(N+2)}.$$

(c) *Comparison.* Which auction format a risk-averse seller would prefer to sell the object if this seller seeks to minimize the variance associated with the payment that he receives?

- The variance of the payment that the seller receives is smaller in the first- than in the second-price auction since

$$\frac{(N-1)^2}{N(N+1)^2(N+2)} < \frac{2(N-1)}{(N+1)^2(N+2)}$$

simplifies to $\frac{N-1}{N} < 2$ that holds. Intuitively, bid shading in the first-price auction "flattens" the distribution of bids, making them less fluctuating than those in the second-price auction. Formally, we can say that the distribution of bids in the first-price auction second-order stochastically dominates that in the second-price auction, since they both have the same mean (expected revenue) but that of the first-price auction lies below that of the second-price auction.

Exercise #6.5: Revenue Equivalence Theorem–Short Proof[B]

6.5 In this exercise we first examine a short proof of the Revenue Equivalence Theorem which does not rely on the use of mechanism design, and thus is often regarded as a more direct proof of the theorem. Afterwards, we analyze a direct implication of our results, which helps us write the expected payment of every bidder i in a compact expression. In this exercise, consider a setting in which bidders' valuations are independent identically distributed (i.i.d.), and bidders are risk-neutral. In addition, assume that every bidder i uses a symmetric and strictly increasing bidding function $b(v_i)$, where v_i denotes bidder i's valuation and $v_i \in [0, 1]$, and that the seller's expected payment from the bidder with the lowest valuation $v_i = 0$ is zero.

(a) Consider two auction formats, A and B. Find bidder i's expected utility from participating in auction A.

- Let us allow bidder i to submit a bid $b(z)$ where $z \neq v_i$, that is, he can submit a bid according to a valuation different from the value he actually assigns to the object. Hence, bidder i wins if his bid is the highest, which occurs with probability

$$\underbrace{F(z) \times \ldots \times F(z)}_{N-1 \text{ times}} = F(z)^{N-1}$$

which, for compactness, we denote as $G(z) \equiv F(z)^{N-1}$. Hence, bidder i's expected payoff in auction A is

$$G(z)v_i - m^A(z)$$

where, upon winning, bidder i obtains the good and enjoys his true valuation v_i; and he must pay a price $m^A(z)$ which depends on how z ranks relative to the other bidders' valuations, but is independent on bidder i's true valuation.

(b) Differentiate bidder i's expected utility with respect to the valuation that he uses to submit his bid.

- Differentiating $G(z)v_i - m^A(z)$ with respect to z yields

$$g(z)v_i - \frac{dm^A(z)}{dz} = 0$$

Intuitively, bidder i increases his reported valuation of the good, z, until the point where his marginal benefit, in terms of higher probability of winning the object, coincides with his marginal cost, i.e., higher payment to the seller $m^A(z)$.

(c) If every bidder i does not have incentives to misreport his valuation, identify the seller's expected payment from bidder i.

- At the equilibrium bidding function, it must be optimal that $z = v_i$. Inserting this result in the above first-order condition, we obtain

$$g(v_i)v_i = \frac{dm^A(v_i)}{dv_i}$$

Exercise #6.5: Revenue Equivalence Theorem–Short Proof[B]

Recall that our objective is to find an expression for the payment to the seller, $m^A(v_i)$. Hence, we rearrange the above equality, $\frac{dm^A(v_i)}{dv_i} = v_i g(v_i)$, and integrate on both sides, which yields

$$m^A(v_i) = m^A(0) + \int_0^{v_i} yg(y)dy$$

and, since the expected payment from the bidder with the lowest valuation is zero by definition, $m^A(0) = 0$, the above equality reduces to

$$m^A(v_i) = \int_0^{v_i} yg(y)dy$$

(d) *Revenue Equivalence.* Argue that the expected payment found in part (c) is unaffected by the particular auction format being considered.

- The right-hand side of the expression found in part (c), $\int_0^{v_i} yg(y)dy$, does not depend on the auction format being used. It only depends on the distribution of bidders' valuations. As a consequence, the seller obtains the same expected payment from bidder i regardless of the auction format (e.g., first-price auction, second-price auction, or all-pay auction) that the seller uses to assign the good.

(e) *Uniform distribution.* Consider that valuations are uniformly distributed, i.e., $v_i \sim U[0, 1]$. Find the payment that bidder i (with privately observed valuation v_i) pays to the seller, $m^A(v_i)$; the expected payment that the seller receives from bidder i; and the expected revenue that the seller receives in this auction.

- *Payment of bidder i.* Since valuations are uniformly distributed, $F(v_i) = v_i$, we have that $G(v_i) \equiv F(v_i)^{N-1} = v_i^{N-1}$, with associated density $g(v_i) = (N-1)v_i^{N-2}$. In this setting, $m^A(0) = 0$ holds, and bidder i's payment becomes

$$m^A(v_i) = \int_0^{v_i} yg(y)dy$$
$$= \int_0^{v_i} y(N-1)y^{N-2}dy$$
$$= (N-1)\int_0^{v_i} y^{N-1}dy$$
$$= (N-1)\left[\frac{y^N}{N}\right]_0^{v_i}$$

$$= (N-1)\left(\frac{v_i^N}{N} - 0\right)$$

$$= \frac{N-1}{N} v_i^N$$

- *Expected payment of bidder i.* The expectation of the payment $m^A(v_i)$ found above is[1]

$$E[m^A(v_i)] = \int_0^1 \frac{N-1}{N} y^N dy$$

$$= \frac{N-1}{N} \int_0^1 y^N dy$$

$$= \frac{N-1}{N} \left[\frac{y^{N+1}}{N+1}\right]_0^{v_i}$$

$$= \frac{N-1}{N} \left(\frac{v_i^{N+1}}{N+1} - 0\right)$$

$$= \frac{N-1}{N} \frac{1}{N+1}$$

- *Expected revenue.* Since there are N bidders competing in the auction, the seller's expected revenue is just N times the expected payment found above, that is,

$$E[R^A] = N \times E[m^A(v_i)]$$

$$= N \times \frac{N-1}{N} \frac{1}{N+1}$$

$$= \frac{N-1}{N+1}$$

which is strictly increasing in the number of bidders, N, because

$$\frac{dE[R^A]}{dN} = \frac{N+1-N+1}{(N+1)^2} = \frac{2}{(N+1)^2} > 0$$

---

[1] Recall that, in the case in which $N=1$, the expectation of a uniformly distributed random variable is $\int_0^1 y \, dy = \frac{1}{2}$; whereas if $N=2$, such expectation becomes $\int_0^1 y^2 dy = \frac{1}{3}$; and similarly for any other N, where $\int_0^1 y^N dy = \frac{1}{N+1}$.

Exercise #6.6: Revenue Equivalence Theorem–Longer Proof^C

6.6 Consider $N \geq 2$ bidders. Every bidder i's valuation for the object, v_i, is drawn according to the cumulative distribution function $F_i(v_i)$ with positive density in all its support; that is, $f_i(v_i) > 0$ for all $v_i \in [0, 1]$ with independent distribution of valuations among the bidders. Every bidder's utility function is linear, $u_i(v_i, m_i) = v_i + m_i$, where $m_i \in \mathbb{R}$ denotes, for generality, the price that the bidder pays (if $m_i < 0$) or the transfer that he receives (if $m_i > 0$). We are now ready to state the Revenue Equivalence Theorem.

Revenue Equivalence Theorem. *If bidding behavior in the BNE of two auction formats yields, for all profiles of bidders' valuations $v = (v_1, \cdots, v_N)$:*
(1) The same assignment rule $(y_1(v), y_2(v), \cdots, y_N(v))$, and
(2) The same expected utility $u_i(0, m_i)$, for the bidder with the lowest valuation, $v_i = 0$.
Then the seller's expected revenue coincides across both auction formats.

Intuitively, this theorem says that, if two auction formats with risk-neutral bidders and independent valuations assign the object to the same bidder (or bidders) and generate the same expected utility for the bidder with the lowest valuation for the object, then they must generate the same expected revenue for the seller. In this exercise, we prove this result in different steps.

(a) We first apply the "Revelation Principle." Consider an auction where every bidder privately observes his valuation for the object, submits a bid b_i, and a winner is chosen given the profile of bids. In this setting, the Revelation Principle says that we can construct a direct revelation mechanism where bidders can choose a valuation, rather than a bid. In this direct revelation mechanism, the equilibrium outcome (winning bidder/s and equilibrium payoffs for each bidder) coincides with the equilibrium outcome in the auction we were considering. Applying the Revelation Principle, express the seller's expected revenue from bidder i as a function of y_i and v_i.

- Applying the Revelation Principle, we can express assignment rule for bidder i as a function of his valuation, $y_i(v_i)$, rather than as a function of the profile of bids. A similar argument applies to his expected payment (negative transfer), $-\bar{t}_i(v_i)$.
- The seller's expected revenue is then given by the sum of expected transfers, i.e., $\sum_{i=1}^{N} E[-\bar{t}_i(v_i)]$. We initially find the expected transfer from bidder i, $E[-\bar{t}_i(v_i)]$, as follows:

$$E[-\bar{t}_i(v_i)] = \int_0^1 -\bar{t}_i(v_i) f_i(x) \, dx$$

Since utility is given by $u_i(v_i) = y_i(v_i)v_i + t_i(v_i)$ in this linear utility environment, we can solve for the expected transfer $\bar{t}_i(v_i)$, which yields $\bar{t}_i(v_i) = u_i(v_i) - \bar{y}_i(v_i)v_i$. Multiplying by -1 on both sides, we obtain $-\bar{t}_i(\theta_i) = \bar{y}_i(\theta_i)\theta_i - u_i(\theta_i)$, which implies that the above expression becomes

$$E[-\bar{t}_i(v_i)] = \int_0^1 \underbrace{[\bar{y}_i(x)x - u_i(x)]}_{-\bar{t}_i(x)} f_i(x)\,dx$$

(b) Apply now Myerson's Characterization Theorem which says that, in a setting where bidders have linear utility functions, every bidder i has incentives to truthfully report his valuation of the object, v_i, when all other bidders are truthfully reporting their valuations, if and only if: (1) every bidder i's expected utility $\bar{u}_i(v_i)$ is increasing in his valuation v_i; and (2) every bidder i's utility, $u_i(v_i)$, can be expressed as $u_i(v_i) = u_i(0) + \int_0^{v_i} \bar{y}(x)\,dx$. In other words, a direct revelation mechanism where every bidder truthfully reports his valuation v_i can be achieved when individuals have linear utility functions satisfying points (1) and (2). Use Myerson's Characterization Theorem to simplify the expected revenue you found in part (a). [*Hint*: After a couple of steps, you may need to apply integration by parts to simplify your results.]

- From Myerson's Characterization Theorem, bidders have incentives to truthfully report their valuations if we can express their utility function as $u_i(v_i) = u_i(0) + \int_0^{v_i} \bar{y}(x)\,dx$. Then, the expected transfer we found in part (a), $E[-\bar{t}_i(v_i)]$, becomes

$$E[-\bar{t}_i(v_i)] = \int_0^1 \left[\bar{y}_i(v_i)v_i - \underbrace{\left(u_i(0) + \int_0^{v_i} y(x)\,dx\right)}_{u_i(v_i)}\right] f_i(x)\,dx$$

Since $u_i(0)$ is a constant, we can take it out of the integral, yielding

$$E[-\bar{t}_i(v_i)] = \int_0^1 \underbrace{\left[\bar{y}_i(v_i)v_i - \int_0^{v_i} y(x)\,dx\right] f_i(x)\,dx}_{\text{Term A}} - u_i(0)$$

Exercise #6.6: Revenue Equivalence Theorem–Longer Proof[C]

Applying integration by parts in term A, we obtain[2]

$$\int_0^1 \left[\int_0^{v_i} \bar{y}_i(x)\,dx\right] f_i(x)\,dx = \int_0^1 \bar{y}_i(x)\,dx - \int_0^1 \bar{y}_i(x) F_i(x)\,dx$$

$$= \int_0^1 \bar{y}_i(x)(1 - F_i(x))\,dx$$

Substituting this result inside $E[-\bar{t}_i(v_i)]$ yields

$$E[-\bar{t}_i(v_i)] = \int_0^1 \left[\bar{y}_i(x)x - \bar{y}_i(x)\frac{1 - F_i(x)}{f_i(x)}\right] f_i(x)\,dx - u_i(0)$$

$$= \int_0^1 \bar{y}_i(x) \left[x - \frac{1 - F_i(x)}{f_i(x)}\right] f_i(x)\,dx - u_i(0)$$

which represents the expected transfer from bidder i. Finally, summing over all N bidders, we find

$$\sum_{i=1}^N E[-\bar{t}_i(v_i)]$$

$$= \underbrace{\int_0^1 \cdots \int_0^1 \sum_{i=1}^N \bar{y}_i(x) \left[x - \frac{1 - F_i(x)}{f_i(x)}\right]}_{\text{Term } B} \underbrace{\prod_{i=1}^N f_i(x)\,dx \cdots dx}_{\text{Term } C} - \underbrace{\sum_{i=1}^N u_i(0)}_{\text{Term } D}$$

where we moved the summation signs inside the integral because valuations $(v_1, \ldots, v_N)$ are independently distributed.

(c) Show that the expected revenue found in part (b) is the same for two auction formats satisfying points (1) and (2) in the Revenue Equivalence Theorem.

- From the expected revenue in part (b), we see that, if the BNE of two different auction formats has: (1) the same probabilities of assigning the object to each bidder, $(\bar{y}_1(v_1), \cdots, \bar{y}_N(v_N))$; and (2) the same expected utility for the bidder with the lowest valuation for the good, $u_i(0)$, for every player i, the expected revenue for the seller coincides.

---

[2] In order to apply integration by parts, a common trick is to first recall the derivative of the product of two functions $h(x)$ and $g(x)$, that is, $(h \cdot g)' = h'g + hg'$, or alternatively $hg' = (h \cdot g)' - h'g$. Integrating both sides yields $\int hg'\,dx = hg - \int h'g\,dx$. In our current example, let $h(x) = \int_0^{v_i} \bar{y}_i(x)\,dx$, $g'(x) = f_i(x)\,dx$, $h'(x) = \bar{y}_i(x)$ and $g(x) = F_i(x)$. Plugging these functions and rearranging yield the above result.

- In particular, term B is constant across two auctions formats by property (1), and so is term D by property (2).
- Finally, term C is unaffected by the rules of the auction since it is just given by the distribution of bidders' valuations for the good, including $F_i(x)$ and $f_i(x)$. This allows for valuations to be drawn from asymmetric probability distributions, $F_1(x)$ and $F_2(x)$.

(d) Apply your results to a first-price and a second-price auction.

- The first-price and second-price satisfy the conditions in this Revenue Equivalence theorem since: (1) the allocation rule in both auctions coincides, i.e., the bidder submitting the highest bid receives the object; and (2) the expected utility of the bidder with the lowest valuation, $u_i(0)$, coincides in both auctions (his utility is zero in both auction formats since this bidder does not win the object). Therefore, the first-price and second-price auctions generate the same expected revenue for the seller.

Common-Value Auctions 7

Keywords

Common-value auctions · Winner's curse · Observed signal · Overestimation · Interdependent values

Introduction

This chapter examines auctions where all bidders share the same (common) value for the object, but none of them observes an accurate signal of the object's true valuation. Instead, every bidder privately observes a noisy signal about the object's value and, based on this signal, submits a bid. This model is typically used to analyze the auctioning of oil leases, as firms exploiting the oil reservoir would earn a similar profit if they win the auction, but they have imprecise and potentially different signals of the amount of oil barrels in the reservoir (as they receive different engineering reports) and, as a consequence, hold different estimates of the oil lease's profitability.

Exercise 7.1 starts by verbally describing the potential for a "winner's curse" result occurring in this type of auction, meaning that one of the bidders, because he received the highest signal, submits the highest bid and wins the auction. However, the bidder then realizes that his signal overestimated the value of the object so much that he ends up paying for more than its true value, so winning becomes a "curse."

Exercise 7.2 then formally analyzes the equilibrium bidding function in this auction assuming, for simplicity, only two bidders. In this setting, equilibrium bids predict a significant bid shading to avoid falling prey of the winner's curse. Exercise 7.3 generalizes our analysis of equilibrium bids to auctions with more than two bidders and examines how the bids are affected by more bidders competing for the object.

Exercise 7.4 considers an auction with two bidders where, again, every bidder receives a signal of the object's true value, and this value is a weighted average of his bid and that of his rival. We first analyze the equilibrium bids when players face a first-price auction (in Exercise 7.4) and when they face a second-price auction (in Exercise 7.5). Finally, Exercise 7.6 evaluates the seller's expected revenue in these two auction formats, showing that, as in the auctions where bidders privately observe their valuation analyzed in the previous chapters, the expected revenues coincide.

Exercise #7.1: The Winner's Curse in Common-Value Auctions–Introduction[A]

7.1 Consider a common-value auction with two bidders. The true value of the object being auctioned is v and is the same for all bidders (common value). Each bidder gets a noisy (inexact) signal of v drawn from the uniform distribution $U[0, 1]$. That is, the cumulative distribution function of bidder i's signal, denoted s_i, is $F(s_i) = s_i$, where $s_i \in [0, 1]$. It is common knowledge that each bidder's signal is independently drawn from $[0, 1]$ according to F.

Finally, we assume that the true value is equal to the average of all bidders' signals, so that

$$v = \frac{1}{2}\left(s_i + s_j\right) \tag{7.1}$$

Bidders participate in a first-price, sealed-bid auction, which means that if bidder i wins, then his realized payoff is $v - b_i$, where b_i is his bid.

(a) Verbally describe why bidding according to your observed signal, $b(s_i) = s_i$, yields an expected negative payoff even if you win the auction.

- If every bidder i submits a bid $b(s_i) = s_i$, this bid is increasing in his own signal s_i, $b'(s_i) > 0$, meaning that the winner is the bidder who received the highest signal. Signals can, however, overestimate the true value of the object, which occurs if $s_i > v$ or

$$s_i > \frac{1}{2}\left(s_i + s_j\right),$$

and it simplifies to $s_i > s_j$. In this context, a winning bidder may be submitting a bid that is actually higher than the true value of the object. In other words, despite winning, the net payoff that the winner earns is negative since he pays for the object more than v. This is the so-called "winner's curse," after Klemperer (1998).

- As we show in Exercise 7.2 with $N = 2$ bidders and in Exercise 7.3, more generally, with $N \geq 2$ bidders, participants in common-value auctions can anticipate the winner's curse and significantly shade their bid to account for the possibility that the estimates they receive are above the true value of the object.

Exercise #7.2: Equilibrium Bidding in Common-Value Auctions with Two Bidders[B]

7.2 Consider the common-value auction in Exercise 7.1. We now seek to find the equilibrium bidding function in this setting. The following steps should help you find this equilibrium bid. In deriving a Bayesian Nash equilibrium (BNE), let us conjecture that the optimal bidding function is linear in the bidder's signal, that is, every bidder i submits a bid

$$b_i = \alpha s_i \text{ where } \alpha \in [0, 1]$$

Bidder i's expected payoff is the probability that he wins (i.e., his bid is higher than all other bids) times his expected payoff, conditional on having submitted the highest bid:

$$\text{Prob}(b_i > b_j) \times \left\{ E\left[v | s_i, b_i > b_j\right] - b_i \right\} \quad (7.2)$$

where $E\left[v | s_i, b_i > b_j\right]$ is bidder i's expected valuation, conditional not only on his signal s_i but also on knowing that he submitted the highest bid, i.e., $b_i > b_j$.

(a) Show that bidder i's expected utility from participating in the auction can be expressed as follows:

$$\frac{b_i}{\alpha}\left[\frac{s_i}{2} + \frac{1}{2}\left(\frac{b_i}{2\alpha}\right) - b_i\right].$$

- Let us use the property that the other bidders use bidding function $b_j = \alpha s_j$. Substituting αs_j for b_j into expression (7.2), we obtain

$$\text{Prob}(b_i > \alpha s_j) \times \left\{ E\left[v | s_i, b_i > \alpha s_j\right] - b_i \right\}$$

$$= \text{Prob}\left(\frac{b_i}{\alpha} > s_j\right) \times \left\{ E\left[v \Big| s_i, \frac{b_i}{\alpha} > s_j\right] - b_i \right\}$$

- Next, inserting $v - \frac{1}{2}(s_i + s_j)$ from expression (7.1):

$$\text{Prob}\left(\frac{b_i}{\alpha} > s_j\right) \times \left\{ E\left[\frac{1}{2}(s_i + s_j) \Big| \frac{b_i}{\alpha} > s_j\right] - b_i \right\}$$

Since bidder i knows his signal, s_i, $E[s_i] = s_i$. However, he does not know s_j, so we obtain

$$\text{Prob}\left(\frac{b_i}{\alpha} > s_j\right) \times \left\{ \frac{s_i}{2} + \frac{1}{2} E\left[s_j \Big| \frac{b_i}{\alpha} > s_j\right] - b_i \right\}.$$

- Using the uniform distribution on s_j, we have

$$\text{Prob}\left(\frac{b_i}{\alpha} > s_j\right) = \frac{b_i}{\alpha}, \text{ and}$$

$$E\left[s_j \middle| \frac{b_i}{\alpha} > s_j\right] = \frac{b_i}{2\alpha}$$

because s_j, which is a uniformly distributed random variable, falls into the range $\left[0, \frac{b_i}{\alpha}\right]$, yielding an expected value of $\frac{b_i}{2\alpha}$. Inserting these results into bidder i's expected payoff, we obtain that

$$\frac{b_i}{\alpha}\left[\frac{s_i}{2} + \frac{1}{2}\left(\frac{b_i}{2\alpha}\right) - b_i\right]$$

as required.

(b) Take first-order conditions with respect to b_i, and rearrange, to find the equilibrium bidding function in this common-value auction.

- Every bidder i chooses his bid b_i to solve

$$\max_{b_i \geq 0} \frac{b_i}{\alpha}\left[\frac{s_i}{2} + \frac{1}{2}\left(\frac{b_i}{2\alpha}\right) - b_i\right]$$

Taking first-order conditions with respect to b_i, we obtain

$$\frac{1}{\alpha}\left(\frac{s_i}{2} + \frac{1-4\alpha}{4\alpha}b_i\right) + \frac{b_i}{\alpha}\left(\frac{1-4\alpha}{4\alpha}\right) = 0$$

Simplifying, we obtain

$$\frac{1-4\alpha}{2\alpha^2}b_i = -\frac{s_i}{2\alpha}$$

Rearranging, and solving for b_i, we find

$$b_i = \frac{\alpha}{4\alpha - 1}s_i. \qquad (7.3)$$

- Recall that we conjecture that the symmetric equilibrium bidding rule is $b_i = \alpha s_i$ for some value of α. Equation (7.3) is indeed linear in s_i, and furthermore, we can now solve for α as follows:

$$\alpha = \frac{\alpha}{4\alpha - 1}$$

Cancelling out s_i on both sides, we have

$$4\alpha - 1 = 1$$

and solving for α, we obtain

$$\alpha = \frac{1}{2}$$

In conclusion, a symmetric BNE has a bidder using the rule

$$b_i(s_i) = \frac{1}{2} s_i$$

which entails that every bidder i submits a bid equal to half of his privately observed signal s_i.

(c) Find the expected payoff from participating in this auction, in equilibrium.

- Since every bidder submits an equilibrium bid of $b_i(s_i) = \frac{1}{2} s_i$, and $\alpha = \frac{1}{2}$, we can insert these results into bidder i's expected payoff, from part (b), to find

$$\frac{\frac{1}{2} s_i}{\frac{1}{2}} \left[\frac{s_i}{2} + \frac{1}{2} \left(\frac{\frac{1}{2} s_i}{2 \frac{1}{2}} \right) - \frac{1}{2} s_i \right]$$

$$= s_i \left[\frac{s_i}{2} + \frac{1}{4} s_i - \frac{1}{2} s_i \right]$$

$$= \frac{1}{4} s_i^2$$

which is strictly positive and increasing in the signal that bidder i receives.

Exercise #7.3: Equilibrium Bidding in Common-Value Auctions with $N \geq 2$ Bidders[C]

7.3 Consider again the common-value auction in Exercise 7.1, but now allow for $N \geq 2$ bidders. In this context, the true value of the object is equal to the average of all bidders' signals:

$$v = \frac{s_1 + s_2 + \ldots + s_N}{N} \tag{7.4}$$

$$= \left(\frac{1}{N}\right)\sum_{j=1}^{N} s_j$$

In deriving a BNE, let us conjecture that the optimal bidding function is linear in a bidder's signal. That is, there is some value for $\alpha \in [0, 1]$ such that

$$b_j = \alpha s_j \qquad (7.5)$$

Bidder i's expected payoff is the probability that he wins (i.e., his bid is higher than all other bids) times his expected payoff, conditional on having submitted the highest bid:

$$\text{Prob}(b_i > b_j \ \forall j \neq i) \times \left\{ E\left[v | s_i, b_i > b_j \ \forall j \neq i\right] - b_i \right\} \qquad (7.6)$$

where $E\left[v | s_i, b_i > b_j \ \forall j \neq i\right]$ is bidder i's expected valuation, conditional not only on his signal s_i but also on knowing that he submitted the highest bid, that is, $b_i > b_j \ \forall j \neq i$.

(a) *Writing expected utility*. Show that bidder i's expected utility from participating in the auction can be expressed as follows:

$$\left(\frac{b_i}{\alpha}\right)^{N-1} \left[\left(\frac{s_i}{N}\right) + \left(\frac{N-1}{N}\right)\left(\frac{b_i}{2\alpha}\right) - b_i \right].$$

- Let us use the property that the other bidders use the bidding rule in expression (7.5). Substitute αs_j for b_j into expression (7.6) to obtain

$$\text{Prob}(b_i > \alpha s_j \ \forall j \neq i) \times \left\{ E\left[v | s_i, b_i > \alpha s_j \ \forall j \neq i\right] - b_i \right\}$$

$$= \text{Prob}\left(\frac{b_i}{\alpha} > s_j \ \forall j \neq i\right) \times \left\{ E\left[v \middle| s_i, \frac{b_i}{\alpha} > s_j \ \forall j \neq i\right] - b_i \right\}$$

- Next, substituting the expression for v from expression (7.4):

$$\text{Prob}\left(\frac{b_i}{\alpha} > s_j \ \forall j \neq i\right)$$

$$\times \left\{ E\left[\left(\frac{1}{N}\right)\left(s_i + \sum_{j \neq i} s_j\right) \middle| \frac{b_i}{\alpha} > s_j \ \forall j \neq i\right] - b_i \right\}$$

$$= \text{Prob}\left(\frac{b_i}{\alpha} > s_j \ \forall j \neq i\right)$$

Exercise #7.3: Equilibrium Bidding in Common-Value Auctions with $N \geq 2$...

$$\times \left\{ \left(\frac{s_i}{N}\right) + \left(\frac{1}{N}\right) E\left[\sum_{j \neq i} s_j \left| \frac{b_i}{\alpha} > s_j \; \forall j \neq i \right.\right] - b_i \right\}$$

$$= \text{Prob}\left(\frac{b_i}{\alpha} > s_j \; \forall j \neq i\right)$$

$$\times \left\{ \left(\frac{s_i}{N}\right) + \left(\frac{1}{N}\right) \sum_{j \neq i} E\left[s_j \left| \frac{b_i}{\alpha} > s_j \right.\right] - b_i \right\}$$

The second line follows from the fact that bidder i knows s_i, so that $E[s_i] = s_i$, but he does not know s_j, and the third line is due to signals being independent random variables.

- Using the uniform distribution on s_j, bidder i's expected payoff becomes

$$\left(\frac{b_i}{\alpha}\right)^{N-1} \left[\left(\frac{s_i}{N}\right) + \left(\frac{N-1}{N}\right)\left(\frac{b_i}{2\alpha}\right) - b_i\right]$$

where[1]

$$E\left[s_j \left| \frac{b_i}{\alpha} > s_j \right.\right] = \frac{b_i}{2\alpha}.$$

because s_j, which is a uniformly distributed random variable, now falls into the range $\left[0, \frac{b_i}{\alpha}\right]$, yielding an expected value of $\frac{b_i}{2\alpha}$.

(b) *Equilibrium bid.* Take first-order conditions with respect to b_i, and rearrange, to find the equilibrium bidding function in this common-value auction.

---

[1] This does not presume that $\frac{b_i}{\alpha} \leq 1$. Since we seek to show that $b_i = \alpha s_i$ for some value of α, it follows that $\frac{b_i}{\alpha} \leq 1$ is equivalent to $\frac{\alpha s_i}{\alpha} \leq 1$, or $s_i \leq 1$, which is true by assumption since $s_i \in [0, 1]$.

- Every bidder i chooses his bid b_i to solve

$$\max_{b_i \geq 0} \left(\frac{b_i}{\alpha}\right)^{N-1} \left[\left(\frac{s_i}{N}\right) + \left(\frac{N-1}{N}\right)\left(\frac{b_i}{2\alpha}\right) - b_i\right]$$

Taking first-order conditions with respect to b_i, we obtain

$$(N-1)\frac{b_i^{N-2}}{\alpha^{N-1}}\left(\frac{s_i}{N} + \frac{N-1-2\alpha N}{2\alpha N}b_i\right)$$

$$+ \left(\frac{b_i}{\alpha}\right)^{N-1} \frac{N-1-2\alpha N}{2\alpha N} = 0$$

Rearranging, we find

$$(N-1)\frac{b_i^{N-2}}{\alpha^{N-1}}\frac{s_i}{N} + \frac{(N-1)(N-1-2\alpha N)}{\alpha^{N-1}(2\alpha N)}b_i^{N-1}$$

$$+ \left(\frac{b_i}{\alpha}\right)^{N-1} \frac{N-1-2\alpha N}{2\alpha N} = 0$$

which yields

$$\frac{N-1}{N}\frac{s_i}{b_i} + (N-1)\frac{N-1-2\alpha N}{2\alpha N} + \frac{N-1-2\alpha N}{2\alpha N} = 0$$

that further simplifies to

$$\frac{1-N+2\alpha N}{2\alpha}b_i = \frac{N-1}{N}s_i$$

Solving for bid b_i yields

$$b_i = \frac{2\alpha(N-1)}{[1+(2\alpha-1)N]N}s_i \quad (7.7)$$

- Recall that we conjecture that the symmetric equilibrium bidding rule is $b_i = \alpha s_i$ for some value of α. Equation (7.7) is indeed linear, and furthermore, we can solve for α, as follows:

$$\alpha = \frac{2\alpha(N-1)}{[1+(2\alpha-1)N]N}$$

Rearranging, we obtain

Exercise #7.3: Equilibrium Bidding in Common-Value Auctions with $N \geq 2$...

$$(2\alpha - 1) N = \frac{N-2}{N}$$

Solving for α, we find

$$\alpha = \frac{(N+2)(N-1)}{2N^2}.$$

In conclusion, a symmetric BNE has every bidder i using the following rule:

$$b_i(s_i) = \frac{(N+2)(N-1)}{2N^2} s_i$$

where $\frac{(N+2)(N-1)}{2N^2} < 1$ reduces to $N^2 - N + 2 > 0$ that holds for all $N \geq 2$.

(c) *Comparative statics.* How is the equilibrium bidding function $b_i(s_i)$ found in part (b) affected by an increase in the number of bidders? Interpret.

- Differentiating the equilibrium bidding function $b_i(s_i)$ with respect to N yields

$$\frac{\partial b_i(s_i)}{\partial N} = \frac{(4-N) s_i}{2N^3}$$

which is positive for all $N < 4$, but negative otherwise. In other words, only when there are $N = 2$ or $N = 3$ bidders competing for the object and a new bidder joining the auction will their equilibrium bids increase. However, when there are more than 4 bidders joining the auction, equilibrium bids decrease.

(d) *Numerical example.* Evaluate the equilibrium bids when $N = 2$, $N = 3$, and $N = 10$ bidders participate in the auction.

- If only two bidders participate in the auction, $N = 2$, bidder i's equilibrium bid becomes

$$b_i(s_i) = \frac{(2+2)(2-1)}{2 \times 2^2} s_i = \frac{1}{2} s_i.$$

- Whereas, if $N = 3$ bidders participate, equilibrium bids increase to

$$b_i(s_i) = \frac{(3+2)(3-1)}{2 \times 3^2} s_i = 0.56 s_i.$$

- Moreover, if $N = 10$ bidders participate in the auction, equilibrium bids decrease to

$$b_i(s_i) = \frac{(10+2)(10-1)}{2 \times 10^2} s_i = 0.54 s_i.$$

- Finally, we are also interested in the asymptotic property of the equilibrium bidding function. When the number of bidders grows to infinity, we obtain

$$\lim_{N \to \infty} b_i(s_i) = \lim_{N \to \infty} \frac{N^2 + N - 2}{2N^2}$$
$$= \lim_{N \to \infty} \frac{2N+1}{4N}$$
$$= \frac{1}{2}$$

where we apply the L'Hôpital's rule to the second and third line. Intuitively, when the number of bidders becomes infinitely large, every bidder i will bid the expected value of the object, which coincides with the expected realization of the signal, i.e., $1/2$.

- For illustration purposes, Fig. 7.1 depicts the equilibrium bidding function evaluated at $N = 2$, $N = 10$, and $N = 100$. As more bidders participate in the auction, the equilibrium bidding function shifts upward, approaching the 45°-line and reducing bid shading.

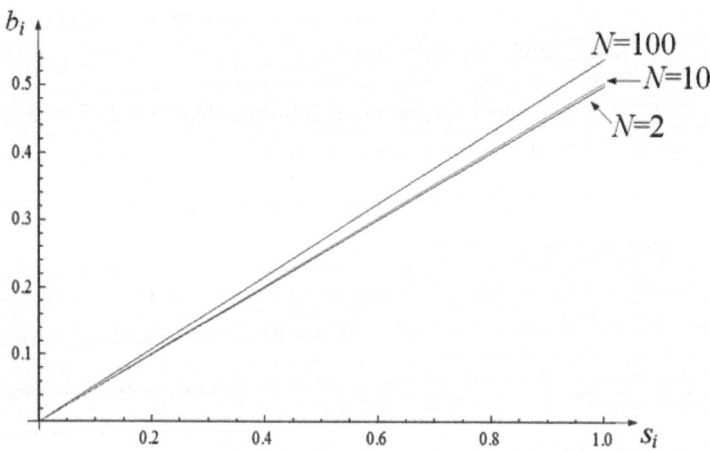

Fig. 7.1 Equilibrium bidding function in a common-value auction—more bidders

Exercise #7.4: First-Price Auction When Bidders Have Interdependent Values[B]

7.4 Consider the following first-price sealed-bid auction with two bidders competing for the object. Players receive private and independent signals, s_1 and s_2, drawn from the uniform distribution $U[0, 1]$. Player i's valuation for the object is defined as

$$v_i = \alpha_i s_i + \alpha_j s_j,$$

where $i \neq j$ and α_i satisfies $\alpha_i \geq 0$ for all $i = \{1, 2\}$. Intuitively, every player's valuation is a function of the signal he receives, s_i, and the signal that his rival receives, s_j. When $\alpha_i = \alpha_j = 1$, bidder i's and j's valuations for the object coincide and both are equal to the sum of their privately observed signals; that is, $v_i = v_j = s_i + s_j$. In that setting, players face a common-value auction as they assign the same value to the object.

In this exercise, we seek to show that bidding function, $b(s_i) = \frac{\alpha_i + \alpha_j}{2} s_i$, for every bidder $i = \{1, 2\}$ is a Bayesian Nash equilibrium of this game. Answer the following questions to confirm that this bidding function is an equilibrium of this auction.

(a) Show that bidder i's probability of winning, given bidding function, $b(s_i) = \frac{\alpha_i + \alpha_j}{2} s_i$, is $\Pr(win) = \frac{2b_i}{\alpha_i + \alpha_j}$.

- Bidder i wins the auction if his bid is larger than that of bidder j, that is, $b_i > \frac{\alpha_i + \alpha_j}{2} s_j$. We can write this probability as follows:

$$\Pr(win) = \Pr\left[b_i > \frac{\alpha_i + \alpha_j}{2} s_j\right]$$

$$= \Pr\left[\frac{2b_i}{\alpha_i + \alpha_j} > s_j\right]$$

where the last inequality solves for signal s_j in $b_i > \frac{\alpha_i + \alpha_j}{2} s_j$, yielding $\frac{2b_i}{\alpha_i + \alpha_j} > s_j$. Since signal s_j is drawn from a uniform distribution, this probability becomes

$$\Pr(win) = \frac{2b_i}{\alpha_i + \alpha_j}.$$

For instance, when $\alpha_i = \alpha_j = 1$, this probability simplifies to b_i.

(b) Show that the expected value of the object for bidder i, if he wins the auction, can be expressed as $v_i = \alpha_i s_i + \frac{\alpha_j b_i}{\alpha_i + \alpha_j}$.

- If bidder i wins the auction, which occurs when $b_i > b_j$, his expected value v_i is

$$E\left[v_i \mid b_i > b_j\right] = E\left[\alpha_i s_i + \alpha_j s_j \mid b_i > b_j\right]$$
$$= \alpha_i s_i + \alpha_j E\left[s_j \mid b_i > b_j\right]$$
$$= \alpha_i s_i + \alpha_j E\left[s_j \mid b_i > \frac{\alpha_i + \alpha_j}{2} s_j\right]$$
$$= \alpha_i s_i + \alpha_j E\left[s_j \mid \frac{2b_i}{\alpha_i + \alpha_j} > s_j\right]$$
$$= \alpha_i s_i + \frac{\alpha_j}{2}\left[\frac{2b_i}{\alpha_i + \alpha_j} - 0\right]$$
$$= \alpha_i s_i + \frac{\alpha_j b_i}{\alpha_i + \alpha_j}.$$

For instance, when $\alpha_i = \alpha_j = 1$, the expected value from the object (conditional on winning) collapses to $s_i + \frac{b_i}{2}$.

(c) Write bidder i's expected utility from participating in the auction. Differentiate with respect to b_i, and show that his equilibrium bidding function is, indeed, $b(s_i) = \frac{\alpha_i + \alpha_j}{2} s_i$.

- The expected utility of bidder i when observing signal s_i is

$$EU_i(b_i \mid s_i) = \Pr(win) \times (\text{Expected value if winning} - b_i).$$

Using the results from parts (a) and (b), this expected utility becomes

$$EU_i(b_i \mid s_i) = \underbrace{\frac{2b_i}{\alpha_i + \alpha_j}}_{\Pr(win)} \times \left(\underbrace{\alpha_i s_i + \frac{\alpha_j b_i}{\alpha_i + \alpha_j}}_{\text{Expected value if winning}} - b_i\right).$$

Differentiating with respect to the bid b_i yields

$$\left(\frac{2}{\alpha_i + \alpha_j}\right)\left(\alpha_i s_i - \frac{\alpha_i b_i}{\alpha_i + \alpha_j}\right) - \left(\frac{2b_i}{\alpha_i + \alpha_j}\right)\left(\frac{\alpha_i}{\alpha_i + \alpha_j}\right) = 0$$

which simplifies to

$$(\alpha_i + \alpha_j)\alpha_i s_i - 2\alpha_i b_i = 0$$

Exercise #7.4: First-Price Auction When Bidders Have Interdependent Values[B]

Solving for b_i, we find that bidder i maximizes his expected utility by submitting the bid

$$b(s_i) = \frac{\alpha_i + \alpha_j}{2} s_i,$$

as required. In other words, if bidder j submits a bid $b(s_j) = \frac{\alpha_i + \alpha_j}{2} s_j$, bidder i does not have incentives to deviate from the bidding function $b(s_i) = \frac{\alpha_i + \alpha_j}{2} s_i$.

(d) *Numerical example.* Evaluate the equilibrium bidding function $b(s_i) = \frac{\alpha_i + \alpha_j}{2} s_i$ at $\alpha_i = \alpha_j = 1$. Then, evaluate it at $\alpha_i = 2$ and $\alpha_j = 1$. Interpret.

- When $\alpha_i = \alpha_j = 1$, this equilibrium bidding function simplifies to

$$b(s_i) = \frac{1+1}{2} s_i = s_i,$$

indicating that, in the Bayesian Nash equilibrium of a first-price auction with interdependent values, every bidder i submits a bid equal to his privately observed signal s_i.

- When $\alpha_i = 2$ and $\alpha_j = 1$, the equilibrium bidding function becomes

$$b(s_i) = \frac{2+1}{2} s_i = \frac{3}{2} s_i,$$

suggesting that every bidder submits more aggressive bids, as signals have a greater impact on the object's value.

(e) How is the equilibrium bidding function affected by a marginal increase in signal s_i?

- If bidder i's signal marginally increases, the equilibrium bidding function $b(s_i) = \frac{\alpha_i + \alpha_j}{2} s_i$ increases by

$$\frac{\partial b(s_i)}{\partial s_i} = \frac{\alpha_i + \alpha_j}{2}.$$

which is unambiguously positive.
- In addition, a marginal increase in s_i increases bidder i's bid by the same amount as a marginal increase in s_j would increase bidder j's bid. Formally, we can say that

$$\frac{\partial b(s_i)}{\partial s_i} = \frac{\partial b(s_j)}{\partial s_j} = \frac{\alpha_i + \alpha_j}{2},$$

so that bids are symmetrically affected by a common increase in signals. This result holds both when weights satisfy $\alpha_i = \alpha_j$ and when $\alpha_i \neq \alpha_j$.

Exercise #7.5: Second-Price Auction When Bidders have Interdependent Values[B]

7.5 Consider the setting in Exercise 7.4, but now assume that bidders participate in a second-price auction. In this exercise, we seek to show that the bidding function, $b(s_i) = (\alpha_i + \alpha_j) s_i$, for every bidder $i = \{1, 2\}$ is a Bayesian Nash equilibrium of this game. Answer the following questions to confirm that this bidding function is indeed an equilibrium of this auction:

(a) Show that bidder i's probability of winning, given the bidding function, $b(s_i) = (\alpha_i + \alpha_j) s_i$, is $\Pr(win) = \frac{b_i}{\alpha_i + \alpha_j}$.

- Bidder i wins the auction if his bid is larger than that of bidder j, that is, $b_i > (\alpha_i + \alpha_j)s_j$. We can write this probability as follows:

$$\Pr(win) = \Pr\left[b_i > (\alpha_i + \alpha_j)s_j\right]$$

$$= \Pr\left[\frac{b_i}{\alpha_i + \alpha_j} > s_j\right]$$

where the last inequality solves for signal s_j in $b_i > (\alpha_i + \alpha_j)s_j$, yielding $\frac{b_i}{\alpha_i + \alpha_j} > s_j$. Since signal s_j is drawn from a uniform distribution, this probability becomes

$$\Pr(win) = \frac{b_i}{\alpha_i + \alpha_j}.$$

For instance, when $\alpha_i = \alpha_j = 1$, this probability simplifies to $\frac{1}{2} b_i$.

(b) Show that the expected value of the object for bidder i, if he wins the auction, can be expressed as $v_i = \alpha_i s_i + \frac{\alpha_j b_i}{2(\alpha_i + \alpha_j)}$.

- If bidder i wins the auction, which occurs when $b_i > b_j$, his expected value v_i is

$$E\left[v_i \,|\, b_i > b_j\right] = E\left[\alpha_i s_i + \alpha_j s_j \,|\, b_i > b_j\right]$$
$$= \alpha_i s_i + \alpha_j E\left[s_j \,|\, b_i > b_j\right]$$
$$= \alpha_i s_i + \alpha_j E\left[s_j \,|\, b_i > (\alpha_i + \alpha_j)s_j\right]$$

Exercise #7.5: Second-Price Auction When Bidders have Interdependent Values

$$= \alpha_i s_i + \alpha_j E\left[s_j \,\bigg|\, \frac{b_i}{\alpha_i + \alpha_j} > s_j\right]$$

$$= \alpha_i s_i + \frac{\alpha_j}{2}\left[\frac{b_i}{\alpha_i + \alpha_j} - 0\right]$$

$$= \alpha_i s_i + \frac{\alpha_j b_i}{2(\alpha_i + \alpha_j)}.$$

For instance, when $\alpha_i = \alpha_j = 1$, this expected value from the object (conditional on winning) collapses to $s_i + \frac{b_i}{4}$.

(c) Show that the expected price that bidder i pays for the object, if winning, is $\frac{b_i}{2}$.

- When bidder i wins, he pays the bid submitted by bidder j, which we fixed in equilibrium all throughout our analysis at $b(s_j) = (\alpha_i + \alpha_j)s_j$. Recall that this is the standard approach to showing that a strategy profile can be sustained as an equilibrium in any game. First, we fix the strategies of player i's rivals at the strategy profile we are testing and then confirm that player i cannot increase his payoff by choosing a different strategy.
- Bidder i does not observe the realization of signal s_j, so he must compute the expected value of this random variable, conditional on bidder i winning, which occurs when his bid satisfies $b_i > (\alpha_i + \alpha_j)s_j$, or in terms of s_j when $\frac{b_i}{\alpha_i + \alpha_j} > s_j$. Therefore, the expected price that bidder i pays when winning the object is

$$E\left[b(s_j) \,\big|\, b_i > b_j\right] = E\left[(\alpha_i + \alpha_j)s_j \,\bigg|\, \frac{b_i}{\alpha_i + \alpha_j} > s_j\right]$$

$$= (\alpha_i + \alpha_j) E\left[s_j \,\bigg|\, \frac{b_i}{\alpha_i + \alpha_j} > s_j\right]$$

$$= \frac{\alpha_i + \alpha_j}{2}\left[\frac{b_i}{\alpha_i + \alpha_j} - 0\right]$$

$$= \frac{b_i}{2}.$$

(d) Write bidder i's expected utility from participating in the auction. Differentiate with respect to b_i, and show that his equilibrium bidding function is, indeed, $b(s_i) = (\alpha_i + \alpha_j)s_i$.

- The expected utility of bidder i when observing signal s_i and paying price $b(s_j)$ is

$$EU_i(b_i|s_i) = \Pr(win) \times \left[\text{Expected value if winning} - b(s_j)\right]$$

because, upon winning, bidder i pays the second-highest bid, $b(s_j)$. Using the results from parts (a), (b), and (c), this expected utility becomes

$$EU_i(b_i|s_i) = \underbrace{\frac{b_i}{\alpha_i + \alpha_j}}_{\Pr(win)} \times \underbrace{\left(\alpha_i s_i + \frac{\alpha_j b_i}{2(\alpha_i + \alpha_j)}\right)}_{\text{Expected value if winning}} - \underbrace{\frac{b_i}{2}}_{\text{Price}}$$

$$= \frac{\alpha_i}{\alpha_i + \alpha_j} s_i b_i + \frac{\alpha_j}{2(\alpha_i + \alpha_j)^2} b_i^2 - \frac{b_i^2}{2(\alpha_i + \alpha_j)}.$$

Differentiating with respect to bid b_i yields

$$\frac{\alpha_i}{\alpha_i + \alpha_j} s_i - \frac{2\alpha_j}{2(\alpha_i + \alpha_j)^2} b_i - \frac{2 b_i}{2(\alpha_i + \alpha_j)} = 0$$

which simplifies to

$$\alpha_i s_i + \frac{\alpha_j}{\alpha_i + \alpha_j} b_i - b_i = 0$$

Solving for b_i, we find that bidder i maximizes his expected utility by submitting the following bid:

$$b(s_i) = (\alpha_i + \alpha_j) s_i$$

as required. In other words, if bidder j submits a bid $b(s_j) = (\alpha_i + \alpha_j) s_j$, bidder i does not have incentives to deviate from the bidding function $b(s_i) = (\alpha_i + \alpha_j) s_i$.

(e) *Numerical example.* Evaluate the equilibrium bidding function $b(s_i) = (\alpha_i + \alpha_j) s_i$ at $\alpha_i = \alpha_j = 1$. Then, evaluate it at $\alpha_i = 2$ and $\alpha_j = 1$. Interpret.

- When $\alpha_i = \alpha_j = 1$, this equilibrium bidding function simplifies to

$$b(s_i) = (1+1) s_i = 2 s_i,$$

indicating that, in the Bayesian Nash equilibrium of a second-price auction with interdependent values, every bidder i submits a bid equal to the double of his privately observed signal s_i.

- When $\alpha_i = 2$ and $\alpha_j = 1$, the equilibrium bidding function becomes

  $$(s_i) = (2+1)s_i = 3s_i,$$

  suggesting that every bidder submits more aggressive bids (three times his privately observed signal). This was expected since signals have a greater effect on the valuation of the object.

(f) *How is the equilibrium bidding function affected by a marginal increase in signal s_i?*

- If bidder i's signal marginally increases, the equilibrium bidding function $b(s_i) = (\alpha_i + \alpha_j)s_i$ increases by

  $$\frac{\partial b(s_i)}{\partial s_i} = \alpha_i + \alpha_j.$$

  Therefore, a marginal increase in s_i increases bidder i's bid by the same amount as a marginal increase in s_j would increase bidder j's bid. Formally, we can say that

  $$\frac{\partial b(s_i)}{\partial s_i} = \frac{\partial b(s_j)}{\partial s_j} = \alpha_i + \alpha_j,$$

  so that bids are symmetrically affected by a common increase in signals. This result holds both when weights satisfy $\alpha_i = \alpha_j$ and when $\alpha_i \neq \alpha_j$.

Exercise #7.6: Revenue Comparison in Auctions When Bidders Have Interdependent Values[B]

7.6 Consider the setting in Exercises 7.4 and 7.5. Answer the following questions:

(a) *First-price auction.* Find the seller's expected revenue in the first-price auction where bidders have interdependent values, as analyzed in Exercise 7.4.

- In the first-price auction of Exercise 7.4, we found that every bidder i submits an equilibrium bid $b(s_i) = \frac{\alpha_i + \alpha_j}{2} s_i$. We can then find the following two results:
  - *Probability of winning.* Bidder i wins if his bid satisfies $b(s_i) > b(s_j)$ or, more explicitly, $\frac{\alpha_i + \alpha_j}{2} s_i > \frac{\alpha_i + \alpha_j}{2} s_j$, which simplifies to $s_i > s_j$. Since signal s_j is uniformly distributed, the probability that bidder i wins is $\Pr(win) = F(s_i > s_j) = s_i$.
  - *Price if winning.* If winning, bidder i must, of course, pay his bid, $b(s_i) = \frac{\alpha_i + \alpha_j}{2} s_i$.

- Hence, his expected payment in the first-price auction is

$$m^1(s_i) = \underbrace{s_i}_{\Pr(win)} \times \underbrace{\frac{\alpha_i + \alpha_j}{2} s_i}_{b(s_i)} = \frac{\alpha_i + \alpha_j}{2} s_i^2.$$

For instance, when $\alpha_i = \alpha_j = 1$, this expected payment simplifies to $m^1(s_i) = s_i^2$.

- *Expected revenue.* We can now obtain the seller's expected revenue in the first-price auction, as the expected payment per bidder, $E[m^1(s_i)]$, times the number of bidders, N, as follows:

$$R^1 = N \times E[m^1(s_i)]$$

$$= 2 \times E\left[\frac{\alpha_i + \alpha_j}{2} s_i^2\right]$$

$$= 2 \times \frac{\alpha_i + \alpha_j}{2} \times E[s_i^2]$$

$$= (\alpha_i + \alpha_j) \times E[s_i^2]$$

and since s_i is uniformly distributed, $E[s_i^2] = \int_0^1 s_i^2 ds_i$, implying that the expected revenue becomes

$$R^1 = (\alpha_i + \alpha_j) \int_0^1 s_i^2 ds_i$$

$$= (\alpha_i + \alpha_j) \left[\frac{s_i^2}{3}\right]_0^1$$

$$= \frac{\alpha_i + \alpha_j}{3}.$$

(b) *Second-price auction.* Find the seller's expected revenue in the second-price auction where bidders have interdependent values, as analyzed in Exercise 7.5.

- In the second-price auction of Exercise 7.5, we found that every bidder i submits an equilibrium bid $b(s_i) = (\alpha_i + \alpha_j)s_i$.
  - *Probability of winning.* Bidder i wins if his bid satisfies $b(s_i) > b(s_j)$ or, more explicitly,

$$(\alpha_i + \alpha_j)s_i > (\alpha_i + \alpha_j)s_j,$$

which simplifies to $s_i > s_j$. Since signal s_j is uniformly distributed, the probability that bidder i wins is $\Pr(win) = F(s_i > s_j) = s_i$.

– *Price if winning.* If winning, bidder i pays bidder j's bid, $b(s_j) = (\alpha_i + \alpha_j)s_j$. This price, however, is a function of signal s_j, which bidder i cannot observe. In any case, bidder i knows that he only pays this price when winning the auction, which occurs when $s_i > s_j$. Then, his expected price is

$$\begin{aligned} E\left[b(s_j)|s_i > s_j\right] &= E[(\alpha_i + \alpha_j)s_j|s_i > s_j] \\ &= (\alpha_i + \alpha_j) E[s_j|s_i > s_j] \\ &= (\alpha_i + \alpha_j) \frac{s_i - 0}{2} \\ &= \frac{(\alpha_i + \alpha_j)s_i}{2}. \end{aligned}$$

For instance, when $\alpha_i = \alpha_j = 1$, this expected price simplifies to s_i.

- Hence, his expected payment in the second-price auction is

$$m^2(s_i) = \underbrace{s_i}_{\Pr(win)} \times \underbrace{\frac{s_i(\alpha_i + \alpha_j)}{2}}_{E[b(s_j)|s_i > s_j]} = \frac{\alpha_i + \alpha_j}{2}s_i^2$$

which coincides with the expected payment in the first-price auction that we found in part (a).

- *Expected revenue.* We can now obtain the seller's expected revenue in the second-price auction, as the expected payment per bidder, $E[m^2(s_i)]$, times the number of bidders, N, as follows:[2]

$$\begin{aligned} R^2 &= N \times E[m^2(s_i)] \\ &= 2 \times E\left[\frac{\alpha_i + \alpha_j}{2}s_i^2\right] \\ &= \frac{\alpha_i + \alpha_j}{3}. \end{aligned}$$

---

[2]Because of the Revenue Equivalence Theorem, we could just claim that the expected revenue in the second-price auction coincides with that in the first-price auction of part (a) in this exercise. However, as a practice, we next describe how to calculate the expected revenue in the second-price auction.

(c) *Comparison.* Show that both auction formats generate the same expected revenue for the seller.

- From parts (a) and (b), it is clear that both auction formats generate the same expected payment from the winning bidder, $m^1(s_i) = m^2(s_i)$, and, as a consequence, the same expected revenue for the seller.

Multi-Unit Auctions 8

Keywords

Single unit · Multiple units · Discriminatory auction · Quadratic equation · Discriminant Conjugate pair · Complex roots · Divide and conquer

Introduction

In this chapter we introduce the reader to auctions where the seller offers more than one unit (multi-unit auctions). For a more detailed analysis, see Krishna (2009, chapters 12–18). Exercise 8.1 considers a "discriminatory auction," which is essentially a first-price auction selling multiple units of the same (homogeneous) good at potentially different prices. Similarly, Exercise 8.2 examines a second-price auction offering several units at the same price, which is often known as "uniform-price auction," while Exercise 8.3 considers this auction format but allowing for each unit to be sold at a different price. In this context, we also evaluate the seller's expected revenue, and compare it to single-unit auction.

Exercises 8.4 and 8.5 analyze whether the seller has incentives to sell each unit at a separate auction, earning the same expected revenue as we found in the previous chapters, or if, instead, selling all units simultaneously in the same auction yields a larger expected revenue. Informally, we examine whether a "divide and conquer" strategy is profitable for the seller in his auction design, showing that in both first- and second-price auction he earns a larger expected revenue from selling the units simultaneously rather than running separate auctions if the number of bidders is relatively low. Otherwise, the seller earns a higher expected revenue separately selling each unit.

Exercise #8.1: First-Price Auction Selling Multiple Units (Discriminatory Auction)[A]

8.1 Consider a seller offering two identical goods for sale in an auction. Every bidder $i \in N$ randomly draws his valuation pair (v_i^1, v_i^2) where v_i^1 is his valuation for the first unit of the good and v_i^2 denotes his valuation for the second unit, where $v_i^1 > v_i^2 \geq 0$ and $v_i^j \in [0, 1]$ for unit $j = \{1, 2\}$. Unit j is assigned to the bidder submitting the highest bid, $b_i^j > \max\left\{b_k^j\right\}$ for every bidder $k \neq i$. In the case of a tie, unit j is randomly assigned among the bidders who submitted the highest bid.

Assume that the seller uses a first-price auction, so the winner of unit j pays his bid b_i^j. The first-price auction selling multiple units is often known as "discriminatory auction," as it can charge different prices to the first and second unit sold. Show that it is not a dominant strategy for bidder i to submit a bid equal to his value for that unit, $b_i^j = v_i^j$.

- If bidder i submits a bid $b_i^j = v_i^j$ for every unit j, his expected payoff is zero. However, shading his bid, $b_i^j < v_i^j$, entails that, if he wins, he earns a positive payoff in every unit j. As a result, submitting a bid equal to his value for that unit is not a dominant strategy (i.e., something that bidder i would do regardless of his rivals' bids).

Exercise #8.2: Second-Price Auction Selling Multiple Units (Uniform-Price Auction)[A]

8.2 Consider the same setting in Exercise 8.1, but now assume that bidders pay the same price for each object, and equal to the second-highest bid for all objects. This is similar to a second-price auction selling multiple units, which is often known as uniform-price auction since all units are sold at the same price. Using the same notation as in Exercise 8.1, assume that every bidder $i \in N$ randomly draws his valuation pair (v_i^1, v_i^2) where $v_i^1 > v_i^2 \geq 0$ and $v_i^j \in [0, 1]$ denotes bidder i's valuation of unit $j = \{1, 2\}$.

Show that it is not a dominant strategy for bidder i to submit a bid equal to his value for every unit, $b_i^j = v_i^j$.

- To prove that it is not a dominant strategy for bidder i to submit a bid $b_i^j = v_i^j$ for every unit j, we must show that, if he did, he could find another bid that constitutes a profitable deviation. Assume that
  - Bidder i submits the highest bid among all units, that is,

$$b_i^1 > \max\left\{b_k^1\right\} \text{ and } b_i^1 > \max\left\{b_k^2\right\} \text{ for every bidder } k \neq i$$

- Bidder $m \neq i$ submits the second-highest bid among all units.
- Bidder i submits the third-highest bid among all units.

According to the auction rules, bidder i wins the first unit, bidder m wins the second unit, and both bidders must pay a price equal to the third-highest bid. However, since bidder i submitted the third-highest bid, he can control how much he pays for the unit he received. In other words, reducing that bid would increase his expected payoff from participating in the auction.

- Therefore, submitting a bid equal to his value for every unit, $b_i^j = v_i^j$, is not a dominant strategy. For this bidding strategy to be dominant, we should find that bidder i does not have incentives to deviate regardless of his valuation for each unit and independently of how other bidders behave.

Exercise #8.3: Second-Price Auction Selling Multiple Units, Allowing for Different Prices[B]

8.3 Consider the setting in Exercise 8.1, but assume that the two units sold in the auction are identical, so each bidder i only cares about acquiring one unit of the good. In this setting, the auction rules change so that each bidder can only submit a bid $b_i \geq 0$, the seller assigns one unit of the object to the two bidders submitting the two highest bids, and each of them pays the third-highest bid. All other bidders pay zero.

(a) Show that every bidder i submitting a bid that coincides with his valuation, $b_i(v_i) = v_i$, can be sustained in equilibrium.

- Since every bidder only cares about winning one unit of the good, we can follow a similar approach as in the proof of the equilibrium bidding function in the second-price auction where the seller offers a single unit for sale (see Exercise 1.2).
  - If every bidder $j \neq i$ submits a bid that coincides with his valuation, $b_j(v_j) = v_j$, and bidder i also submits a bid $b_i(v_i) = v_i$, he wins one unit of the object if $v_i > v_i^{[3]}$, where $v_i^{[3]}$ is the third-order statistic[1] of the valuation profile $v \equiv (v_1, v_2, \ldots, v_N)$, enjoying a utility of $v_i - v_i^{[3]}$. (Note that bidder i's valuation does not need to satisfy $v_i >$

---

[1] For more details on the third-order statistic of a valuation profile, see Exercise 2.10, which analyzes how to find the cumulative distribution function of the kth-order statistic, its associated density function, its application to the case in which valuations are uniformly distributed, and how to obtain the expected value and variance of the kth-order statistic.

$v_i^{[2]}$, where $v_i^{[2]}$ is the second-order statistic[2] of the valuation profile v, as in the single-unit second-price auction of Chap. 1, since there are two identical units of the good in this context.)

- If bidder i lowers his bid to $b_i(v_i) < v_i$, he wins one unit of the object if $b_i(v_i) > v_i^{[3]}$, which is less likely than when submitting a bid $b_i(v_i) = v_i$, since $v_i > b_i(v_i) > v_i^{[3]}$, but still earns the same payoff $v_i - v_i^{[3]}$ if winning. Therefore, bidder i does not have strict incentives to lower his bid.
- If bidder i increases his bid to $b_i(v_i) > v_i$, he wins one unit of the object if $b_i(v_i) > v_i^{[3]}$, which is more likely than when submitting a bid $b_i(v_i) = v_i$, since $b_i(v_i) > v_i > v_i^{[3]}$, but still earns the same payoff $v_i - v_i^{[3]}$ if winning. Therefore, bidder i does not have strict incentives to raise his bid. Overall, every bidder i has no incentives to deviate from submitting a bid $b_i(v_i) = v_i$.
- We can then claim that submitting a bid equal to bidder i's valuation, $b_i(v_i) = v_i$, is a weakly dominant strategy.

(b) *Uniformly distributed valuations.* Find the density function of the third-highest valuation $g(v)$ (that is, density function of the third-order statistic), assuming that $v \sim U[0, 1]$. Use this result to find the seller's expected revenue from this auction, $R = 2 \times \int_0^1 vg(v)dv$, and explain how this revenue changes when there are more bidders participating in the auction.

- From part (b) of Exercise 2.10, the probability density function of the third-order statistic for the valuation profile v is

$$g(v) = N \binom{N-1}{2} [1 - F(v)]^{N-3} F(v)^2 f(v)$$

$$= \frac{N(N-1)(N-2)}{2} [1 - F(v)]^{N-3} F(v)^2 f(v)$$

In this context, the seller's expected revenue from the multi-unit auction is

$$R = 2 \int_0^1 vg(v)dv$$

$$= 2 \int_0^1 v \left[\frac{N(N-1)(N-2)}{2} [1 - F(v)]^{N-3} F(v)^2 f(v) \right] dv$$

---

[2]See Exercise 2.9 for more details about the second-order statistic, its cumulative distribution function, density function, and its application to the case in which valuations are uniformly distributed.

Exercise #8.3: Second-Price Auction Selling Multiple Units, Allowing for...

$$= N(N-1)(N-2) \int_0^1 vf(v) [1-F(v)]^{N-3} F(v)^2 \, dv$$

Since valuations are uniformly distributed in [0, 1], we have that $F(v) = v$ and $f(v) = 1$, which simplifies the above expected revenue to

$$R = N(N-1)(N-2) \int_0^1 [1-v]^{N-3} v^3 \, dv$$

where the integral is

$$\int_0^1 [1-v]^{N-3} v^3 \, dv = -\frac{1}{N-2} \int_0^1 v^3 d(1-v)^{N-2}$$

$$= -\frac{1}{N-2} \left[v^3 (1-v)^{N-2} \right]_0^1$$

$$+ \frac{3}{N-2} \int_0^1 (1-v)^{N-2} v^2 \, dv$$

$$= -\frac{1}{N-2} [1 \times 0 - 0 \times 1]$$

$$- \frac{3}{(N-1)(N-2)} \int_0^1 v^2 d(1-v)^{N-1}$$

$$= \frac{3}{(N-1)(N-2)}$$

$$\times \left[2 \int_0^1 (1-v)^{N-1} v \, dv - \left[(1-v)^{N-1} v^2 \right]_0^1 \right]$$

$$= \frac{6}{N(N-1)(N-2)}$$

$$\times \left[\int_0^1 (1-v)^N \, dv - \left[(1-v)^N v \right]_0^1 \right]$$

$$= \frac{6}{N(N-1)(N-2)} \left[-\frac{(1-v)^{N+1}}{N+1} \right]_0^1$$

$$= \frac{6}{N(N-1)(N-2)(N+1)}$$

Therefore, the seller's expected revenue becomes

$$R = N(N-1)(N-2) \int_0^1 [1-v]^{N-3} v^3 \, dv$$

$$= N(N-1)(N-2) \frac{6}{N(N-1)(N-2)(N+1)}$$

$$= \frac{6}{N+1}$$

- This expected revenue is decreasing in the number of competing bidders since

$$\frac{\partial R}{\partial N} = -\frac{6}{(N+1)^2} < 0$$

In contrast with single-unit auction where $R^2 = \frac{N-1}{N+1}$ (see Exercise 1.7) is monotonically increasing in the number of bidders, expected revenue in this multi-unit auction decreases as the auction becomes more competitive since both the highest-valued and second-highest-valued bidder can purchase one unit of the object.

(c) *Exponentially distributed valuations.* Consider now that individual valuations are drawn from an exponential distribution, $F(v) = 1 - \exp(-\lambda v)$ where $v \in [0, \infty)$. Find the seller's expected revenue from this auction. How does the expected revenue change with the number of bidders N and parameter λ? Interpret your results.

- The seller's expected revenue from the multi-unit auction is

$$R = N(N-1)(N-2) \int_0^\infty vf(v)[1-F(v)]^{N-3} F(v)^2 \, dv$$

$$= N(N-1)(N-2) \int_0^\infty v\underbrace{\lambda \exp(-\lambda v)}_{f(v)} \times \underbrace{[\exp(-\lambda v)]^{N-3}}_{1-F(v)}$$

$$\times \underbrace{[1-\exp(-\lambda v)]^2}_{F(v)} dv$$

$$= N(N-1)(N-2)\lambda$$

$$\times \int_0^\infty \left\{ \underbrace{v[\exp(-\lambda v)]^{N-2}}_{\text{Term A}} - \underbrace{2v[\exp(-\lambda v)]^{N-1}}_{\text{Term B}} + \underbrace{v[\exp(-\lambda v)]^N}_{\text{Term C}} \right\} dv$$

Consider term A, where

$$\int_0^\infty v[\exp(-\lambda v)]^{N-2} \, dv$$

$$= -\frac{1}{\lambda(N-1)} \int_0^\infty vd[\exp(-\lambda v)]^{N-1}$$

$$= -\frac{1}{\lambda(N-1)} \left[v \exp(-\lambda v)^{N-1} \right]_0^\infty$$

$$+ \frac{1}{\lambda(N-1)} \int_0^\infty \exp[(-\lambda v)]^{N-1} dv$$

$$= -\frac{1}{\lambda^2 N(N-1)} \int_0^\infty d\exp[(-\lambda v)]^N$$

$$= \frac{1}{\lambda^2 N(N-1)}$$

Consider now term B, where

$$\int_0^\infty -2v \left[\exp(-\lambda v)\right]^{N-1} dv$$

$$= \frac{2}{\lambda N} \int_0^\infty v d\left[\exp(-\lambda v)\right]^N$$

$$= \frac{2}{\lambda N} \left[v \exp(-\lambda v)^N \right]_0^\infty - \frac{2}{\lambda N} \int_0^\infty \left[\exp(-\lambda v)\right]^N dv$$

$$= \frac{2}{\lambda^2 N(N+1)} \int_0^\infty d\left[\exp(-\lambda v)\right]^{N+1}$$

$$= -\frac{2}{\lambda^2 N(N+1)}$$

Consider now term C, where

$$\int_0^\infty v\left[\exp(-\lambda v)\right]^N dv$$

$$= -\frac{1}{\lambda(N+1)} \int_0^\infty v d\left[\exp(-\lambda v)\right]^{N+1}$$

$$= -\frac{1}{\lambda(N+1)} \left[v \exp(-\lambda v)^{N+1} \right]_0^\infty$$

$$+ \frac{1}{\lambda(N+1)} \int_0^\infty \left[\exp(-\lambda v)\right]^{N+1} dv$$

$$= -\frac{1}{\lambda^2(N+1)(N+2)} \int_0^\infty d\left[\exp(-\lambda v)\right]^{N+2}$$

$$= \frac{1}{\lambda^2(N+1)(N+2)}$$

Adding up the terms, the seller's expected revenue becomes

$$R = N(N-1)(N-2)\lambda$$
$$\times \left[\underbrace{\frac{1}{\lambda^2 N(N-1)}}_{\text{Term A}} - \underbrace{\frac{2}{\lambda^2 N(N+1)}}_{\text{Term B}} + \underbrace{\frac{1}{\lambda^2(N+1)(N+2)}}_{\text{Term C}} \right]$$

$$= \frac{N(N-1)(N-2)\lambda}{N(N-1)(N+1)(N+2)\lambda^2}$$
$$\times [(N+1)(N+2) - 2(N-1)(N+2) + N(N-1)]$$

$$= \frac{6(N-2)}{(N+1)(N+2)\lambda}$$

- This expected revenue is decreasing in the number of bidders N since

$$\frac{\partial R}{\partial N} = \frac{6}{\lambda} \frac{N^2 + 3N + 2 - (N-2)(2N+3)}{(N+1)^2(N+2)^2}$$
$$= -\frac{6(N^2 - 4N - 8)}{\lambda(N+1)^2(N+2)^2}$$

which is negative (so R is decreasing in N) when we have $N \geq 6$ bidders participating in this auction.

- This expected revenue is decreasing in parameter λ since

$$\frac{\partial R}{\partial \lambda} = -\frac{6(N-2)}{(N+1)(N+2)\lambda^2} < 0$$

since bidders' valuation is more likely to be low for higher values of λ.

(d) How does the seller's revenue compare with the single-unit second-price auction in Exercise 1.7? For simplicity, you can assume that the cost to produce every unit is zero.

- *Uniform distribution.* The seller earns a higher revenue by selling two units when N satisfies

$$R = \frac{6}{N+1} > \frac{N-1}{N+1} = R^2$$

which simplifies to

$$6 > N - 1$$

that holds when $N < 7$. Since $N \geq 3$, as required (remembering that every winning bidder pays the third-highest price for the object), when there are no more than 7 bidders participating in the auction, the seller can generate a higher revenue by selling two identical units instead of just one unit of the object. For example, when $N = 4$, revenue satisfies

$$R = \frac{6}{5} > \frac{3}{5} = R^2$$

but when we have $N = 8$ bidders, we obtain

$$R = \frac{2}{3} < \frac{7}{9} = R^2$$

- *Exponential distribution.* The seller earns a higher revenue by selling two units when N satisfies

$$R = \frac{6(N-2)}{(N+1)(N+2)\lambda} > \frac{1}{2\lambda} = R^2$$

which simplifies to

$$N^2 - 9N + 26 < 0$$

that does not hold for any values of N since the discriminant[3] of the above quadratic equation is

$$\Delta = (-9)^2 - 4 \times 26 = -23 < 0$$

so that the function is always above the horizontal axis. In this context, the seller should only sell one unit of the object.

---

[3] Recall that, in the case of a quadratic equation $ax^2 + bx + c = 0$, where coefficients a, b, and c are real numbers (not necessarily positive), the discriminant is $\Delta = b^2 - 4ac$. When the discriminant is positive, $\Delta > 0$, the roots of the quadratic equation are real numbers and different from each other. Graphically the quadratic equation crosses the x-axis at two different points. However, when the discriminant is zero, $\Delta = 0$, roots are real numbers and at least two of them coincide. Finally, if $\Delta < 0$, there is a conjugate pair of complex roots, graphically meaning that the quadratic equation lies above the x-axis.

Exercise #8.4: Divide Bidders and Conquer in a First-Price Auction[B]

8.4 Consider the setting in Exercise 8.3, but now assume that the seller evenly separates the bidders into two separate rooms. For simplicity, assume that the total number of bidders, N, is even, so that both rooms have the same number of bidders, $N/2$, and valuations are uniformly distributed in $[0, 1]$. Once bidders are separated, the seller offers a single unit in each room using a standard first-price auction. In this exercise, we investigate if this "divide and conquer" strategy generates a larger expected revenue for the seller than simultaneously offering both units to all bidders, as examined in Exercise 8.3.

(a) *Equilibrium bids.* Find the equilibrium bidding function that every bidder i uses in each room.

- In a standard first-price auction with uniformly distributed valuations, we know from Exercise 2.3 that the equilibrium bidding function is

$$b_i(v_i) = \frac{N-1}{N} v_i$$

when N bidders participate in the auction. Since only $\frac{N}{2}$ bidders participate in this auction room, the equilibrium bidding function becomes

$$b_i(v_i) = \frac{\frac{N}{2} - 1}{\frac{N}{2}} v_i = \frac{N-2}{N} v_i$$

(b) *Expected revenue.* Find the seller's expected revenue in each room.

- In a standard first-price auction with uniformly distributed valuations, we know from Exercise 3.1 that the seller's expected revenue is

$$R^1 = \frac{N-1}{N+1}$$

when N bidders participate in the auction. Since there are only $\frac{N}{2}$ bidders participating in this auction room, the expected revenue in each room is

$$R^1 = \frac{\frac{N}{2} - 1}{\frac{N}{2} + 1} = \frac{N-2}{N+2}$$

(c) *Revenue comparison.* Does the seller increase its expected revenue by dividing bidders into two rooms (as in this exercise), relative to his expected revenue when he sells both units simultaneously to all N bidders (as analyzed in Exercise 8.3)?

- In the auction of Exercise 8.3, where the seller simultaneously sells both units to all bidders, we found that his expected revenue is $R = \frac{6}{N+1}$. In contrast, now the seller only earns $\frac{N-2}{N+2}$ in each auction room, for a total of $\frac{2(N-2)}{N+2}$. Therefore, the expected revenue from simultaneously selling both units is larger if

$$\frac{6}{N+1} > \frac{2(N-2)}{N+2}$$

  which simplifies to $N^2 - 4N - 8 < 0$. Solving for N, we find two roots, $N > -1.46$ and $N < 5.46$. Since the number of bidders satisfies $N > 2$ by definition, we obtain that the seller earns a higher expected revenue by selling both units simultaneously when relatively few bidders compete for these units ($N < 5.46$, which entails $N \leq 5$), but otherwise earns a higher expected revenue separately selling each unit.
- Since the number of bidders in each room, $\frac{N}{2}$, is even by definition, condition $N < 5$ implies that the only case in which running two separate auctions generates a larger expected revenue than simultaneously selling both units is when exactly two bidders participate in each auction room ($N = 4$ bidders in total). Otherwise, the seller earns a larger expected revenue by simultaneously selling both units.

Exercise #8.5: Divide Bidders and Conquer in a Second-Price Auction[A]

8.5 Consider the setting in Exercise 8.4, but assume that the seller uses a second-price auction in each room.

(a) *Equilibrium bids.* Find the equilibrium bidding function that every bidder i uses in each room.

- In a standard second-price auction, we know from Exercise 1.2 that the equilibrium bidding function is $b_i(v_i) = v_i$, so every bidder i bids his valuation for the object. This result is unaffected by how many bidders are in the auction room or by whether values are distributed according to a uniform distribution or other distribution.

(b) *Expected revenue.* Find the seller's expected revenue in each room.

- In the standard second-price auction with uniformly distributed valuations, we know from Exercise 1.7 that the seller's expected revenue is

$$R^2 = \frac{N-1}{N+1}$$

when N bidders participate in the auction. Since only $\frac{N}{2}$ bidders participate in his auction room, the expected revenue becomes

$$R^2 = \frac{\frac{N}{2} - 1}{\frac{N}{2} + 1} = \frac{N-2}{N+2}$$

This result could be directly obtained by the Revenue Equivalence Principle, so R^2 coincides with the expected revenue from the first-price auction in Exercise 8.4, part (b).

(c) *Revenue comparison.* Does the seller increase his expected revenue by dividing bidders into two rooms (as in this exercise), relative to his expected revenue when he sells both units simultaneously to all N bidders (as analyzed in Exercise 8.4)?

- By the same argument as in Exercise 8.4, the seller earns a higher expected revenue selling both units of the good simultaneously (as in Exercise 8.3) than by dividing bidders into two rooms and running a second-price auction when $N \leq 5$ bidders. Otherwise, the seller earns a higher expected revenue separately selling each unit.

Mechanism Design 9

Keywords

Mechanism design · Reported valuation · Over-reporting · Under-reporting · Incentive compatibility · Assignment rule · Allocation rule · Truthtelling · Several units · Direct revelation mechanism · Vickrey-Clarke-Groves (VCG) mechanism · Generalized second-price auction · Keyword auctions · Clicks

Introduction

Previous chapters consider different auction formats, understood as mechanisms to allocate an object among different individuals (bidders). Each of them is, essentially, characterized by an allocation rule (who gets the object) and a payment rule (how much each bidder has to pay when winning the object and otherwise). In this chapter, we take a more general approach by considering a richer set of mechanisms, seeking to: (1) allocate the object to the individual with the highest valuation (efficiency); (2) maximize the seller's expected revenue; (3) maximize the social planner's welfare function; or (4) a combination of these objectives.

Exercise 9.1 examines a direct revelation mechanism, where every bidder is asked to report his valuation for the object, rather than bidding for it. Every bidder privately observes his valuation and can truthfully report it to the seller or not. We show that, when the winner of this allocation mechanism pays his announced valuation, he would have incentives to underreport it. However, when the winner pays the second-highest announced valuation, players have incentives to truthfully report their valuations.

In Exercise 9.2 (9.3), we study the first-price (second-price) auction and how it can be transformed into a direct revelation mechanism where players truthfully report their valuation for the object.

Exercises 9.4 and 9.5 analyze the Vickrey–Clarke–Groves (VCG) mechanism to assign one, or several, unit of a good, respectively, and show that players have incentives to truthfully report their valuation. Finally, Exercise 9.6 introduces "keyboard auctions," also known as generalized second-price auctions, and compares them against the VCG mechanism. For an introduction to the theory of mechanism design, see Börgers (2015).

Exercise #9.1: Incentives to Truthfully Reveal Valuations[A]

9.1 Consider a setting with two bidders. Unlike in previous chapters where every bidder i submits a bid, we consider that he can only report a valuation $\widehat{v}_i \in [0, 1]$ for the object to the seller, where $\widehat{v}_i$ can coincide with his valuation, $\widehat{v}_i = v_i$ (which we denote as "truthful reporting" or "truthtelling"), or not, $\widehat{v}_i \neq v_i$ (which we refer to as "misreporting" his valuation). The seller observes the reported valuations from each bidder, $\widehat{v} = (\widehat{v}_i, \widehat{v}_j)$, but cannot observe their true valuation for the object, $v = (v_i, v_j)$.

Before choosing his reported valuation, $\widehat{v}_i$, every bidder i knows the rules of this game: the seller will assign the object to one bidder (the individual who reported the highest valuation) and ask every bidder i to pay a price (that is normalized to zero for the bidders who did not win the object and a positive price to the bidder who won the object). Answer the following questions assuming that the seller assigns the object to the individual reporting the highest valuation, that is, bidder i wins the object if his report, $\widehat{v}_i$, satisfies $\widehat{v}_i > \widehat{v}_j$ for all $j \neq i$.

(a) Consider that the seller asks the winner to pay his reported valuation, $\widehat{v}_i$. Show that bidders do not have incentives to truthfully report their valuations in the Bayesian Nash equilibrium of the game.

- If bidder j truthfully reports $\widehat{v}_j = v_j$, and bidder i also truthfully reports his valuation, $\widehat{v}_i = v_i$, bidder i gains a utility $v_i - \widehat{v}_i = 0$ if he wins the object (which occurs when $\widehat{v}_i > \widehat{v}_j$), or a zero utility if he loses.
- Bidder i, then, has incentives to underreport his true valuation (i.e., $\widehat{v}_i < v_i$) so his payoff from winning, $v_i - \widehat{v}_i$, becomes positive, while his payoff from losing remains zero. Therefore, we found that players have incentives to unilaterally deviate from the strategy profile $(\widehat{v}_i, \widehat{v}_j) = (v_i, v_j)$, implying that it cannot be sustained in equilibrium.

(b) Assume now that the seller asks the winner to pay the second-highest reported valuation that, in a two-bidder setting, is the reported valuation of the losing bidder, $\widehat{v}_j$. Show that bidders have incentives to truthfully report their valuations in the Bayesian Nash equilibrium of the game.

- If bidder j behaves as prescribed, truthfully reporting his valuation $\widehat{v}_j = v_j$, bidder i has incentives to truthfully report his valuation too, $\widehat{v}_i$. By doing so, he may win, paying the lowest reported valuation, $\widehat{v}_j$, and thus

gaining a utility $\widehat{v}_i - \widehat{v}_j = v_i - v_j$. We now show that bidder i has no incentives to deviate:
- Bidder i could deviate to over-report his valuation to $\widehat{v}_i$, where $\widehat{v}_i > v_i$. This deviation does not increase his payoff from winning, which is still $v_i - v_j$, so that he does not have incentives to over-report his valuation.
- Similarly, bidder i could deviate to underreport his valuation to $\widehat{v}_i$, where $\widehat{v}_i < v_i$. However, this deviation does not increase his payoff from winning either, which is still $v_i - v_j$, but he may lose the object if $\widehat{v}_i < v_j < v_i$, so that he does not have incentives to underreport his true valuation of the object.

- In this auction format, every bidder finds truthtelling to be a *weakly dominant* strategy, that is, every bidder i has no strict incentives to deviate from $\widehat{v}_i = v_i$ regardless of the reporting strategy that his opponent selects (i.e., both when $\widehat{v}_j = v_j$ and when $\widehat{v}_j \neq v_j$).

Exercise #9.2: First-Price Auction as a Direct Revelation Mechanism[B]

9.2 Consider a first-price auction with $N \geq 2$ bidders, each of them privately observing his valuation of the object, v_i, where $v_i \in [0, 1]$. It is common knowledge that valuations are independent and identically distributed with $F_i(v_i)$ and positive support in $[0, 1]$. Let us define a direct revelation mechanism (DRM) as a game where every bidder i, rather than submitting a bid, reports a valuation $\widehat{v}_i$ to the seller (which may coincide with his true valuation of the object, $\widehat{v}_i = v_i$, or not, $\widehat{v}_i \neq v_i$). The seller observes the profile of reported valuations $\widehat{v} \equiv (\widehat{v}_1, \widehat{v}_2, \ldots, \widehat{v}_N)$ and responds by announcing a winner and a payment for each bidder i, $t_i(\widehat{v})$.

Can you construct a DRM in which the object is assigned to the same individual as in the standard first-price auction, and where the payment that each bidder makes to the seller also coincides with that in the first-price auction?

- In the first-price auction, every bidder i submits a bid $b_i(v_i)$, which is increasing in his valuation, v_i, and exhibits bid shading, $b_i(v_i) < v_i$. For instance, in a setting where all bidders draw their valuations from the same cumulative distribution function $F(v_i)$, Exercise 2.4 showed that the equilibrium bidding function in the first-price auction is

$$b_i(v_i) = v_i - \underbrace{\frac{\int_0^{v_i} F(x)^{N-1} dx}{F(v_i)^{N-1}}}_{\text{bid shading}}.$$

In this setting, the winner is the individual submitting the highest bid, who pays his own bid, $b_i(v_i) \equiv \max\{b_1(v_1), b_2(v_2), \ldots, b_N(v_N)\}$, while all other bidders pay zero.

- We can construct a DRM with the following assignment rule:

$$y_i(v) = \begin{cases} 1 & \text{if } v_i > b_i(v_i), \text{ and} \\ 0 & \text{otherwise.} \end{cases}$$

And the following payment rule:

$$t_i(v) = y_i(v)v_i - \int_0^{v_i} y_i(x, v_{-i})dx \quad \text{for every bidder } i.$$

Intuitively, assignment rule $y_i(v)$ says that the object goes to bidder i if his valuation is higher than the highest bid, which is his own. The payment rule can be interpreted by separately looking at each term. The first term says that, if the individual is the winner, $y_i(v) = 1$, the first term collapses to his valuation, v_i, but if he does not win the object, $y_i(v) = 0$, the first term is nil. The second term in the payment rule captures the expected probability that bidders with a lower valuation than bidder i are assigned the object. Given the above assignment rule, the second term of the payment rule is zero for all individuals losing the auction.

- We need to show that this DRM: (1) satisfies incentive compatibility (i.e., every bidder i has incentives to truthfully report his valuation v_i to the seller); and (2) produces the same equilibrium results as the first-price auction (i.e., the same winner and payment from every bidder).
  - *Incentive compatibility.* To check if the above DRM is incentive compatible, we need to check if it satisfies Myerson's characterization theorem (as stated in Exercise 6.6). In particular, assignment rule $y_i(v)$ is non-decreasing in bidder i's valuation, v_i, implying that bidder i's expected assignment $\bar{y}_i(v) \equiv E_{v_{-i}}(y_i(v_i, v_{-i}))$ is also non-decreasing in v_i. In addition, bidder i's expected payment, $\bar{t}_i(v) \equiv E_{v_{-i}}(t_i(v_i, v_{-i}))$, can be expressed as

$$\bar{t}_i(v) = -\bar{y}_i(0) + \bar{y}_i(v)v_i - \int_0^{v_i} \bar{y}_i(x)dx$$

where $\bar{y}_i(0) = 0$ is for the individual with the lowest valuation, so that

$$\bar{t}_i(v) = \bar{y}_i(v)v_i - \int_0^{v_i} \bar{y}_i(x)dx$$

as required by Myerson's characterization theorem. Then, the above DRM is incentive compatible.

- *Same equilibrium results as the first-price auction.* As described above, for a given valuation profile, v, the assignment rule gives the object to the individual with the highest valuation, $y_i(v) = 1$ if his valuation satisfies $v_i > b_i(v_i)$. This individual pays

$$t_i(v) = \overbrace{y_i(v)}^{1} v_i - \int_0^{v_i} y_i(x, v_{-i}) dx$$

$$= v_i - \int_{b_i(v_i)}^{v_i} 1 dx$$

$$= v_i - (v_i - b_i(v_i)) = b_i(v_i)$$

so that the winner of the DRM pays a price equal to the bid he submitted in the first-price auction. All other bidders pay zero since

$$t_i(v) = \overbrace{y_i(v)}^{0} v_i - \int_0^{v_i} y_i(x, v_{-i}) dx$$

$$= 0 - \int_0^{v_i} 0 dx = 0.$$

Therefore, the winner of the DRM coincides with that in the first-price auction, and every bidder i pays the same amount ($b_i(v_i)$ or zero) as they did in the first-price auction.

- **Remark:** This result holds regardless of whether valuations are drawn from symmetric or asymmetric distributions, independently on the information that the seller has about these distributions, and robust to the risk preferences of the seller and the bidders.

Exercise #9.3: Second-Price Auction as a Direct Revelation Mechanism[A]

9.3 Consider the setting in Exercise 9.2. If bidders face a second-price auction, can you construct a DRM where the object is assigned to the same individual as in the standard first-price auction, and where the payment that each bidder makes to the seller coincides with that in the second-price auction?

- In the second-price auction, every bidder i submits a bid equal to his valuation for the object, $b_i(v_i) = v_i$. In this setting, the highest bidder wins the object and pays the second-highest bid (which coincides with the second-highest valuation), $t_i(v) = h_i$, where $h_i = \max\{v_j | j \neq i\}$, while all other bidders $j \neq i$ pay zero, $t_j(v) = 0$.
- We can construct a DRM with the following assignment rule:

$$y_i(v) = \begin{cases} 1 & \text{if } v_i > h_i, \text{ and} \\ 0 & \text{otherwise.} \end{cases}$$

And the following payment rule:

$$t_i(v) = y_i(v)v_i - \int_0^{v_i} y_i(x, v_{-i})dx \text{ for every bidder } i.$$

Intuitively, assignment rule $y_i(v)$ says that the object goes to bidder i if his valuation is higher than the second-highest valuation, implying that he is the individual with the highest valuation. This payment rule can be interpreted by separately looking at each term. The first term says that, if the individual is the winner, $y_i(v) = 1$, the first term collapses to his valuation, v_i, but if he does not win the object, $y_i(v) = 0$, the first term is nil. The second term in the payment rule captures the expected probability that bidders with a lower valuation than bidder i will be assigned the object. Given the above assignment rule, the second term of the payment rule is zero for all individuals losing the auction.

We need to show that this DRM: (1) satisfies incentive compatibility (i.e., every bidder i has incentives to truthfully report his valuation v_i to the seller); and (2) produces the same equilibrium results as the second-price auction (i.e., the same winner and payment from every bidder).

- *Incentive compatibility.* To check if the above DRM is incentive compatible, we need to check if it satisfies Myerson's characterization theorem (as stated in Exercise 6.6). In particular, assignment rule $y_i(v)$ is non-decreasing in bidder i's valuation, v_i, implying that bidder i's expected assignment $\bar{y}_i(v) \equiv E_{v_{-i}}(y_i(v_i, v_{-i}))$ is also non-decreasing in v_i. In addition, bidder i's expected payment, $\bar{t}_i(v) \equiv E_{v_{-i}}(t_i(v_i, v_{-i}))$, can be expressed as

$$\bar{t}_i(v) = \bar{y}_i(0) + \bar{y}_i(v)v_i - \int_0^{v_i} \bar{y}_i(x)dx$$

where $\bar{y}_i(0) = 0$ is for the individual with the lowest valuation, so that

$$\bar{t}_i(v) = \bar{y}_i(v)v_i - \int_0^{v_i} \bar{y}_i(x)dx$$

as required by Myerson's characterization theorem. Then, the above DRM is incentive compatible.

- *Same equilibrium results as the second-price auction.* As described above, for a given valuation profile, v, the assignment rule gives the object to the individual with the highest valuation, $y_i(v) = 1$ if his valuation satisfies $v_i > h_i$. This individual pays

$$t_i(v) = \overbrace{y_i(v)}^{1} v_i - \int_0^{v_i} y_i(x, v_{-i}) dx$$

$$= v_i - \int_{h_i}^{v_i} 1 \, dx$$

$$= v_i - (v_i - h_i) = h_i$$

so that the winner of the DRM pays a price equal to the second-highest valuation, h_i. All other bidders pay zero since

$$t_i(v) = \overbrace{y_i(v)}^{0} v_i - \int_0^{v_i} y_i(x, v_{-i}) dx$$

$$= 0 - \int_0^{v_i} 0 \, dx = 0.$$

Therefore, the winner of the DRM coincides with that in the second-price auction, and every bidder i pays the same amount (h_i or zero) as he did in the second-price auction.

- **Remark:** Finally, note that this result holds regardless of whether valuations are drawn from symmetric or asymmetric distributions, independently on the information that the seller has about these distributions, and robust to the risk preferences of the seller and the bidders.

Exercise #9.4: VCG Mechanism Selling a Single Unit[B]

9.4 Consider an auction with $N \geq 2$ bidders, each privately observing his valuation v_i for the object, which is drawn from a distribution function $F(v_i)$ with positive density in all its support, $v_i \in [0, 1]$. The seller seeks to implement a direct revelation mechanism where every bidder i announces his valuation $v_i \in [0, 1]$. The seller assigns the object to the bidder who announced the highest valuation, and every bidder i pays a price according to the following transfer function (often known as the "VCG mechanism" or Vickrey–Clarke–Groves mechanism, after Vickrey (1961), Clarke (1971), and Groves (1973)):

$$t_i(v) = \sum_{j \neq i} u_j\left(k(v), v_j\right) - \sum_{j \neq i} u_j\left(k_{-i}(v_{-i}), v_j\right)$$

where $v_{-i} \equiv (v_1, \ldots v_{i-1}, v_{i+1}, \ldots, v_N)$ denotes the valuations of all other bidders $j \neq i$, and $v \equiv (v_1, \ldots, v_N)$ represents the valuations of all bidders (including i).

Intuitively, the first term represents the utility that all $j \neq i$ bidders obtain when the seller considers all players' preferences (including i) in allocating the object according to the allocation rule $k(v)$. The second term, in contrast, describes the utility that they enjoy when the seller ignores player i's preferences, so the allocation becomes $k_{-i}(v_{-i})$. Therefore, the price that bidder i pays (receives) is a function of the negative (positive) externality that his preference for the object imposes (generates) on the other bidders.

Consider five individuals participating in this auction, with the following valuations for a good, $v_1 = 20$, $v_2 = 15$, $v_3 = 12$, $v_4 = 10$, and $v_5 = 6$.

(a) Assume that every bidder has incentives to truthfully report his valuation of the object. Find the winner of the auction and the price that every bidder pays.

- *Winning bidder.* If every individual truthfully reveals his valuation, bidder 1 reports the highest value, $v_1 = 20$, winning the auction, and paying a price

$$t_1(v) = \sum_{j \neq 1} u_j \left(k(v), v_j\right) - \sum_{j \neq 1} u_j \left(k_{-1}(v_{-1}), v_j\right)$$
$$= 0 - 15 = -\$15$$

  - In the first term, the allocation rule considers the reported valuations of all bidders, v. Then, the object would be assigned to bidder 1 (the individual with the highest valuation), entailing a value of 0+0+0+0=0 to the other $j \neq 1$ bidders.
  - The second term, in contrast, ignores bidder 1's preferences (valuation), thus assigning the object to bidder 2 (as he is now the player with the highest valuation). Bidder 2's utility from receiving the good is 15, implying that the sum of valuations is now $15+0+0+0 = 15$.
  - The difference between the two terms yields a transfer of $t_1(v) = 0-15 = -\$15$, thus indicating that player 1 pays \$15, i.e., the second-highest valuation. Intuitively, the price that player 1 pays coincides with the negative externality that his presence generates on player 2, since the latter would receive the object should the former not participate in the auction.
- *Losing bidders.* A similar argument applies to all other players. However, since their valuations are lower than that of player 1, their transfers become

$$t_i(v) = (20+0+0+0) - (20+0+0+0) = \$0 \text{ for every player } i \neq 1.$$

In other words, the object is assigned to player 1 in both settings, thus yielding the same profile of utilities when individual $i \neq 1$ participates, $20+0+0+0$, and when he does not, $20+0+0+0$. In particular, player 1 obtains 20 as he receives the object, and all other individuals $i \neq 1$ do not receive the object.

Intuitively, their decision to participate in the mechanism does not produce an externality on other players, thus yielding a nil transfer $t_i(v) = 0$ for every valuation vector v and all $i \neq 1$.

(b) Show that, in the strategy profile analyzed in part (a), every bidder has incentives to truthfully report his valuation of the object.

- The winning bidder's utility in the strategy profile we considered in part (a) is $20 - 15 = 5$. If, instead, he reports a higher valuation for the object, he still wins the auction and pays a price of $15, enjoying the same utility. If, instead, he reports a lower valuation, he may lose the auction, decreasing his utility from 5 to 0. Therefore, the winning bidder has no incentives to misreport his valuation.
- Every losing bidder has a zero utility in the above strategy profile.
  - If, instead, bidder $i \neq 1$ increases its reported valuation (e.g., bidder 2) from v_i to v_i' where $v_i' < v_1$, he still loses the auction. However, if he over-reports $v_i' \geq v_1$, he can win the auction but pays the second-highest valuation, $20, for the object, ultimately making a loss.
  - Bidder i does not have incentives to underreport his valuation either, since he keeps losing the auction, making a zero utility. Overall, losing bidders do not have incentives to over- or underreport their valuations.
- As a consequence, the strategy profile in part (a) is a weakly dominant strategy profile for every player.

(c) Verbally discuss the similarities between the VCG mechanism and a second-price auction.

- The VCG mechanism leads to the same outcome (the object is allocated to the bidder with highest valuation) and transfer profile (the individual receiving the object pays a transfer equal to the valuation of the individual with the second-highest valuation, while everyone else pays zero) as in the second-price auction.

Exercise #9.5: VCG Mechanism Selling Several Units[B]

9.5 Consider the same auction as in Exercise 9.4, with the same set of bidders. However, assume now that the seller offers 3 identical units of the object, with every bidder willing to buy only one unit.

(a) If every bidder truthfully reports his valuation, find the winning bidder and the price that every bidder pays.

- *Assignment.* If every bidder truthfully reports his valuation, one unit of the object is assigned to bidder 1, one unit to bidder 2, and one unit to

bidder 3, as they reported the three highest valuations. Bidders 4 and 5 receive no units.
- *Payments.* In this context, the price that bidder 1 pays is

$$t_1(v) = \sum_{j \neq 1} u_j\left(k(v), v_j\right) - \sum_{j \neq 1} u_j\left(k_{-1}(v_{-1}), v_j\right)$$
$$= (15 + 12) - (15 + 12 + 10)$$
$$= -\$10.$$

  – The first term indicates that, when the valuation profiles of all bidders are considered, the three available units are assigned to the bidders with the highest valuations: bidders 1, 2, and 3. The first term, however, measures the utility that players $j \neq 1$ obtain from such an allocation, i.e., the valuations of players 2 and 3, $(15 + 12)$.
  – In the second term, we still measure the utility of bidders $j \neq 1$ but ignoring player 1's preferences when allocating the three units. In this case, the three units go to the three remaining bidders with the highest valuations (bidders 2, 3, and 4) yielding a total utility of $(15 + 12 + 10)$.
  – As a result, the transfer that bidder 1 has to pay is $-\$10$, indicating that, if his preferences were considered, he would impose a negative externality of -10 on the remaining players. Specifically, this externality captures the utility loss of 10 that player 4 suffers as he would get one unit when bidder 1's preferences are ignored but does not receive the object when the preference of bidder 1 is considered.
- Similarly, the price that bidder 2 pays is

$$t_2(v) = \sum_{j \neq 2} u_j\left(k(v), v_j\right) - \sum_{j \neq 2} u_j\left(k_{-2}(v_{-2}, v_j)\right)$$
$$= (20 + 12) - (20 + 12 + 10)$$
$$= -\$10$$

  which, as expected, coincides with bidder 4's valuation since the participation of bidder 2 means that bidder 4 no longer gets one unit of the object.
- Lastly, the price that bidder 3 pays is

$$t_3(v) = \sum_{j \neq 3} u_j\left(k(v), v_j\right) - \sum_{j \neq 3} u_j\left(k_{-3}(v_{-3}), v_j\right)$$
$$= (20 + 15) - (20 + 15 + 10)$$
$$= -\$10$$

which, again, coincides with the price that bidders 1 and 2 pay, since bidder 4 is outbid by bidder 3's participation in the auction.
- Finally, every losing bidder $i = \{4, 5\}$ pays zero since

$$t_i(v) = \sum_{j \neq i} u_j\left(k(v), v_j\right) - \sum_{j \neq i} u_j\left(k_{-i}(v_{-i}), v_j\right)$$
$$= (20 + 15 + 12) - (20 + 15 + 12)$$
$$= \$0,$$

which holds for both bidders 4 and 5. Intuitively, their presence or absence in the auction does not change the assignment of the three units, thus implying that these bidders do not impose a positive or negative externality on the other three bidders.

(b) Argue that your results from part (a) coincide with those in a second-price auction.

- The assignment rule coincides with that of the second-price auction (the object is allocated to bidders with the highest valuation), and transfer profile coincides too (bidders receiving the object pay a transfer equal to the highest valuation of the individual who does not win the object, while everyone else pays zero).

Exercise #9.6: VCG Mechanism and the Generalized Second-Price Auction[B]

9.6 In this exercise, we provide a brief introduction to generalized second-price auctions (GSPs) and how they differ from the VCG. GSPs are often used in "keyword auctions," where sponsored search slots are sold by search engines, such as Google or Yahoo! As an example, type some keywords in your favorite search engine and note that, at the top of the results, there are often a few websites showing up as sponsored search results (slots).

The rules of a GSP are analogous to those of the second-price auction, namely, the highest bidder receives the first slot of sponsored search results and pays the second-highest bid; the second-highest bidder receives the second slot and pays the third-highest bid, and similarly for other bidders. Formally, ordering bids as follows $b_1 \geq b_2 \geq \ldots \geq b_N$, where $N \geq 2$ denotes the number of bidders, and the bidder who submitted bid b_i receives slot i and pays b_{i+1} for it.

Consider a multi-unit auction with $N = 3$ bidders, A, B, and C, and two slots, 1 and 2. Slot 1 gets 100 clicks per hour, while slot 2 only gets 70 clicks per hour. Bidder A's value per click is $v_A = \$10$, bidder B's is $v_B = \$8$, and

C's is $v_C = \$5$. This information is common knowledge among players, so they interact in a complete information game.

(a) If the seller uses a GSP and bidders bid truthfully, what is his total revenue from the auction?

- If bidders submit their bids truthfully, their bids coincide with their valuations, that is, $b_A = \$10$, $b_B = \$8$, and $b_C = \$5$. Ranking these bids, the seller assigns slot 1 to the highest bidder (bidder A), who pays the second-highest bid, $\$8$; and slot 2 to the second-highest bidder (bidder B), who pays the third-highest bid, $\$5$.
- Since slot 1 (2) gets 100 (70) clicks per hour, bidder A pays $\$8 \times 100 = \$800$ and bidder B pays $\$5 \times 70 = \$350$ to the seller, yielding a total revenue of

$$\$800 + \$350 = \$1{,}150.$$

(b) If the seller uses, instead, a VCG mechanism, which is his total revenue from the auction?

- Bidder B in this setting still pays $\$5$ per click (the externality that he imposes on bidder C, who does not win a slot because of bidder B's presence in the auction). Therefore, bidder B's payment is still $\$5 \times 70 = \$350$.
- However, bidder A's price per click now changes, as we have to measure the externality that he imposes on bidders B and C. In particular:
  - *Bidder B.* If bidder A was absent, bidder B would win slot 1, with 100 clicks, rather than slot 2, with only 70 clicks. In other words, bidder A's presence reduces the number of clicks that bidder B is allocated by $100 - 70 = 30$. Since bidder B's value per click is $v_B = \$8$, we can say that bidder A's externality on bidder B is $\$8 \times 30 = \$240$.
  - *Bidder C.* If bidder A was absent, bidder C would win slot 2, with 70 clicks. Since he values each click at $v_C = \$5$, bidder A's externality on C is $\$5 \times 70 = \$350$.

  Overall, the sum of the above two externalities on bidders B and C is

$$\$240 + \$350 = \$590,$$

which implies that the seller's total revenue in this case is $\$350$ from bidder B and $\$590$ from bidder A, adding up to $\$940$. This amount, however, falls below $\$1{,}150$, what the seller would earn by running the GSP of part (a).

(c) The results in parts (a) and (b) suggest that GSP generates a larger revenue for the seller if bidders submit their bids truthfully. Show, however, that truthtelling is not a dominant strategy in the GSP. [*Hint*: Consider bidder A finding a profitable deviation from $b_A = \$10$.]

Exercise #9.6: VCG Mechanism and the Generalized Second-Price Auction   249

- If all players bid truthfully, bidder A's payoff when receiving slot 1 is

$$(v_A - p) \times 100 = (\$10 - \$8) \times 100 = \$200.$$

However, if he unilaterally deviates to a lower bid, such as $7, bids become $b_B = \$8$, $b_A = \$7$, and $b_C = \$5$, implying that bidder B submits the highest bid, $b_B = \$8$, winning slot 1; bidder A submits the second-highest bid, winning slot 2; and bidder C submits the lowest bid, winning no slot. In this setting, bidder A's payoff becomes

$$(v_A - p) \times 70 = (\$10 - \$5) \times 70 = \$350$$

which exceeds his payoff from submitting a bid that coincides with his valuation, $200. Intuitively, by shading his bid, bidder A loses 30 clicks (moving from slot 1 to 2) but captures a larger margin per click that, overall, yields a larger payoff. In summary, we found one profitable deviation from bidding $b_A = \$10$, which strictly increases bidder A's payoff, implying that truthtelling is not a dominant strategy in the GSP.
- For a more formal presentation of this type of auctions, along with several industry examples, see the book by Nisan et al. (2007), especially chapter 28, and the articles by Edelman et al. (2007) and Caragiannis et al. (2015).

(d) Assume now that bidders' valuations satisfy $v_A > v_B > v_C > 0$ and that slot 1 gets α_1 clicks, while slot 2 gets α_2, where $\alpha_1 > \alpha_2$. Reproduce parts (a)–(b) and find that the GSP yields a larger revenue than the VCG for all parameter conditions. Then argue that truthtelling is not a dominant strategy in the GSP for certain parameter conditions.

- *GSP.* If bidders bid truthfully, their bids are $b_A = v_A$, $b_B = v_B$, and $b_C = v_C$. Ranking these bids, the seller assigns slot 1 to the highest bidder (bidder A), who pays the second-highest bid, v_B; and slot 2 to the second-highest bidder (bidder B) who pays the third-highest bid, v_C. Since slot 1 (2) gets α_1 (α_2) clicks per hour, bidder A pays $v_B \alpha_1$ and bidder B pays $v_C \alpha_2$ to the seller, yielding a total revenue of $v_B \alpha_1 + v_C \alpha_2$.
- *VCG.* Bidder B in this setting still pays v_C per click (the externality that he imposes on bidder C, who does not win a slot because of bidder B's presence in the auction). Therefore, bidder B's payment is still $v_C \alpha_2$.
- However, bidder A's price per click now changes, as we have to measure the externality that he imposes on bidders B and C. Specifically:
    - *Bidder B.* If bidder A was absent, bidder B would win slot 1, with α_1 clicks, rather than slot 2, with only α_2 clicks. In other words, bidder A's presence reduces by $\alpha_1 - \alpha_2$ the number of clicks that bidder B is allocated. Since bidder B's value per click is v_B, we can say that bidder A's externality on bidder B is $v_B(\alpha_1 - \alpha_2)$.

- *Bidder C*. If bidder A was absent, bidder C would win slot 2, with α_2 clicks. Since he values each click at v_C, bidder A's externality on C is $v_C \alpha_2$.

  Overall, the sum of the above two externalities on bidders B and C is

  $$v_B(\alpha_1 - \alpha_2) + 2v_C\alpha_2.$$

- *Revenue comparison.* Comparing the total revenue that the seller earns in the GSP, $v_B\alpha_1 + v_C\alpha_2$, and that in the VCG, $v_B(\alpha_1 - \alpha_2) + 2v_C\alpha_2$, we find that

  $$v_B\alpha_1 + v_C\alpha_2 > v_B(\alpha_1 - \alpha_2) + 2v_C\alpha_2$$

  holds since this inequality simplifies to $v_B > v_C$ that is satisfied. In summary, the GSP generates a larger total revenue than the VCG when bidder B has a higher valuation than bidder C and for any clicks that each of the two slots get.

- *Truthtelling.* If all players bid truthfully, bidder A's payoff when receiving slot 1 is $(v_A - v_B)\alpha_1$ since his valuation is v_A but pays a price v_B. However, if he unilaterally deviates to a lower bid, such as $b_A = v'_A$, where $v_C < v'_A < v_B$, bids become $b_B = v_B$, $b_A = v'_A$ and $b_C = v_C$, implying that bidder B submits the highest bid, $b_B = v_B$, winning slot 1; bidder A submits the second-highest bid, winning slot 2; and bidder C submits the lowest bid, winning no slot. In this setting, bidder A's payoff becomes

  $$(v_A - v_C)\alpha_2$$

  which exceeds his payoff from submitting a bid that coincides with his valuation, $(v_A - v_B)\alpha_1$, if

  $$(v_A - v_C)\alpha_2 > (v_A - v_B)\alpha_1$$

  or, after rearranging,

  $$\frac{v_A - v_C}{v_A - v_B} > \frac{\alpha_1}{\alpha_2}.$$

  The left-hand side represents the margin that bidder A earns when deviating to a bid $b_A = v'_A$, $v_A - v_C$, relative to the margin he earns when telling the truth, $v_A - v_B$. The right-hand side measures how more attractive is slot 1 relative to slot 2. When the margin increases, as captured by the left-hand ratio, exceeds the appeal of slot 1, as measured by the right-hand ratio, bidder A has incentives to deviate from truthtelling. Otherwise, he does not strictly improve his payoff.

Procurement Auctions 10

Keywords

Procurement auctions · Town mayor · Water distribution · Water production · Reservation utility · Welfare function · Procurer · Contract · Distortionary taxes · Shadow cost · Public funds · Partial derivative · Quasilinear utility function · Voluntary participation · Individual rationality · Single-crossing property · Spence-Mirrlees sorting condition · Unconstrained problem · Marginal virtual cost · Information rent · Truthful revelation · No distortion at the top · External effects · Positive externalities · Negative externalities · Perfect monitoring · Imperfect monitoring · Monitoring cost · Monitoring probability · Conservation project · Migratory birds · Biodiversity value · Acreage of farmland · Know-how · Monotonicity · Under-investment · Flat rate payment scheme

Introduction

This chapter examines procurement auctions, such as those that utility companies use to deliver water or garbage collection services, under different settings. In these auctions, firms compete to be awarded the service delivery contract, privately observe their costs, and bid to a social planner who seeks to award the contract to the most efficient firm (lowest cost). First, Exercise 10.1 studies the standard procurement auction where firms' decisions do not produce any external effects on their rivals' costs. As a benchmark, this exercise assumes that bidder's efficiency is observable, so the procurer (e.g., government agency) assigns output levels among bidders in a context of complete information. Exercise 10.2, however, considers that each bidder privately observes his efficiency and analyzes the inefficiencies that arise due to the procurer's incomplete information.

Exercise 10.3 then allows for these external effects and compares our results with those in Exercise 10.2. Finally, Exercise 10.4 (10.5) considers a procurement auction where the social planner can perfectly (imperfectly) observe the output decisions of the winning firm once the contract has been awarded, showing how imperfect monitoring affects the initial decision to allocate the contract among firms.

Exercise #10.1: Procurement Auctions Under Complete Information[A]

10.1 Consider a town mayor inviting N firms to bid in a procurement contract that will allocate to the selected firm the right of water distribution for town residents. The efficiency in implementing the project is $\theta_i \in [0, 1]$, so bidders are regarded as more efficient when their efficiency parameter, θ_i, increases. In this exercise, we consider that all players can observe every bidder's efficiency, while in the next exercise we relax this assumption, allowing bidder i to privately observe his efficiency parameter.

The cost of bidder i to implement the contract is $C_i(q_i, \theta_i)$, which is increasing and convex in output q_i, decreasing and convex in bidder i's efficiency θ_i, and satisfies $\frac{\partial^2 C_i(q_i, \theta_i)}{\partial q_i \partial \theta_i} \leq 0$. Each bidder has a quasilinear utility function,

$$U(q_i, \theta_i) = t_i(q_i) - C_i(q_i, \theta_i),$$

where $t_i(q_i)$ represents the transfer that the bidder receives from the procurer when the bidder produces q_i units of output (e.g., gallons of water). For simplicity, assume that bidders earn a zero reservation utility if they choose to not participate in the auction.

The procurer's welfare function is

$$V(q_i) - (1 + \lambda) t_i(q_i)$$

where $V(q_i)$ denotes the value that the procurer assigns to q_i units of output, while λ captures the shadow cost of raising public funds (as the procurer needs to raise distortionary taxes in order to pay the transfer $t_i(q_i)$ to bidder i).

(a) Interpret the sign of the cross-partial derivative, $\frac{\partial^2 C_i(q_i, \theta_i)}{\partial q_i \partial \theta_i}$.

- This negative cross-partial derivative means that the more efficient the bidder i is, the lower is his marginal cost of production. That is, $\frac{\partial C_i(q_i, \theta_i)}{\partial q_i}$ decreases in θ_i.

(b) Set up the procurer's program that induces participation of the bidders.

- The procurer chooses the output–transfer pair, (q_i, t_i), for each bidder i to maximize

Exercise #10.1: Procurement Auctions Under Complete Information

$$\max_{q_i, t_i(q_i)} V(q_i) - (1+\lambda) t_i(q_i)$$

subject to the individual rationality condition

$$U_i(q_i(\theta_i), \theta_i) \geq 0 \text{ for all } \theta_i \in [0, 1]$$

Intuitively, the procurer maximizes social welfare generated from all bidders, subject to the voluntary participation of bidders of all types θ_i.

(c) Solve for the socially optimal output of bidder i.

- Using bidder i's utility function, $U(q_i, \theta_i) = t_i(q_i) - C_i(q_i, \theta_i)$, we can solve for transfer $t_i(q_i)$ to obtain $t_i(q_i) = C_i(q_i, \theta_i)$ since the individual rationality condition must be binding, i.e., $U_i(q_i(\theta_i), \theta_i) = 0$. Inserting it into the procurer's objective function yields

$$\max_{q_i \geq 0} V(q_i) - (1+\lambda) C_i(q_i, \theta_i)$$

which has only one choice variable, q_i. Differentiating with respect to q_i, we obtain

$$\frac{\partial V(q_i^{CI})}{\partial q_i} - (1+\lambda) \frac{\partial C_i(q_i^{CI}, \theta_i)}{\partial q_i} = 0$$

where q_i^{CI} denotes the socially optimal output under complete information. Rearranging this first-order condition yields

$$\underbrace{\frac{\partial V(q_i^{CI})}{\partial q_i}}_{MB_i} = \underbrace{(1+\lambda) \frac{\partial C_i(q_i^{CI}, \theta_i)}{\partial q_i}}_{MC_i}$$

Intuitively, the procurer increases water production until its marginal benefit (MB_i, in the left-hand side of the above equality) coincides with its associated marginal cost (MC_i, in the right-hand side). Since benefit function $V(q_i)$ is increasing and concave, its derivative lies in the positive quadrant but decreases in q_i, as depicted in Fig. 10.1. Similarly, because the production cost $C_i(q_i, \theta_i)$ is increasing and convex, its derivative lies in the positive quadrant and increases in q_i. The crossing point between the marginal benefit and cost functions entails $MB_i = MC_i$, yielding a socially optimal output.

If water produces a larger marginal benefit, the MB_i function shifts upward, increasing the socially optimal output q_i^{CI}. In contrast, an increase in the marginal cost of production, $\frac{\partial C_i(q_i, \theta_i)}{\partial q_i}$, or in the shadow

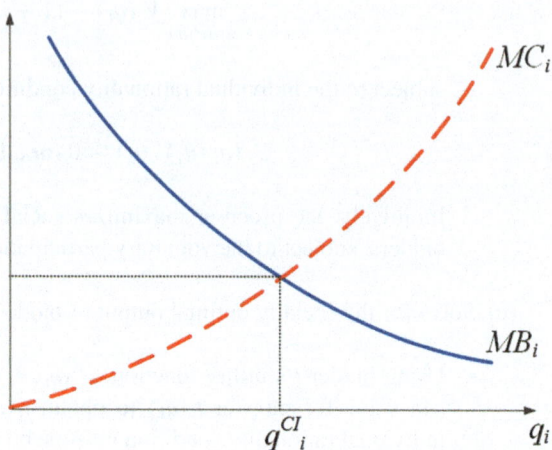

Fig. 10.1 Optimal output under complete information

cost of raising public funds, λ, yields an upward shift in the MC_i function, ultimately reducing the socially optimal output q_i^{CI} that the procurer implements.

(d) *Parametric example.* Let us now assume a parametric form for the value and cost functions in a setting with two bidders. In particular, assume that the cost function of bidder i is

$$C_i(q_i, \theta_i) = \frac{q_i^2}{1 + 2\theta_i}$$

Furthermore, the value that the procurer assigns to the output of bidder i is $V(q_i) = q_i$, and $\lambda = \frac{1}{10}$. Solve for the optimal output and transfer of bidder i.

- In this setting, the marginal benefit is $\frac{\partial V(q_i^*)}{\partial q_i} = 1$, the marginal cost of production is given by $\frac{\partial C_i}{\partial q_i} = \frac{2q_i}{1+2\theta_i}$, and the single-crossing property holds because

$$\frac{\partial^2 C_i}{\partial q_i \partial \theta_i} = -\frac{4q_i}{(1+2\theta_i)^2} < 0,$$

that is, bidder i's marginal cost decreases in its efficiency parameter θ_i. The optimal output solves $MB_i = MC_i$, which in this parametric setting entails

$$1 = \left(1 + \frac{1}{10}\right)\frac{2q_i}{1 + 2\theta_i}$$

Exercise #10.1: Procurement Auctions Under Complete Information

Simplifying the above expression yields

$$5 = \frac{11 q_i}{1 + 2\theta_i}.$$

- Solving for output q_i, we obtain the socially optimal output under complete information

$$q_i^{CI} = \frac{5(1 + 2\theta_i)}{11}$$

which is increasing in bidder i's efficiency parameter, θ_i, as depicted in Fig. 10.2.

- Substituting q_i^{CI} into the transfer function, we obtain

$$t_i^{CI} = C_i(q_i, q_j, \theta_i)$$

$$= \frac{1}{1 + 2\theta_i} \left[\frac{5(1 + 2\theta_i)}{11} \right]^2$$

$$= \frac{25(1 + 2\theta_i)}{121}.$$

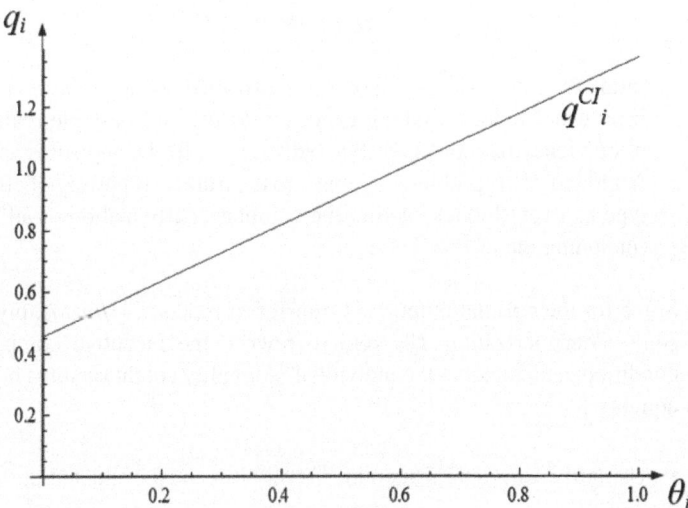

Fig. 10.2 Socially optimal output, q_i^{CI}, as a function of θ_i

Exercise #10.2: Procurement Auctions Under Incomplete Information[B]

10.2 Consider the procurement auction in Exercise 10.1, but assume that every bidder i's efficiency of implementing the project, θ_i, is privately observable by bidder i. Efficiency θ_i is uniformly distributed, $U[0, 1]$, which is common knowledge among all players.

(a) Set up the procurer's program that induces participation and revelation of the bidders.

- The procurer chooses the output–transfer pair, (q_i, t_i), for each bidder i to maximize

$$\max_{\{q_i, t_i(q_i)\}_{i=1}^N} \sum_{i=1}^N E_{\theta_i}\left[V(q_i) - (1+\lambda) t_i(q_i)\right]$$

subject to incentive compatibility:

$$U_i(q_i(\theta_i), \theta_i) \geq U_i\left(q_i(\widehat{\theta}_i), \widehat{\theta}_i\right) \quad \text{for every } \theta_i, \text{ where } \widehat{\theta}_i \neq \theta_i \tag{IC_i}$$

and individual rationality:

$$U_i(q_i(0), 0) \geq 0 \tag{IR_i}$$

Intuitively, the procurer maximizes expected social welfare generated from every bidder i, taking expectations over all possible realizations of efficiency parameter for this bidder, $\theta_i \in [0, 1]$, and sums across all N bidders. This problem is subject to the truthful reporting of efficiency type by every bidder i and to the voluntary participation of all bidders (including the least efficient one).

(b) Solve for the optimal output and transfer of bidder i. [*Hint*: Apply Myerson's Characterization Theorem to rewrite the incentive compatibility condition, and note that the individual rationality condition must hold with equality.]

- Using the Envelope Theorem, $\frac{\partial U_i(q_i(\widehat{\theta}_i), \widehat{\theta}_i)}{\partial q_i(\widehat{\theta}_i)} = 0$ evaluated at $\widehat{\theta}_i = \theta_i$, we obtain

$$dU_i(q_i, \theta_i) = -\frac{\partial C_i(q_i, \theta_i)}{\partial \theta_i} d\theta_i$$

Exercise #10.2: Procurement Auctions Under Incomplete Information[B]

Applying Myerson's Characterization Theorem (Myerson 1981) to the incentive compatibility condition (IC_i), we obtain the following differential equation for all possible realizations of the efficiency parameter $\theta_i \in [0, 1]$.

$$U_i(q_i, \theta_i) = U_i(q_i(0), 0) - \int_0^{\theta_i} \frac{\partial C_i(q_i, \widetilde{\theta}_i)}{\partial \theta_i} d\widetilde{\theta}_i$$

In addition, the individual rationality condition (IR_i) must hold with equality. Otherwise, the procurer could further reduce bidder i's residual utility and still induce his participation. When IR_i binds, $U_i(q_i(0), 0) = 0$, so that

$$U_i(q_i, \theta_i) = - \int_0^{\theta_i} \frac{\partial C_i(q_i, \widetilde{\theta}_i)}{\partial \theta_i} d\widetilde{\theta}_i \tag{10.1}$$

Substituting expression (10.1) into the utility function of bidder i yields

$$- \int_0^{\theta_i} \frac{\partial C_i(q_i, \widetilde{\theta}_i)}{\partial \theta_i} d\widetilde{\theta}_i = t_i(q_i) - C_i(q_i, \theta_i)$$

Solving for the transfer of bidder i, we have

$$t_i(q_i) = C_i(q_i, \theta_i) - \int_0^{\theta_i} \frac{\partial C_i(q_i, \widetilde{\theta}_i)}{\partial \theta_i} d\widetilde{\theta}_i$$

- Inserting this transfer, the welfare maximization program of the procurer simplifies to the following unconstrained problem:

$$\max_{\{q_i\}_{i=1}^N} \sum_{i=1}^N E_{\theta_i} \left[V(q_i) - (1+\lambda) \underbrace{\left(C_i(q_i, \theta_i) - \int_0^{\theta_i} \frac{\partial C_i(q_i, \widetilde{\theta}_i)}{\partial \theta_i} d\widetilde{\theta}_i \right)}_{t_i(q_i)} \right]$$

Conducting integration by parts on the right-hand side of the integral yields

$$E_{\theta_i} \left[\int_0^{\theta_i} \frac{\partial C_i(q_i, \widetilde{\theta}_i)}{\partial \theta_i} d\widetilde{\theta}_i \right]$$

$$= \int_0^1 \int_0^{\theta_i} \frac{\partial C_i(q_i, \widetilde{\theta}_i)}{\partial \theta_i} d\widetilde{\theta}_i d\theta_i$$

$$= \left[\theta_i \int_0^{\theta_i} \frac{\partial C_i(q_i, \widetilde{\theta}_i)}{\partial \theta_i} d\widetilde{\theta}_i \right]_0^1 - \int_0^1 \theta_i \frac{\partial C_i(q_i, \theta_i)}{\partial \theta_i} d\theta_i$$

$$= \int_0^1 (1-\theta_i) \frac{\partial C_i(q_i,\theta_i)}{\partial \theta_i} d\theta_i$$

$$= E_{\theta_i}\left[(1-\theta_i)\frac{\partial C_i(q_i,\theta_i)}{\partial \theta_i}\right]$$

Then, the welfare maximization program of the procurer can be simplified to

$$\max_{\{q_i\}_{i=1}^N} \sum_{i=1}^N E_{\theta_i}\left[V(q_i) - (1+\lambda)\left(C_i(q_i,\theta_i) - (1-\theta_i)\frac{\partial C_i(q_i,\theta_i)}{\partial \theta_i}\right)\right]$$

The procurer maximizes welfare by taking first-order condition with respect to q_i.

$$\underbrace{\frac{\partial V(q_i^*)}{\partial q_i}}_{MB_i} = (1+\lambda)\underbrace{\left[\frac{\partial C_i(q_i^*,\theta_i)}{\partial q_i} - \overbrace{(1-\theta_i)\frac{\partial^2 C_i(q_i^*,\theta_i)}{\partial q_i \partial \theta_i}}^{\text{Information rent}}\right]}_{MVC_i}$$

In words, the procurer increases water production until the point at which its marginal benefit (MB_i, in the left-hand side of the above equality) coincides with its associated marginal virtual cost (MVC_i, in the right-hand side). This cost embodies not only firm i's marginal production cost (first term on the right-hand side) but also the information rent that the procurer needs to provide in order to induce bidder i report his type truthfully (last term on the right-hand side).

- From the welfare maximization program above, we can evaluate bidder i's transfer at the optimal output q_i^*, to obtain the optimal transfer to bidder i as follows:

$$t_i(q_i^*) = C_i(q_i^*,\theta_i) - (1-\theta_i)\frac{\partial C_i(q_i^*,\theta_i)}{\partial \theta_i}.$$

(c) *Comparison.* Compare your results against those in the complete information setting of Exercise 10.1. Interpret.

- Under a complete information setting, the last term in MVC_i (information rent) was absent, as shown in Exercise 10.1. Figure 10.3 superimposes MVC_i on Fig. 10.1 to facilitate our comparison. Since the cross-partial derivative $\frac{\partial^2 C_i(q_i^*,\theta_i)}{\partial q_i \partial \theta_i}$ is negative, we obtain that $MVC_i \geq MC_i$, as depicted in the figure.

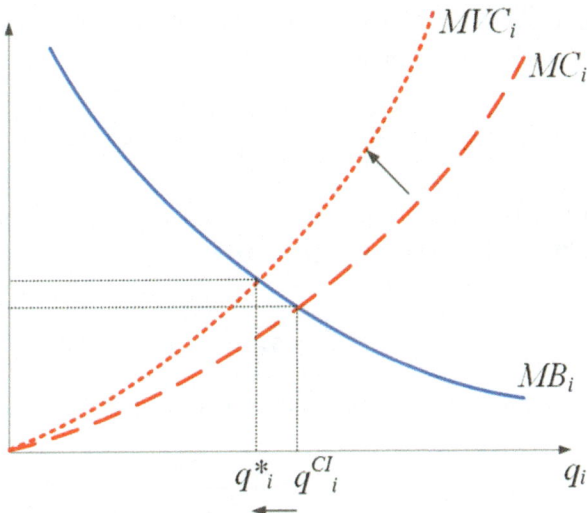

Fig. 10.3 Socially optimal output under complete and incomplete information

- Therefore, the socially optimal output under complete information is larger than that under incomplete information, $q_i^{CI} \geq q_i^*$. Intuitively, the procurer must pay an information rent to all bidders to induce truthful revelation of their types, incurring more costs to implement the auction than under complete information, ultimately inducing lower output levels. This is commonly referred in the literature as "downward distortion" for all bidders with efficiency levels $\theta_i \neq 1$.
- However, the output of the bidder with the highest efficiency, $\theta_i = 1$, suffers no distortion when moving from a complete to an incomplete information context. Indeed, MVC_i simplifies to MC_i when evaluated at $\theta_i = 1$, so first-order conditions across information contexts coincide, and $q_i^{CI} = q_i^*$. Intuitively, the most efficient bidder has no incentives to underreport his valuation at $\theta_i = 1$. This result is known as "no distortion at the top."

(d) *Parametric example.* Consider the same parametric forms as in Exercise 10.1(d). Solve for the optimal output and transfer of bidder i. Compare your results with those in Exercise 10.1(d).

- The optimal output solves $MB_i = MVC_i$, which in this parametric setting entails

$$1 = \left(1 + \frac{1}{10}\right)\left[\frac{2q_i}{1 + 2\theta_i} + (1 - \theta_i)\frac{4q_i}{(1 + 2\theta_i)^2}\right]$$

since $\frac{\partial V(q_i^*)}{\partial q_i} = 1$, $\lambda = \frac{1}{10}$, marginal cost is given by $\frac{\partial C_i}{\partial q_i} = \frac{2q_i}{1+2\theta_i}$, and the single-crossing property holds because

$$\frac{\partial^2 C_i}{\partial q_i \partial \theta_i} = -\frac{4q_i}{(1+2\theta_i)^2} < 0,$$

that is, bidder i's marginal cost decreases in its efficiency parameter θ_i. Simplifying the above expression yields

$$1 = \frac{33 q_i}{5(1+2\theta_i)^2}.$$

- Solving for output q_i, we obtain the optimal output

$$q_i^* = \frac{5(1+2\theta_i)^2}{33}$$

which is increasing in bidder i's efficiency parameter, θ_i, as depicted in Fig. 10.4. To facilitate our comparisons, the figure also includes the socially optimal output under complete information, $q_i^{CI} = \frac{5(1+2\theta_i)}{11}$, which clearly lies above q_i^* for all efficiency levels $\theta_i \neq 1$ but coincides at exactly $\theta_i = 1$ (no distortion at the top), since $q_i^{CI} \geq q_i^*$ entails

$$\frac{5(1+2\theta_i)}{11} \geq \frac{5(1+2\theta_i)^2}{33}$$

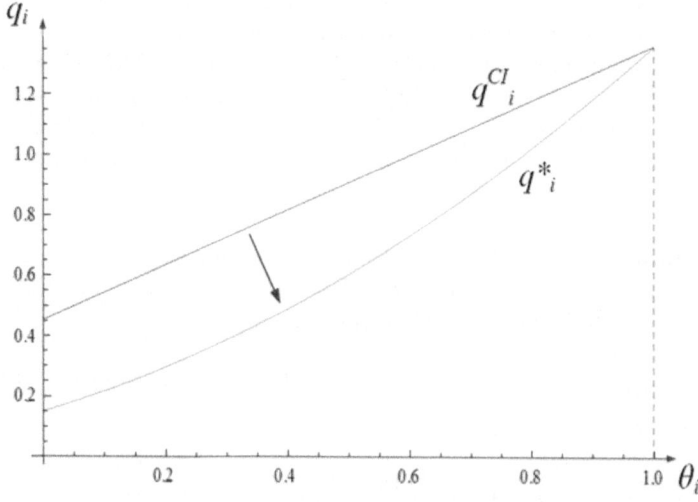

Fig. 10.4 Socially optimal output under incomplete information as a function of θ_i

simplifies to

$$3 \geq 1 + 2\theta_i$$

that holds for all $\theta \in [0, 1]$.

- Inserting this optimal level into the transfer function $t_i(q_i)$ yields

$$\begin{aligned}
t_i^* &= C_i(q_i, \theta_i) - (1 - \theta_i) \frac{\partial C_i(q_i, \theta_i)}{\partial \theta_i} \\
&= \frac{q_i^2}{1 + 2\theta_i} + (1 - \theta_i) \frac{2q_i^2}{(1 + 2\theta_i)^2} \\
&= \frac{3q_i^2}{(1 + 2\theta_i)^2} \\
&= \frac{25(1 + 2\theta_i)^2}{363}
\end{aligned}$$

which is increasing in bidder i's efficiency parameter, θ_i, but falls below the transfer under complete information, t_i^{CI}, since

$$\frac{25(1 + 2\theta_i)^2}{363} < \frac{25(1 + 2\theta_i)}{121}$$

simplifies to $3 > 1 + 2\theta_i$ that is true for all $\theta_i \in [0, 1]$. This happens since the procurer reduces the output of bidder i under incomplete information that reduces production cost more than the information rent that bidder i seeks to truthfully report his type, ultimately reducing the transfer that the procurer offers to bidder i.

Exercise #10.3: Procurement Auctions with External Effects, Based on Choi et al. (2018) ^{C}

10.3 Consider the procurement auction in Exercise 10.2 but assume that bidder i's cost depends on the output of other bidders. Specifically, consider $C_i(q_i, q_{-i}, \theta_i)$, where vector $q_{-i} = (q_1, \cdots, q_{i-1}, q_{i+1}, \cdots, q_N)$ can be understood as the externalities that other bidders impose on bidder i, which can be positive, if bidder i's costs satisfy $\frac{\partial C_i(q_i, q_{-i}, \theta_i)}{\partial q_{-i}} < 0$, or negative, if his costs satisfy $\frac{\partial C_i(q_i, q_{-i}, \theta_i)}{\partial q_{-i}} > 0$.

(a) Write down the welfare maximization program of the procurer. You may assume that the IC_i and IR_i conditions still hold so that you need not repeat them again.

- The procurer's problem is given by

$$\max_{\{q_i\}_{i=1}^N} \sum_{i=1}^N E_{\theta_i} \left[V(q_i) - (1+\lambda)\left(C_i(q_i, q_{-i}, \theta_i)\right.\right.$$
$$\left.\left. - (1-\theta_i) \frac{\partial C_i(q_i, q_{-i}, \theta_i)}{\partial \theta_i}\right)\right]$$

(b) Solve for the optimal output and transfer of bidder i.

- The procurer maximizes welfare by taking first-order condition with respect to q_i, which yields

$$\underbrace{\frac{\partial V(q_i^*)}{\partial q_i}}_{MB_i} = \underbrace{(1+\lambda) \left[\frac{\partial C_i(q_i^*, q_{-i}^*, \theta_i)}{\partial q_i} - (1-\theta_i) \frac{\partial^2 C_i(q_i^*, q_{-i}^*, \theta_i)}{\partial q_i \partial \theta_i} \right]}_{MVC_i'}$$

(10.2)

$$+ \underbrace{\sum_{j \neq i}(1+\lambda) \left[\frac{\partial C_j(q_j^*, q_{-j}^*, \theta_j)}{\partial q_i} - (1-\theta_j) \frac{\partial^2 C_j(q_j^*, q_{-j}^*, \theta_j)}{\partial q_i \partial \theta_j} \right]}_{VEC_i}$$

In other words, the procurer balances the marginal benefit, MB_i, against the marginal virtual cost of bidder i, MVC_i', and the virtual external cost bidder i imposes on other bidders, VEC_i.

- Also, from the welfare maximization program above, the optimal transfer to bidder i is

$$t_i(q_i^*) = C_i(q_i^*, q_{-i}^*, \theta_i) - (1-\theta_i) \frac{\partial C_i(q_i^*, q_{-i}^*, \theta_i)}{\partial \theta_i} \quad (10.3)$$

(c) Compare the results between Exercises 10.2(b) and 10.3(b), and describe how equilibrium output is affected. For simplicity, consider that, with positive externalities, we have that $\frac{\partial^2 C_i(q_i, q_{-i}, \theta_i)}{\partial q_{-i} \partial \theta_i} \geq 0$ such that the efficiency of bidder i (higher θ_i) attenuates the effect of positive externalities that other bidders impose on his costs. Similarly, with negative externalities, $\frac{\partial^2 C_i(q_i, q_{-i}, \theta_i)}{\partial q_{-i} \partial \theta_i} \leq 0$, so that bidder i's efficiency ameliorates the negative effects that other bidders impose on i's costs.

- We compare the optimal output of bidder i in Exercises 10.2(b) and 10.3(b) by sequentially presenting the impact of externalities as follows:

- *No Externalities.* This is the setting analyzed in Exercise 10.2(b), where the procurer balances the marginal benefit, MB_i, with the marginal virtual cost of bidder i, MVC_i, which generates the optimal output q_i^0.
- *Externalities from other bidders on bidder i.* This is the setting analyzed in Exercise 10.3(b), where the positive (negative) externalities from other bidders decrease (increase) the marginal virtual cost of bidder i, that is, $MVC_i' < MVC_i$ $(MVC_i' > MVC_i$, respectively). Therefore, the right-hand side of expression (10.2) becomes smaller (larger).

  Then, by the concavity of the value function $V(q_i)$, a higher (lower) output emerges. Intuitively, in the case of positive (negative) externalities, it becomes less (more) costly for bidder i to produce additional units of output, so he should produce more (less) output than in the absence of external effects.
- *Externalities from bidder i on other bidders.* Furthermore, if bidder i imposes positive (negative) externalities on other bidders, then the virtual cost of other bidders would decrease (increase) such that VEC_i would be negative (positive). Therefore, the right-hand side of (2) decreases (increases), yielding a higher (lower) output.

  Intuitively, if bidder i imposes a positive (negative) externality on other bidders, he makes it less (more) costly for them to produce additional units of output, implying that he should produce more (less) output, thus generating a larger (smaller) externality on its rivals.

(d) *Parametric example.* Let us now consider a similar parametric example as in Exercises 10.1 and 10.2 but with the following cost function to allow for cost externalities:

$$C_i(q_i, q_j, \theta_i) = \frac{q_i^2}{1 + 2\theta_i} - \alpha q_j \quad (10.4)$$

where parameter α captures the degree of externality that bidder j imposes on bidder i. We normalize this parameter, $\alpha \in [-1, 1]$, such that: (i) if $\alpha \in (0, 1]$, externality is positive, thus reducing the cost of bidder i; (ii) if $\alpha = 0$, no externality on bidder i; and (iii) if $\alpha \in [-1, 0)$, the externality is negative, thus increasing the cost of bidder i.

As in Exercise 10.2, continue to assume that $V(q_i) = q_i$ and $\lambda = \frac{1}{10}$. Find the optimal output and transfer of bidder i in the following three cases: (1) $\alpha = 0$, (2) $\alpha = \frac{1}{3}$, and (3) $\alpha = -\frac{1}{3}$. Compare and interpret your results.

- *No external effects.* The case where bidders' actions entail no external effects, $\alpha = 0$, coincides with Exercise 10.2(d).

- *External effects.* When $\alpha > 0$ ($\alpha < 0$), bidder j's output reduces (increases) bidder i's costs. In this context, optimal output solves $MB_i = MVC_i + VEC_i$, which in this parametric setting entails

$$\underbrace{1}_{MB_i} = \underbrace{\left(1 + \frac{1}{10}\right)\left[\frac{2q_i}{1+2\theta_i} + (1-\theta_i)\frac{4q_i}{(1+2\theta_i)^2}\right]}_{MVC_i} + \underbrace{\left(1+\frac{1}{10}\right)(-\alpha)}_{VEC_i}$$

where $\frac{\partial V(q_i^*)}{\partial q_i} = 1$, $\lambda = \frac{1}{10}$, marginal cost is given by $\frac{\partial C_i}{\partial q_i} = \frac{2q_i}{1+2\theta_i}$, and the single-crossing property holds because

$$\frac{\partial^2 C_i}{\partial q_i \partial \theta_i} = -\frac{4q_i}{(1+2\theta_i)^2} < 0,$$

and bidder j's cost decreases after one-unit increase in bidder i's output since $\frac{\partial C_j}{\partial q_i} = -\alpha$.
- Rearranging, the above expression simplifies to

$$1 = \frac{11}{10}\left(\frac{6q_i}{(1+2\theta_i)^2} - \alpha\right)$$

Solving for output q_i, we obtain that the optimal output under externalities is

$$q_i^E(\theta_i) = \frac{10+11\alpha}{66}(1+2\theta_i)^2$$

where the superscript E denotes externalities. This output is, as expected, increasing in bidder i's efficiency (as captured by the parameter θ_i). It also increases (decreases) in the positive (negative) externalities that bidder j imposes on bidder i (as captured by the parameter α), since

$$\frac{\partial q_i^E(\theta_i)}{\partial \alpha} = \frac{(1+2\theta_i)^2}{6} > 0$$

Figure 10.5 evaluates the output function at $\alpha = 0$ (no externalities), which yields the same optimal output as in Exercise 10.2, $q_i^* = \frac{5(1+2\theta_i)^2}{33}$; $q_i^{PE} = \frac{41(1+2\theta_i)^2}{198}$ under $\alpha = \frac{1}{3}$ (positive externalities, PE, in the superscript) that yields optimal output above q_i^*; and $q_i^{NE} = \frac{19(1+2\theta_i)^2}{198}$ under $\alpha = -1/3$ (negative externalities, NE, in the superscript) that yields optimal output below q_i^*.
- Inserting this optimal level into the transfer function $t_i(q_i)$ yields

Exercise #10.3: Procurement Auctions with External Effects, Based on Choi...

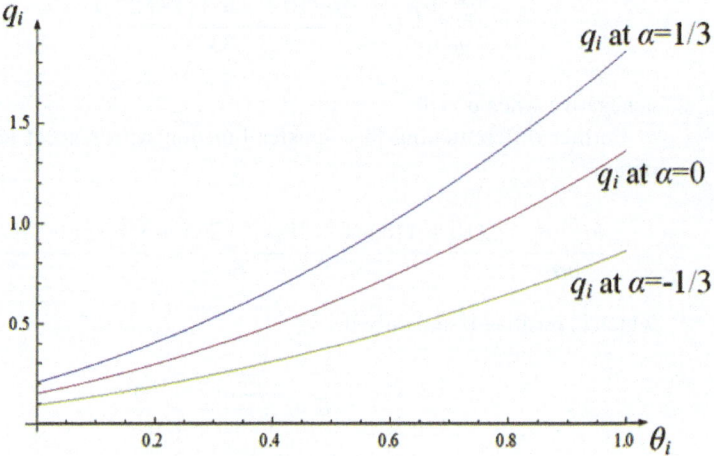

Fig. 10.5 Optimal output with and without externalities

$$t_i^E(\theta) = C_i\left(q_i^E, q_j^E, \theta_i\right) - (1 - \theta_i)\frac{\partial C_i\left(q_i^E, q_j^E, \theta_i\right)}{\partial \theta_i}$$

$$= \frac{3(q_i^E)^2}{(1 + 2\theta_i)^2} - \alpha\left(q_j^E\right).$$

Substituting $q_i^E = \frac{10 + 11\alpha}{66}(1 + 2\theta_i)^2$ and $q_j^E = \frac{10 + 11\alpha}{66}(1 + 2\theta_j)^2$, we find

$$t_i^E(\theta) = \frac{3}{(1 + 2\theta_i)^2}\left(\frac{10 + 11\alpha}{66}\right)^2(1 + 2\theta_i)^4$$

$$- \frac{\alpha(10 + 11\alpha)}{66}(1 + 2\theta_j)^2$$

$$= \frac{10 + 11\alpha}{1452}\left[(10 + 11\alpha)(1 + 2\theta_i)^2 - 22\alpha(1 + 2\theta_j)^2\right].$$

which is increasing in bidder i's efficiency θ_i because

$$\frac{\partial t_i^E(\theta)}{\partial \theta_i} = \frac{(10 + 11\alpha)^2(1 + 2\theta_i)}{363} > 0$$

but decreasing in bidder j's efficiency θ_j under positive externalities since

$$\frac{\partial t_i^E(\theta)}{\partial \theta_j} = -\frac{2\alpha(10 + 11\alpha)(1 + 2\theta_j)}{33}$$

is negative when $\alpha > 0$.

Further differentiating this transfer function with respect to α, we obtain

$$\frac{\partial t_i^E(\theta)}{\partial \alpha} = \frac{(10 + 11\alpha)(1 + 2\theta_i)^2 - 2(5 + 11\alpha)(1 + 2\theta_j)^2}{66}$$

which is positive if and only if

$$\frac{1 + 2\theta_i}{1 + 2\theta_j} > \overline{\alpha} \equiv \sqrt{\frac{2(5 + 11\alpha)}{10 + 11\alpha}}$$

where

$$\frac{d\overline{\alpha}}{d\alpha} = \frac{55(10 + 11\alpha)^{-\frac{3}{2}}}{\sqrt{2(5 + 11\alpha)}} > 0$$

indicating that as positive externality increases, bidder i seeks a larger transfer from the procurer when the first-order effect of output increase more than offsets the second-order effect of cost decrease of producing all the inframarginal units, which happens when bidder i is more efficient relative to bidder j, i.e., $\frac{1+2\theta_i}{1+2\theta_j} > \overline{\alpha}$.

Accordingly, transfer under positive externalities, evaluated at $\alpha = 1/3$, becomes

$$t_i^{PE}(\theta) = \frac{10 + \frac{11}{3}}{1452} \left[\left(10 + \frac{11}{3}\right)(1 + 2\theta_i)^2 - \frac{22}{3}(1 + 2\theta_j)^2 \right]$$

$$= \frac{41}{13068} \left[41(1 + 2\theta_i)^2 - 22(1 + 2\theta_j)^2 \right]$$

Similarly, transfer under negative externalities, evaluated at $\alpha = -1/3$, becomes

$$t_i^{NE}(\theta) = \frac{10 - \frac{11}{3}}{1452} \left[\left(10 - \frac{11}{3}\right)(1 + 2\theta_i)^2 + \frac{22}{3}(1 + 2\theta_j)^2 \right]$$

$$= \frac{19}{13068} \left[19(1 + 2\theta_i)^2 + 22(1 + 2\theta_j)^2 \right]$$

- Let us finally compare optimal output and transfers in Exercises 10.2 and 10.3 (without and with externalities, respectively):

- For the optimal output, we obtain that $q_i^{PE}(\theta_i) > q_i^* > q_i^{NE}(\theta_i)$ for all $\theta_i, \theta_j \in [0, 1]$ because the procurer asks bidder i to produce more (less) output when it reduces (increases) the production cost of bidder j.
- For the optimal transfer, we found that $t_i^{PE}(\theta) > t_i^* > t_i^{NE}(\theta_i)$ for all θ_i, θ_j satisfying $\frac{1+2\theta_i}{1+2\theta_j} > \overline{\alpha}$ when the increase (decrease) in bidder i's cost to produce more (less) output more than offsets of the decrease (increase) in bidder i's cost due to the positive (negative) externalities imposed by bidder j.

Exercise #10.4: Procurement Auctions with Perfect Monitoring[B]

10.4 Consider N bidders bidding for a procurement contract, for example, a conservation project that restores the wetland to provide a habitat for migratory birds. Each bidder i is endowed with a certain acreage of farmland available for conservation, where $i \in \{1, \cdots, N\}$. We summarize the expertise of bidder i in implementing the conservation project and the biodiversity value of his farmland into a unidimensional measure of efficiency, θ_i, which is observable to bidder i but not to the other bidders or the procurer. This efficiency parameter is uniformly distributed, $\theta_i \sim U[0, 1]$, and is common knowledge among bidders.

Bidder i's cost of implementing the project is determined by the acreage of farmland dedicated to wetland conservation q_i (herein denoted as the "input") and his efficiency θ_i, that is,

$$C_i(q_i, \theta_i) = \frac{(q_i)^2}{2(1+\theta_i)}$$

where the cost of conserving zero units of land is $C_i(0, \theta_i) = 0$, for we assume away fixed costs. The cost of conservation is decreasing in the efficiency parameter, θ_i.

Each bidder i has a quasilinear utility function,

$$U(q_i, \theta_i) = t_i(q_i) - C_i(q_i, \theta_i),$$

where $t_i(q_i)$ represents the transfer he receives from the procurer. For simplicity, we assume a zero reservation utility for bidder i. The procurer's valuation of bidder i's conservation is $V(q_i) = A \ln q_i$, where $A > 0$ indicates the intensity of the procurer's valuation.

(a) First, we examine the properties of the cost function $C_i(q_i, \theta_i)$. Show that conservation cost is: (1) increasing and convex in the input q_i; (2) decreasing and convex in efficiency θ_i; and (3) satisfying the Spence–Mirrlees sorting condition (also known as the single-crossing condition).

- We first check that $C_i(q_i, \theta_i)$ is increasing and convex in the land that the bidder conserves

$$\frac{\partial C_i(q_i, \theta_i)}{\partial q_i} = \frac{q_i}{1+\theta_i} \geq 0$$

$$\frac{\partial^2 C_i(q_i, \theta_i)}{\partial q_i^2} = \frac{1}{1+\theta_i} \geq 0$$

Therefore, conservation cost is increasing at an increasing rate in input q_i. Intuitively, the more acres of farmland to be converted, the more expensive it is to use the less efficient resources at the margin.
- We next show that the conservation cost is decreasing and convex in efficiency θ_i.

$$\frac{\partial C_i(q_i, \theta_i)}{\partial \theta_i} = -\frac{1}{2}\left(\frac{q_i}{1+\theta_i}\right)^2 \leq 0$$

$$\frac{\partial^2 C_i(q_i, \theta_i)}{\partial \theta_i^2} = \frac{(q_i)^2}{(1+\theta_i)^3} \geq 0$$

Therefore, conservation cost is decreasing at a decreasing rate in the efficiency parameter θ_i. In words, the more efficient bidder i becomes, the less costly it is for him to apply his know-how in converting his farmland into a wetland. However, the contribution of his know-how at the margin diminishes as he becomes more experienced.
- Last, we evaluate the Spence–Mirrlees sorting condition. In particular, this property holds since

$$\frac{\partial^2 C_i(q_i, \theta_i)}{\partial q_i \partial \theta_i} = -\frac{q_i}{(1+\theta_i)^2} \leq 0$$

In words, the marginal cost of converting land decreases as the bidder becomes more efficient (higher θ_i).

(b) Second, we investigate the properties of the procurer's valuation function $V(q_i)$. Show that the valuation function is increasing and concave in input q_i.

- We confirm that $V(q_i)$ is increasing and concave in q_i,

$$\frac{\partial V(q_i)}{\partial q_i} = \frac{A}{q_i} \geq 0$$

Exercise #10.4: Procurement Auctions with Perfect Monitoring[B]

$$\frac{\partial^2 V(q_i)}{\partial q_i^2} = -\frac{A}{q_i^2} \leq 0$$

Therefore, output is increasing in q_i at a decreasing rate. Intuitively, the more acres of farmland to be converted, the higher will be the total conservation benefits generated, but the marginal benefit diminishes as additional acres of farmland are converted.

(c) *Case 1. Perfect monitoring and observable efficiency.* In Case 1, assume that both efficiency and input are observable. Write down the individual rationality constraint for bidder i to participate. Let λ, where $\lambda \geq 0$, be the shadow cost of raising public funds. What is the procurer's welfare function? Solve for the optimal input and transfer of bidder i.

- The procurer's welfare from the land conserved by bidder i is the procurer's valuation from this land minus the transfer paid to bidder i, that is,

$$W_i(q_i) = V_i(q_i) - (1+\lambda) t_i(q_i)$$

Hence, the procurer maximizes aggregate welfare by choosing the optimal profile of land and transfer (q_i, t_i) for each bidder i that solves

$$\max_{\{q_i, t_i\}_{i=1}^N} \sum_{i=1}^N W_i(q_i)$$

subject to individual rationality:

$$U_i(q_i(\theta_i), \theta_i) \geq 0 \quad \text{for all } \theta_i \in [0,1] \qquad (IRC_i^1)$$

and

$$\frac{\partial q_i(\theta_i)}{\partial \theta_i} \geq 0. \qquad \text{(Monotonicity)}$$

Note that we do not need to impose an incentive compatibility condition, since the procurer observes every bidder's efficiency θ_i in this setting.
- Since the procurer can reduce the residual utility of bidder i while keeping his participation, the individual rationality condition IRC_i^1 holds with equality, that is, $U_i(q_i(\theta_i), \theta_i) = 0$ such that

$$t_i(q_i) = C_i(q_i, \theta_i)$$

which compensates bidder i exactly for the conservation cost he incurs. Therefore, the procurer's welfare maximization problem is simplified to

$$\max_{\{q_i\}_{i=1}^N} \sum_{i=1}^N [V(q_i) - (1+\lambda) C_i(q_i, \theta_i)]$$

$$= \sum_{i=1}^N \left[A \ln q_i - \frac{1+\lambda}{1+\theta_i} \frac{(q_i)^2}{2} \right]$$

subject to the monotonicity condition. (As in similar problems, we next ignore the monotonicity constraint and later check that it holds once we find the optimal input and transfer, q_i and t_i.)

- Taking first-order condition with respect to q_i, we obtain

$$\frac{A}{q_i} = \frac{1+\lambda}{1+\theta_i} q_i$$

Simplifying the above expression, the optimal input in Case 1, q_i^1, is

$$q_i^1 = \sqrt{\frac{A(1+\theta_i)}{1+\lambda}}$$

Substituting the optimal input, q_i^1, into the transfer function, the optimal transfer, t_i^1, becomes

$$t_i^1 = C_i\left(q_i^1, \theta_i\right) = \frac{1}{2(1+\theta_i)} \frac{A(1+\theta_i)}{1+\lambda} = \frac{A}{2(1+\lambda)}$$

Last, we need to check that the above optimal input satisfies the (so far ignored) monotonicity constraint, as follows:

$$\frac{\partial q_i^1}{\partial \theta_i} = \frac{1}{2} \sqrt{\frac{A}{(1+\lambda)(1+\theta_i)}} \geq 0$$

Hence, we verify that the monotonicity constraint holds.

(d) *Case 2. Perfect monitoring, unobservable efficiency.* In Case 2, assume that input is observable but efficiency is not. That is, the procurer can perfectly observe the amount of land being conserved by every bidder but cannot observe each bidder's efficiency parameter. Write down the procurer's problem in this setting, and solve for the optimal input and transfer of bidder i.

Exercise #10.4: Procurement Auctions with Perfect Monitoring[B]

- Since in this context efficiency is unobservable, the procurer needs to induce bidder i to truthfully report his efficiency parameter θ_i rather than a different efficiency $\widehat{\theta}_i \neq \theta_i$ for all $\theta_i, \widehat{\theta}_i \in [0, 1]$, that is,

$$U_i(q_i(\theta_i), \theta_i) \geq U_i\left(q_i(\widehat{\theta}_i), \widehat{\theta}_i\right)$$

In addition, the procurer needs to induce the least efficient bidder i, with parameter $\theta_i = 0$, to participate by ensuring him a participation utility that does not fall below his reservation utility, such that all other bidders have incentives to participate, that is,

$$U_i(q_i(0), 0) \geq 0$$

We can now write down the procurer's problem. Since the procurer does not observe the realization of efficiency profile $\theta \equiv (\theta_1, \ldots, \theta_N) \in \Theta$, he takes the expected welfare that each bidder i generates, $E_\theta[W_i(q_i)]$, and sums over all bidders, choosing (q_i, t_i) for each bidder i to solve

$$\max_{\{q_i, t_i\}_{i=1}^N} \sum_{i=1}^N EW(q)$$

$$= \sum_{i=1}^N E_\theta[W_i(q_i)]$$

subject to

$$U_i(q_i(\theta_i), \theta_i) \geq U_i\left(q_i(\widehat{\theta}_i), \widehat{\theta}_i\right) \qquad (BIC_i^2)$$

$$U_i(q_i(0), 0) \geq 0 \qquad (IRC_i^2)$$

and monotonicity.

- We first simplify our above problem. First, note that, as in Case 1, since the procurer can reduce the residual utility of the least efficient bidder i while keeping his participation, IRC_i^2, holds with equality, we have $U_i(q_i(0), 0) = 0$. Next, applying the Envelope Theorem to IRC_i^2, $\frac{\partial U_i(q_i(\widehat{\theta}_i), \widehat{\theta}_i)}{\partial q_i(\theta_i)} = 0$ when evaluated at $\widehat{\theta}_i = \theta_i$, yielding $dU_i(q_i, \theta_i) = -\frac{\partial C_i(q_i, \theta_i)}{\partial \theta_i} d\theta_i$. Therefore, by Myerson's Characterization Theorem (Myerson 1981), we obtain the following differential equation for all possible realizations of the efficiency parameter, $\theta_i \in [0, 1]$.

$$U_i(q_i, \theta_i) = U_i(q_i(0), 0) - \int_0^{\theta_i} \frac{\partial C_i(q_i, \widetilde{\theta}_i)}{\partial \theta_i} d\widetilde{\theta}_i = -\int_0^{\theta_i} \frac{\partial C_i(q_i, \widetilde{\theta}_i)}{\partial \theta_i} d\widetilde{\theta}_i$$

Substituting the above expression into the utility function of bidder i, we obtain

$$-\int_0^{\theta_i} \frac{\partial C_i(q_i, \tilde{\theta}_i)}{\partial \theta_i} d\tilde{\theta}_i = t_i(q_i) - C_i(q_i, \theta_i)$$

and solving for the transfer of bidder i yields

$$t_i(q_i) = C_i(q_i, \theta_i) - \int_0^{\theta_i} \frac{\partial C_i(q_i, \tilde{\theta}_i)}{\partial \theta_i} d\tilde{\theta}_i$$

which compensates bidder i not only for his actual conservation cost (first part on the right-hand side) but also the information rent to report efficiency θ_i instead of under-reporting at $\widehat{\theta}_i < \theta_i$ to exaggerate his conservation cost (second part on the right-hand side). Therefore, the procurer's welfare maximization program simplifies to

$$\max_{\{q_i\}_{i=1}^N} \sum_{i=1}^N E_\theta \left[V(q_i) - (1+\lambda) t_i(q_i) \right]$$

$$= \sum_{i=1}^N E_{\theta_i} \left[V(q_i) - (1+\lambda) \left(C_i(q_i, \theta_i) - \int_0^{\theta_i} \frac{\partial C_i(q_i, \tilde{\theta}_i)}{\partial \theta_i} d\tilde{\theta}_i \right) \right]$$

subject to monotonicity. We can further simplify the above objective function. Conducting integration by parts on the integral (last term),

$$E_{\theta_i} \left[\int_0^{\theta_i} \frac{\partial C_i(q_i, \tilde{\theta}_i)}{\partial \theta_i} d\tilde{\theta}_i \right]$$

$$= \int_0^1 \int_0^{\theta_i} \frac{\partial C_i(q_i, \tilde{\theta}_i)}{\partial \theta_i} d\tilde{\theta}_i \, dF_i(\theta_i)$$

$$= \left[\theta_i \int_0^{\theta_i} \frac{\partial C_i(q_i, \tilde{\theta}_i)}{\partial \theta_i} d\tilde{\theta}_i \right]_0^1 - \int_0^1 \theta_i \frac{\partial C_i(q_i, \theta_i)}{\partial \theta_i} d\theta_i$$

$$= \int_0^1 (1 - \theta_i) \frac{\partial C_i(q_i, \theta_i)}{\partial \theta_i} d\theta_i$$

$$= E_{\theta_i} \left[(1 - \theta_i) \frac{\partial C_i(q_i, \theta_i)}{\partial \theta_i} \right]$$

where the third line uses the fact that $F(\theta_i) = \theta_i$ for $\theta_i \sim U[0, 1]$. We can then rewrite the procurer's program to

$$\max_{\{q_i\}_{i=1}^N} \sum_{i=1}^N E_{\theta_i} \left[V(q_i) - (1+\lambda) \left(C_i(q_i, \theta_i) - (1-\theta_i) \frac{\partial C_i(q_i, \theta_i)}{\partial \theta_i} \right) \right]$$

$$= \sum_{i=1}^N E_{\theta_i} \left[A \ln q_i - (1+\lambda) \left[\frac{(q_i)^2}{2(1+\theta_i)} + (1-\theta_i) \frac{(q_i)^2}{2(1+\theta_i)^2} \right] \right]$$

$$= \sum_{i=1}^N E_{\theta_i} \left[A \ln q_i - (1+\lambda) \left(\frac{q_i}{1+\theta_i} \right)^2 \right]$$

subject to monotonicity.
- We next ignore the monotonicity constraint and later on check that it does hold once we find the optimal input and transfer, q_i and t_i. Taking first-order condition with respect to q_i,

$$\frac{A}{q_i} = \frac{2(1+\lambda)}{(1+\theta_i)^2} q_i$$

Simplifying the above expression, the optimal input in Case 2, q_i^2, is

$$q_i^2 = \sqrt{\frac{A}{2(1+\lambda)}} (1+\theta_i)$$

Substituting the optimal input, q_i^2, into the transfer function, the optimal transfer, t_i^2, is

$$t_i^2 = \tilde{C}_i\left(q_i^2, \theta_i\right) = \frac{1}{(1+\theta_i)^2} \frac{A}{2(1+\lambda)} (1+\theta_i)^2 = \frac{A}{2(1+\lambda)}$$

Last, we check the monotonicity constraint, which holds since q_i^2 is increasing in the bidder's efficiency,

$$\frac{\partial q_i^2}{\partial \theta_i} = \sqrt{\frac{A}{2(1+\lambda)}} \geq 0.$$

Exercise #10.5: Procurement Auctions with Imperfect Monitoring[C]

10.5 Consider the procurement auction analyzed in Exercise 10.4. However, we will now allow for monitoring to be imperfect, that is, the procurer cannot perfectly observe the amount of land they conserve (i.e., whether bidder i fully implements the contract by conserving q_i or a smaller amount $\widehat{q}_i < q_i$).

In this context, the procurer needs to monitor every bidder i to deter him from investing less than the contracted level of input, or not investing at all. In particular, let α_i be the probability of monitoring, where $0 \leq \alpha_i \leq 1$. For simplicity, we assume that bidder i will be detected with 100% certainty if he is monitored and will not be detected if not monitored. Hence, if bidder i cheats, he is detected with probability α_i, receiving a transfer $t_i(\widehat{q}_i)$, which is a function of his level of under investment $\widehat{q}_i$, where $\widehat{q}_i < q_i$.

Last, assume that the monitoring cost is $m_i(q_i) = \frac{\gamma}{2}(q_i)^2$, where $\gamma \in [0, 1]$, implying that the cost of monitoring is increasing and convex in q_i, that is, $\frac{\partial m_i}{\partial q_i} = \gamma q_i \geq 0$ and $\frac{\partial^2 m_i}{\partial (q_i)^2} = \gamma \geq 0$.

(a) *Case 3: Imperfect monitoring, observable of efficiency.* Consider the opposite scenario of Case 2 in Exercise 10.4, that is, the procurer can observe bidders' efficiency but cannot perfectly observe the amount of land they conserve.

1. What is the expected utility of bidder i if he cheats?

    - When bidder i is detected of cheating, his utility is $t_i(\widehat{q}_i) - C_i(\widehat{q}_i, \theta_i)$, which is the transfer he obtains from the procurer of investing at $\widehat{q}_i$ less the conservation cost. When bidder i is *not* detected of cheating, his utility is $t_i(q_i) - C_i(\widehat{q}_i, \theta_i)$, which is the transfer he obtains as if investing at q_i less the conservation cost. As a result, the expected utility of bidder i becomes

    $$E[U_i(\widehat{q}_i, q_i, \theta_i)] = \alpha_i [t_i(\widehat{q}_i) - C_i(\widehat{q}_i, \theta_i)]$$
    $$+ (1 - \alpha_i)[t_i(q_i) - C_i(\widehat{q}_i, \theta_i)]$$
    $$= \alpha_i t_i(\widehat{q}_i) + (1 - \alpha_i) t_i(q_i) - C_i(\widehat{q}_i, \theta_i)$$

2. What is the expected utility of bidder i if he does not cheat?

    - If bidder i does not cheat by investing at q_i, he receives the transfer, $t_i(q_i)$, whether or not he is monitored. Intuitively, if the bidder does not cheat, the procurer pays him a transfer $t_i(q_i)$ even if the procurer were to monitor the bidder's performance. Therefore, his expected utility becomes

    $$E[U_i(q_i, q_i, \theta_i)] = t_i(q_i) - C_i(q_i, \theta_i)$$

3. What is the incentive compatibility condition for bidder i not to cheat?

    - For bidder i not to cheat, his expected utility from not cheating must be weakly larger than that from cheating, that is, $E[U_i(q_i, q_i, \theta_i)] \geq E[U_i(\widehat{q}_i, q_i, \theta_i)]$, which entails

Exercise #10.5: Procurement Auctions with Imperfect Monitoring[C]

$$t_i(q_i) - C_i(q_i, \theta_i) \geq \alpha_i t_i(\widehat{q}_i) + (1-\alpha_i) t_i(q_i) - C_i(\widehat{q}_i, \theta_i)$$

$$\alpha_i [t_i(q_i) - t_i(\widehat{q}_i)] \geq C_i(q_i, \theta_i) - C_i(\widehat{q}_i, \theta_i)$$

$$\alpha_i \geq \frac{C_i(q_i, \theta_i) - C_i(\widehat{q}_i, \theta_i)}{t_i(q_i) - t_i(\widehat{q}_i)}$$

Intuitively, the bidder does not cheat if the probability of being detected, α_i, is sufficiently high.

4. What is the individual rationality condition for bidder i not to refrain from participation?

- To induce the participation of bidder i, the expected utility from cheating should be weakly larger than from non-participation, that is, $E[U_i(\widehat{q}_i, q_i, \theta_i)] \geq 0$, or

$$\alpha_i t_i(\widehat{q}_i) + (1-\alpha_i) t_i(q_i) - C_i(\widehat{q}_i, \theta_i) \geq 0$$

which, solving for probability α_i, yields

$$\alpha_i \leq \frac{t_i(q_i) - C_i(\widehat{q}_i, \theta_i)}{t_i(q_i) - t_i(\widehat{q}_i)}$$

5. What conditions do we need on probability α_i so that bidder i participates and does not cheat?

- Combining the above inequalities, the feasible range of monitoring probability must satisfy

$$\frac{C_i(q_i, \theta_i) - C_i(\widehat{q}_i, \theta_i)}{t_i(q_i) - t_i(\widehat{q}_i)} \leq \alpha_i \leq \frac{t_i(q_i) - C_i(\widehat{q}_i, \theta_i)}{t_i(q_i) - t_i(\widehat{q}_i)} \quad (SMC_i)$$

6. Write down the procurer's welfare maximization problem and the constraints.

- The procurer chooses the triplet (q_i, t_i, α_i) for each bidder i to solve

$$\max_{\{q_i, t_i, \alpha_i\}_{i=1}^N} \sum_{i=1}^{N} [V(q_i) - (1+\lambda) t_i(q_i) - \alpha_i m_i(q_i)]$$

subject to SMC_i and the monotonicity of q_i.

7. What is the transfer function of the procurer? And, what is the optimal level of bidder i's cheating?

   - Same as Case 1 (part a of Exercise 10.4). Since efficiency is observable, the transfer that the procurer pays to bidder i compensates for his cost but not his information rent; that is, $t_i(q_i) = C_i(q_i, \theta_i)$ if he honestly invests at q_i or is not detected of cheating at $\widehat{q}_i$, whereas the transfer is $t_i(\widehat{q}_i) = C_i(\widehat{q}_i, \theta_i)$ if the bidder is detected of cheating at $\widehat{q}_i$.
   - Knowing that he will be compensated for the conservation cost, bidder i's expected utility from cheating becomes

   $$E[U_i(\widehat{q}_i, q_i, \theta_i)] = \alpha_i t_i(\widehat{q}_i) + (1 - \alpha_i) t_i(q_i) - C_i(\widehat{q}_i, \theta_i)$$
   $$= \alpha_i C_i(\widehat{q}_i, \theta_i) + (1 - \alpha_i) C_i(q_i, \theta_i) - C_i(\widehat{q}_i, \theta_i)$$
   $$= (1 - \alpha_i)[C_i(q_i, \theta_i) - C_i(\widehat{q}_i, \theta_i)]$$
   $$= (1 - \alpha_i) \int_{\widehat{q}_i}^{q_i} \frac{\partial C_i(\widetilde{q}_i, \theta_i)}{\partial q_i} d\widetilde{q}_i$$

   In words, the above expected utility indicates that, when bidder i increases his cheating (i.e., he decreases his conservation level $\widehat{q}_i$, thus slacking), he saves in his conservation cost, as indicated by the positive anti-derivative. Therefore, all levels of cheating, $0 < \widehat{q}_i < q_i$, are dominated by $\widehat{q}_i = 0$. Intuitively, if bidder i ever cheats, he prefers to not conserve any land at all, $\widehat{q}_i = 0$. These incentives are, of course, anticipated by the procurer.

8. What is the procurer's optimal monitoring probability α_i?

   - Substituting our above result about the optimal cheating, $\widehat{q}_i = 0$, and the transfer $t_i(q_i) = C_i(q_i, \theta_i)$ into the SMC_i condition, we obtain

   $$\frac{C_i(q_i, \theta_i)}{C_i(q_i, \theta_i)} \le \alpha_i \le \frac{C_i(q_i, \theta_i)}{C_i(q_i, \theta_i)}$$

   which simplifies to $1 \le \alpha_i \le 1$, ultimately implying that $\alpha_i = 1$. Therefore, the procurer must monitor bidder i at all times to deter him from converting less than the contracted acreage of farmland q_i.

9. Using your above results, solve for the optimal input and transfer of bidder i.

Exercise #10.5: Procurement Auctions with Imperfect Monitoring

- The procurer solves

$$\max_{\{q_i\}_{i=1}^N} \sum_{i=1}^N [V(q_i) - (1+\lambda) C_i(q_i, \theta_i) - m_i(q_i)]$$

$$= \sum_{i=1}^N \left[A \ln q_i - \frac{1+\lambda}{1+\theta_i} \frac{(q_i)^2}{2} - \gamma \frac{(q_i)^2}{2} \right]$$

subject to the monotonicity of q_i.

- We next ignore the monotonicity constraint and later on check that it indeed holds once we find the optimal input and transfer, q_i and t_i. Taking first-order condition with respect to q_i,

$$\frac{A}{q_i} = \left(\frac{1+\lambda}{1+\theta_i} + \gamma \right) q_i$$

Simplifying the above expression, the optimal input in Case 3, q_i^3, is

$$q_i^3 = \sqrt{\frac{A}{\frac{1+\lambda}{1+\theta_i} + \gamma}}$$

Substituting the optimal input, q_i^3, into the transfer function, the optimal transfer, t_i^3, is

$$t_i^3 = C_i\left(q_i^3, \theta_i\right)$$

$$= \frac{1}{2(1+\theta_i)} \frac{A}{\frac{1+\lambda}{1+\theta_i} + \gamma}$$

$$= \frac{A}{2} \frac{1}{(1+\lambda) + \gamma(1+\theta_i)}$$

Last, we confirm that the optimal input, q_i^3, satisfies the monotonicity constraint, since

$$\frac{\partial q_3^1}{\partial \theta_i} = \frac{1+\lambda}{2} \sqrt{A} [(1+\lambda) + \gamma(1+\theta_i)]^{-\frac{3}{2}} (1+\theta_i)^{-\frac{1}{2}} \geq 0$$

(b) *Case 4. Imperfect monitoring, unobservable efficiency.* We finally assume that both efficiency and input are not observable by the procurer.

  1. Write down the procurer's welfare maximization problem and the constraints.

- The procurer chooses the triplet (q_i, t_i, α_i) for each bidder i to solve

$$\max_{\{q_i, t_i, \alpha_i\}_{i=1}^{N}} \sum_{i=1}^{N} [V(q_i) - (1+\lambda) t_i(q_i) - \alpha_i m_i(q_i)]$$

subject to BIC_i, SMC_i, and the monotonicity of q_i.
- Regarding the transfer function, note that, it coincides with that in Case 2 of Exercise 10.4. Since efficiency is unobservable, the transfer that the procurer pays to bidder i compensates for his conservation cost in addition to his information rent, that is, $t_i(q_i) = \widetilde{C}_i(q_i, \theta_i)$ if he invests q_i, or he is not detected cheating, whereas the transfer is $t_i(\widehat{q}_i) = \widetilde{C}_i(\widehat{q}_i, \theta_i)$ if he is detected cheating at $\widehat{q}_i$.

2. What is the optimal level of bidder i's cheating?

- Knowing that he will be compensated for the *virtual* cost, bidder i's expected utility from cheating is

$$\begin{aligned} E[U_i(\widehat{q}_i, q_i, \theta_i)] &= \alpha_i t_i(\widehat{q}_i) + (1 - \alpha_i) t_i(q_i) - C_i(\widehat{q}_i, \theta_i) \\ &= \alpha_i \widetilde{C}_i(\widehat{q}_i, \theta_i) + (1 - \alpha_i) \widetilde{C}_i(q_i, \theta_i) - C_i(\widehat{q}_i, \theta_i) \\ &= (1 - \alpha_i)[C_i(q_i, \theta_i) - C_i(\widehat{q}_i, \theta_i)] \\ &\quad - \frac{1 - F(\theta_i)}{f(\theta_i)} \left[\alpha_i \frac{\partial C_i(\widehat{q}_i, \theta_i)}{\partial \theta_i} \right. \\ &\quad \left. + (1 - \alpha_i) \frac{\partial C_i(q_i, \theta_i)}{\partial \theta_i} \right] \\ &= (1 - \alpha_i) \int_{\widehat{q}_i}^{q_i} \frac{\partial C_i(\widetilde{q}_i, \theta_i)}{\partial q_i} d\widetilde{q}_i \\ &\quad - \frac{1 - F(\theta_i)}{f(\theta_i)} \frac{\partial C_i(q_i, \theta_i)}{\partial \theta_i} \\ &\quad + \alpha_i \frac{1 - F(\theta_i)}{f(\theta_i)} \int_{\widehat{q}_i}^{q_i} \frac{\partial^2 C_i(\widetilde{q}_i, \theta_i)}{\partial q_i \partial \theta_i} d\widetilde{q}_i \end{aligned}$$

In words, when bidder i cheats more (i.e., $\widehat{q}_i$ decreases), two effects arise: on the one hand, he saves in the conservation cost, as indicated by the positive first anti-derivative; on the other hand, cheating reduces the information rent that bidder i enjoys, as indicated by the negative second anti-derivative.

Overall, if bidder i can save more in conservation cost than the foregone information rent, then this bidder has incentives to cheat more. In this case, all levels of cheating, $0 < \widehat{q}_i < q_i$, are dominated

Exercise #10.5: Procurement Auctions with Imperfect Monitoring[C]

by $\widehat{q}_i = 0$. Intuitively, if bidder i ever cheats, he prefers to convert zero acreage of farmland.

3. What is the procurer's optimal monitoring probability?

   - Substituting the optimal cheating $\widehat{q}_i = 0$ and the transfer $t_i(q_i) = \widetilde{C}_i(q_i, \theta_i)$ into SMC_i, the feasible range of monitoring probability becomes

   $$\frac{C_i(q_i, \theta_i)}{\widetilde{C}_i(q_i, \theta_i)} \leq \alpha_i \leq 1$$

   Since monitoring is costly, the procurer would choose the minimum feasible monitoring probability at $\underline{\alpha}_i$ in order to deter bidder i from cheating at the minimal cost, where

   $$\begin{aligned}\underline{\alpha}_i &\equiv \frac{C_i(q_i, \theta_i)}{\widetilde{C}_i(q_i, \theta_i)} \\ &= \frac{(q_i)^2}{2(1+\theta_i)} \frac{(1+\theta_i)^2}{(q_i)^2} \\ &= \frac{1+\theta_i}{2}.\end{aligned}$$

4. Find the optimal input and transfer of bidder i.

   - The procurer now solves

   $$\max_{\{q_i\}_{i=1}^N} \sum_{i=1}^N \left[V(q_i) - (1+\lambda) \widetilde{C}_i(q_i, \theta_i) - \underline{\alpha}_i m_i(q_i) \right]$$

   $$= \sum_{i=1}^N \left[A \ln q_i - (1+\lambda) \left(\frac{q_i}{1+\theta_i}\right)^2 - \frac{1+\theta_i}{2} \frac{\gamma}{2} (q_i)^2 \right]$$

   $$= \sum_{i=1}^N \left[A \ln q_i - \left[\frac{2(1+\lambda)}{(1+\theta_i)^2} + \frac{\gamma(1+\theta_i)}{2}\right] \frac{(q_i)^2}{2} \right]$$

   subject to the monotonicity of q_i.
   - As usual, we next ignore the monotonicity constraint and later on check that it indeed holds once we find the optimal input and transfer, q_i and t_i. Taking first-order condition with respect to q_i,

   $$\frac{A}{q_i} = \left[\frac{2(1+\lambda)}{(1+\theta_i)^2} + \frac{\gamma(1+\theta_i)}{2}\right] q_i$$

Simplifying the above expression, the optimal input in Case 4, q_i^4, is

$$q_i^4 = \sqrt{\frac{A}{\frac{2(1+\lambda)}{(1+\theta_i)^2} + \frac{\gamma(1+\theta_i)}{2}}}$$

Substituting the optimal input, q_i^4, into the transfer function, the optimal transfer, t_i^4, is

$$t_i^4 = \tilde{C}_i\left(q_i^4, \theta_i\right)$$

$$= \frac{A}{\frac{2(1+\lambda)}{(1+\theta_i)^2} + \frac{\gamma(1+\theta_i)}{2}} \frac{1}{(1+\theta_i)^2}$$

$$= \frac{2A}{4(1+\lambda) + \gamma(1+\theta_i)^3}$$

Last, we confirm that the optimal input, q_i^4, satisfies the monotonicity constraint, since

$$\frac{\partial q_4^1}{\partial \theta_i} = \frac{\sqrt{A}}{2}\left(\frac{2(1+\lambda)}{(1+\theta_i)^2} + \frac{\gamma(1+\theta_i)}{2}\right)^{-\frac{3}{2}}\left(\frac{4(1+\lambda)}{(1+\theta_i)^3} - \frac{\gamma}{2}\right)$$

In particular, the second parenthesis is positive since $\frac{4(1+\lambda)}{(1+\theta_i)^3} \geq \frac{\gamma}{2}$ simplifies to

$$\gamma \leq \left(\frac{2}{1+\theta_i}\right)^3 (1+\lambda).$$

Indeed, since $0 \leq \gamma \leq 1$, the above inequality is satisfied because $\frac{2}{1+\theta_i} \geq 1$ and $1+\lambda \geq 1$. Hence, the monotonicity constraint holds.

(c) *Numerical example.* Consider parameter values $A = 3$, $\lambda = \frac{1}{10}$, and $\gamma = \frac{1}{20}$. Solve for the optimal input, transfer, and stochastic monitoring probability in Cases 1–4 when efficiency is low at $\theta_i = \frac{1}{4}$ and high at $\theta_i = \frac{3}{4}$, respectively. Compare your results.

- Inserting these parameter values in our above results yields Table 10.1, which summarizes our equilibrium results when efficiency is low, $\theta_i = \frac{1}{4}$ (top panel), and when efficiency is high, $\theta_i = \frac{3}{4}$ (bottom panel). We separately discuss each case below.

Exercise #10.5: Procurement Auctions with Imperfect Monitoring^C

Table 10.1 Equilibrium results with low efficiency (top panel) and high efficiency (bottom panel)

	Input q_i	Transfer t_i	Monit. α_i
Case 1	1.846	1.364	N/A
Case 2	1.460	1.364	N/A
Case 3	1.796	1.290	1
Case 4	1.444	1.334	0.625

	Input q_i	Transfer t_i	Monit. α_i
Case 1	2.185	1.364	N/A
Case 2	2.044	1.364	N/A
Case 3	2.103	1.263	1
Case 4	1.984	1.285	0.875

- In Cases 1 and 2, when input is observable, the procurer devises a flat rate payment scheme for all bidders because a more efficient bidder i would convert more acres of farmland at a lower cost per acre, rendering the same conservation cost across all bidders. In Case 2 when efficiency is no longer observable, on the one hand it becomes more expensive to induce land conservation due to the payment of information rent, but on the other hand the contracted acreage of land conservation is reduced, such that these two effects offset each other to give exactly the same transfer as in Case 1.
- In Case 3, when input is not observable, the procurer has to spare some resources to monitor the input of bidder i, such that it becomes more expensive to implement land conservation, yielding a reduced acreage of land conservation and also the transfer to bidder i as compared to Case 1. Nonetheless, compared to Case 2, input is larger but transfer is lower because the cost of monitoring is relatively less significant than the cost of information for the procurer.
- In Case 4, when both efficiency and input are not observable, the procurer pays information rent to and also conducts costly monitoring of bidder i. Therefore, it becomes the most expensive to implement land conservation, resulting in the lowest input among all cases. Compared to Case 3, despite reduced acreage of land conservation, transfer is higher because the procurer needs to pay information rent to bidder i to induce his truthful reporting of efficiency θ_i.
- Lastly, evaluating across the two tables, efficiency increases input in all cases. Other than the flat rate payment scheme that is independent of efficiency in Cases 1 and 2, a higher efficiency implies a lower conservation cost so that transfer is reduced despite a higher acreage of land conservation in Cases 3 and 4. Nevertheless, in Case 4, as a more extensive area of land is under conservation, monitoring rate is higher when efficiency increases.

Game Theory Appendix

We next provide a list of basic game theory tools that are used throughout this book. For a more detailed presentation, see Tadelis (2013) and Munoz-Garcia and Toro-Gonzalez (2020).

Background

Consider a setting with $N \geq 2$ players (e.g., bidders in an auction), each choosing a strategy s_i from a strategy set S_i. In an auction, strategy s_i represents the specific bid that bidder i submits, so that $S_i = \mathbb{R}_+$ implies $s_i = b_i \geq 0$. If bidder i faces a budget constraint $w_i > 0$, then his strategy set is $S_i \in [0, w_i]$, intuitively representing the set of affordable bids. Similarly, let s_j represents player j's strategy, where $j \neq i$, from his strategy set S_j, which may differ from player i's, S_i, if each bidder can afford to submit different bids; otherwise, $S_i = S_j = \mathbb{R}_+$.

For compactness, we often use (s_i, s_{-i}) to denote a strategy profile where player i chooses s_i while his rivals select s_{-i}, which is defined as

$$s_{-i} \equiv (s_1, s_2, \ldots, s_{i-1}, s_{i+1}, \ldots, s_N).$$

In an auction context, for instance, strategy profile (s_i, s_{-i}) is denoted as (b_i, b_{-i}) to represent bidder i's bid, b_i, and the vector of his rivals' bids, b_{-i}.

The following subsections present different solution concepts that seek to predict how players behave (e.g., which specific bid each of them submits).

Dominated Strategies

In this first solution concept, rather than focusing on which specific strategy each player chooses to maximize his payoff, we seek to delete those strategies that a rational player (one choosing actions that maximize his payoff) would never select.

Strictly Dominated Strategy. *Strategy s_i strictly dominates strategy $s'_i \neq s_i$ if player i's utility satisfies*

$$u_i(s_i, s_{-i}) > u_i(s'_i, s_{-i})$$

for every strategy profile s_{-i} of player i's rivals.

Intuitively, strategy s_i yields a higher payoff than s'_i *regardless* of the strategy chosen by his rivals. In this setting, we say that s_i strictly dominates s'_i and that s'_i is strictly dominated by s_i. In an auction, a strictly dominated bid, b'_i, yields a strictly lower payoff than bid b_i regardless of the bid that i's rivals choose. Therefore, we should expect a rational bidder to never choose such a strategy. We can, essentially, delete strictly dominated strategies (e.g., bids) from bidder i's strategy set, reducing the list of potential bids that this bidder considers.

If player i finds that strategy s_i strictly dominates all his other available strategies, we can then say that s_i is a *strictly dominant* strategy, and we would expect him choosing that strategy, as defined below.

Strictly Dominant Strategy. *Strategy s_i is strictly dominant if player i's utility satisfies*

$$u_i(s_i, s_{-i}) > u_i(s'_i, s_{-i})$$

for every strategy $s'_i \neq s_i$, and every strategy profile s_{-i} of player i's rivals.

Intuitively, when player i chooses a strategy $s'_i \neq s_i$, he obtains a strictly lower payoff than with s_i, and this happens *regardless* of the strategy profile that his rivals select (i.e., for all s_{-i}). This implies that every strategy $s'_i \neq s_i$ is strictly dominated by s_i, meaning that we can then delete every $s'_i \neq s_i$, leaving him with only one surviving strategy, s_i. This strategy, then, provides player i with an unambiguously higher payoff than any of his available strategies regardless of the strategies that his opponents select.

Iterative Deletion of Strictly Dominated Strategies. From rationality, we know that a player would never use strictly dominated strategies, so we can delete them from his strategy set. We can then proceed by using the definition of "common knowledge of rationality," which in this context entails that every player i can put in his opponent's shoes, identify all strictly dominated strategies for his opponent (player j), and delete them from j's strategy set. Next, player i can find which strategies survived this iterated deletion of strictly dominated strategies (IDSDS) and then remove any strategies that he finds to be strictly dominated at this point. This process continues until no player can identify any further strictly dominated strategies to delete. The remaining strategies are referred to as the strategy profile(s) surviving IDSDS.

In some games, player i may find that his payoff from two strategies, s_i and s'_i, coincides, which yields the following definition, which is less restrictive than strictly dominated strategies.

Weakly Dominated Strategies. *Strategy s_i weakly dominates another strategy $s'_i \neq s_i$ if player i's utility satisfies*

$$u_i(s_i, s_{-i}) \geq u_i(s'_i, s_{-i})$$

for every strategy profile s_{-i} of player i's rivals, and

$$u_i(s_i, s_{-i}) > u_i(s'_i, s_{-i})$$

for at least one strategy profile s_{-i}.

In an auction, this definition entails that bidder i finds that the bid b_i weakly dominates another bid b'_i if submitting b_i provides him with a strictly higher payoff than submitting bid b'_i for at least one of his rivals' bidding profiles, b_{-i}, but provides at least the same payoff as b'_i for the remaining bids of his rivals. This is the case in the second-price auction of Exercise 1.2.

Nash Equilibrium

In this section, we present a solution concept that helps us refine IDSDS. Fewer strategy profiles can, then, emerge as equilibria of the game. This solution concept, known as Nash equilibrium after Nash (1950), builds upon the notion that every player finds his best response to each of his rivals' strategies. We, hence, start defining best responses.

Best Response. *Strategy s_i is player i's best response to strategy profile s_{-i}, if*

$$u_i(s_i, s_{-i}) \geq u_i(s'_i, s_{-i})$$

for every strategy $s'_i \neq s_i$.

Intuitively, in an auction with two bidders, bid b_i is bidder i's best response to bidder j's bid, b_j, if b_i yields a weakly higher payoff than any other bid b'_i against b_j. In other words, when bidder j submits b_j, bidder i maximizes his payoff by responding with bid b_i than with any other available bid.

We next use the concept of best response to define a Nash equilibrium.

Nash Equilibrium (NE). *A strategy profile (s_i^*, s_{-i}^*) is a Nash equilibrium if every player chooses a best response given his rivals' strategies.*

In an auction with two bidders, bid b_i^* is bidder i's best response to his opponent's equilibrium bid, b_j^*, and, similarly, bid b_j^* is bidder j's best response to bidder i's equilibrium bid b_i^*. Therefore, a bidding profile is a NE if it is a *mutual* best response: bidder i's bid is a best response to bidder j's, and vice versa. As a result, no bidder has incentives to unilaterally deviate since doing so would lower his payoff or keep it unchanged.

Mixed-Strategy Nash Equilibrium

The above definition of NE assumes that players choose one of their strategies with certainty (100% probability), which are often known as "pure strategies." In some games, however, restricting players to choose pure strategies may entail no NE. If we allow players to randomize over their available strategies, that is, mixing between at least two strategies, we can then find equilibria. Since players in this setting choose mixed strategies, this solution concept is referred to as "mixed-strategy Nash equilibrium." Before defining this solution concept, we provide a more formal definition of mixed strategy.

Mixed Strategy. *Consider a discrete strategy set $S_i = \{s_1, s_2, \ldots, s_m\}$, where $m \geq 2$ denotes the number of pure strategies. Player i's mixed strategy $\sigma_i = \{\sigma_i(s_1), \sigma_i(s_2), \ldots, \sigma_i(s_m)\}$ is a probability distribution over the pure strategies in S_i, with the property that probability $\sigma_i(s_k)$ satisfies:*

1. *$\sigma_i(s_k) \geq 0$ for every pure strategy s_k, and*
2. *$\sum_{k=1}^{m} \sigma_i(s_k) = 1$.*

Therefore, probability $\sigma_i(s_k)$ assigns a number in the interval $[0, 1]$ to the pure strategy s_k, indicating how frequently player i chooses pure strategy s_k and taking into account that the sum of his probabilities over all his available pure strategies must add up to 1, $\sum_{k=1}^{m} \sigma_i(s_k) = 1$. For instance, in a setting with only two pure strategies, H and L, the above conditions entail that $\sigma_i(H) \geq 0$, $\sigma_i(L) \geq 0$, and $\sigma_i(H) + \sigma_i(L) = 1$, so player i distributes probability weights between H and L.

Examples include $\sigma_i(H) = \sigma_i(L) = \frac{1}{2}$, or $\sigma_i(H) = \frac{2}{3}$ and $\sigma_i(L) = \frac{1}{3}$, among others.[1]

As a remark, the above definition can be applied to games where players choose their strategies from a continuous strategy space, such as bid in an auction $b_i \in \mathbb{R}_+$. In this context, player i's probability distribution over his pure strategies (bids) in $\mathbb{R}_+$ can be represented with a cumulative distribution function $F : \mathbb{R}_+ \to [0, 1]$ mapping every bid $b_i \in \mathbb{R}_+$ into a cumulative probability. In addition, if the cumulative distribution function $F(b_i)$ has a density function $f(b_i)$, we can interpret $f(x)$ as representing the probability that bidder i submits a bid of $b_i = \$x$.

Using the above definition of mixed strategy as probability distribution over the set of pure strategies, we can adapt the definition of the NE solution concept to a setting where players can randomize. In this context, let us first define a player's best response.

Best Response with Mixed Strategies. *Player i's mixed strategy σ_i is a best response to his opponents' mixed strategy σ_{-i} if and only if his expected utility from σ_i satisfies*

$$EU_i(\sigma_i, \sigma_{-i}) \geq EU_i(\sigma'_i, \sigma_{-i}) \text{ for all } \sigma'_i \neq \sigma_i.$$

Intuitively, mixed strategy σ_i is player i's best response to his opponents' mixed-strategy profile σ_{-i} if no other randomization σ'_i (including pure or mixed strategies) yields a higher expected utility than σ_i does. Note that player i needs to compute his expected utility from randomizing over his pure strategies and his rivals also randomizing over their pure strategies.

Using best responses in a context where players can randomize, we now define a mixed-strategy Nash equilibrium, as follows.

Mixed Strategy Nash Equilibrium (msNE). *A strategy profile $\left(\sigma_i^*, \sigma_{-i}^*\right)$ is a mixed-strategy Nash equilibrium if and only if $\sigma_i^* = BR_i(\sigma_{-i}^*)$ for every player i.*

Therefore, when player i chooses his equilibrium strategy σ_i^*, he is best responding to his opponents' equilibrium strategies, σ_{-i}^*. In other words, players are choosing their mutual best responses and, thus, have no incentives to unilaterally deviate, which is analogous to our definition of NE using pure strategies.

---

[1] The definition of mixed strategy also embodies pure strategies as a special case (that is, $\sigma_i(H) = 1$ and $\sigma_i(L) = 0$, or vice versa), which entails that the probability distribution is degenerated, assigning all probability weights to just one pure strategy.

Subgame Perfect Equilibrium

In sequential-move games, the NE solution concept can identify equilibria that are not *sequentially rational*, that is, strategies that do not maximize a player's payoff given: (1) the stage of the game when he is called to move, and (2) given the information he observes at that point. (For a more detailed discussion about this point, see Tadelis (2013, section 8.2).) We then need a new solution concept that guarantees sequential rationality in all stages of the game where players interact.

Before introducing this new solution concept, we need to define subsets of the entire game tree, where we will require players to behave optimally. Specifically, a *subgame* is a tree structure defined by a node and all its successors (not breaking any information sets).[2] We are now ready to use the definition of subgame, as a part of the game tree, to characterize a new solution concept in sequential-move games.

Subgame Perfect Equilibrium (SPE). *A strategy profile (s_i^*, s_{-i}^*) is a subgame perfect equilibrium if it specifies a NE in each subgame.*

To find SPEs in a sequential-move game, we apply the notion of "backward induction," namely, we start at the last stage of the game, finding optimal actions for the player(s) called to move in the last stage. Then, we move to the second-to-last mover who, anticipating equilibrium behavior in the last subgame, chooses his optimal action. We can repeat this process by moving one step closer to the first stage, successively finding equilibrium behavior in subgames as each player can anticipate how players behave in the subgames that unfold in all subsequent stages.

For an example, consider Exercise 3.11 where, in the first stage, the seller chooses an entry fee $E \geq 0$ that every bidder must pay before entering into the auction. In the second stage, every bidder i observes the entry fee E that the seller chose in the first period and responds with its equilibrium bid $b_i(E)$. As that exercise describes, we start analyzing the second stage, finding the equilibrium bid $b_i(E)$ as a function of any entry fee E chosen in the first period. We then move to examine the first stage, where the seller can anticipate the equilibrium bid $b_i(E)$ that bidder i responds with in the second stage. Anticipating this equilibrium bid, the seller chooses the entry fee E^* that maximizes his expected payoff. The pair $(E^*, b_i(E))$ reports the SPE of this two-stage sequential-move game.

---

[2] An information set connects different nodes in a game tree, and the player called on to move cannot distinguish among these connected nodes. For instance, a player does not observe the action chosen by the player acting before him, thus not knowing in which specific node along the game tree he is located.

Bayesian Nash Equilibrium

In some contexts, players interact in games where at least one of them is uninformed about some relevant information, such as his rival's valuation for the object being sold. In an auction, every bidder i observes his valuation, v_i, where $v_i \in \mathbb{R}_+$, but does not observe his rival's valuation, v_j. Bidders, however, know the probability distribution over valuations, e.g., bidder i knows that his rival's valuation is either $v_j = H$ with probability q or $v_j = L$ with probability $1 - q$, and this information is a common knowledge, where $q \in [0, 1]$. Therefore, in an auction where bidders submit their bids simultaneously, every bidder i's bid must be a function of his observed valuation, v_i, and is denoted as $b_i(v_i)$.

We are now ready to adapt the definition of a NE to this incomplete information setting, by first defining a best response in this context. For compactness, let

$$v_{-i} = (v_1, v_2, \ldots, v_{i-1}, v_{i+1}, \ldots, v_N)$$

denote the valuations of bidder i's rivals.

Best Response Under Incomplete Information. *Bid b_i is a best response to bid profile b_{-i}, if bidder i's utility satisfies*

$$EU_i(b_i(v_i), b_{-i}(v_{-i})) \geq EU_i(b'_i(v_i), b_{-i}(v_{-i}))$$

for every strategy $b'_i(v_i) \neq b_i(v_i)$ and every valuation v_i.

Intuitively, in an auction with two bidders, bidder i finds bid $b_i(v_i)$ to be a best response to bidder j's, $b_j(v_j)$, if $b_i(v_i)$ yields a weakly higher *expected* payoff than any other bid $b'_i(v_i)$ against $b_j(v_j)$. In other words, when bidder j submits $b_j(v_j)$, bidder i maximizes his payoff by responding with $b_i(v_i)$.

We next apply the concept of best response to define a Nash equilibrium in this incomplete information context.

Bayesian Nash Equilibrium (BNE). *A bidding profile $\left(b_i^*(v_i), b_{-i}^*(v_{-i})\right)$ is a Bayesian Nash equilibrium if every bidder i chooses a best response (in expectation) given his rivals' strategies.*

Therefore, in an auction with two bidders, a bidding profile is a BNE if it is a *mutual* best response, thus being analogous to the definition of NE: the bid that bidder i chooses is a best response to that selected by his rival, bidder j, and vice versa. As a result, no bidder has unilateral incentives to deviate.

For a detailed analysis of how to apply BNE in simultaneous-move games, see Munoz-Garcia and Toro-Gonzalez (2020, Chapter 7).

References

Augenblick, N. (2016). The sunk-cost fallacy in penny auctions. *The Review of Economic Studies, 83*(1), 58–86.
Baye, M. R., Kovenock, D., & De Vries, C. G. (1996). The all-pay auction with complete information. *Economic Theory, 8*(2), 291–305.
Blume, A., & Heidhues, P. (2004). All equilibria of the Vickrey auction. *Journal of Economic Theory, 114*(1), 170–177.
Börgers, T. (2015). *An introduction to the theory of mechanism design.* Oxford: Oxford University Press.
Bulow, J., & Klemperer, P. (1999). The generalized war of attrition. *American Economic Review, 89*(1), 175–189.
Caragiannis, I., Kaklamanis, C., Kanellopoulos, P., Kyropoulou, M., Lucier, B., Leme, R. P., et al. (2015). Bounding the inefficiency of outcomes in generalized second price auctions. *Journal of Economic Theory, 156*, 343–388.
Che, Y. K., & Gale, I. (1998). Standard auctions with financially constrained bidders. *The Review of Economic Studies, 65*(1), 1–21.
Choi, P-S, Espínola-Arredondo A., & Muñoz-García, F. (2018). Conservation Procurement Auctions with Bidirectional Externalities. *Journal of Environmental Economics and Management, 92*, 559–579.
Clarke, E. (1971). Multipart Pricing of Public Goods. *Public Choice, 11*(1), 17–33.
Edelman, B., Ostrovsky, M., & Schwarz, M. (2007). Internet advertising and the generalized second-price auction: Selling billions of dollars worth of keywords. *American Economic Review, 97*(1), 242–259.
Fudenberg, D., & Tirole, J. (1986). A theory of exit in duopoly. *Econometrica: Journal of the Econometric Society, 54*(4), 943–960.
Graham, D. A., & Marshall, R. C. (1987). Collusive bidder behavior at single-object second-price and English auctions. *Journal of Political Economy, 95*(6), 1217–1239.
Groves, T. (1973). Incentives in Teams. *Econometrica, 41*(4), 617–631.
Haeringer, G. (2018). *Market design: Auctions and matching.* Cambridge, MA: MIT Press.
Hendricks, K., & Porter, R. H. (1989). Collusion in auctions. *Annales d'Éonomie et de Statistique, 15/16*, 217–230.
Karpowicz, M. (2010). Designing auctions: A historical perspective. *Journal of Telecommunications and Information Technology, 3*, 114–122.
Kim, J., & Che, Y.-K. (2004). Asymmetric information about rivals' types in standard auctions. *Games and Economic Behavior, 46*(2), 383–397.
Klemperer, P. (1998). Auctions with almost common values: The 'Wallet Game' and its applications. *European Economic Review, 42*(3–5), 757–769.
Klemperer, P. (2004). *Auctions: Theory and practice.* Princeton, NJ: Princeton University Press.
Krishna, V. (2009). *Auction theory* (2nd ed.). Cambridge, MA: Academic Press.

Nisan, N., Roughgarden, T., Tardos, E., & Vazirani, V. V. (Eds.). (2007). *Algorithmic game theory*. Cambridge: Cambridge University Press.

Lucking-Reiley, D. (2000). Vickrey auctions in practice: From nineteenth-century Philately to twenty-first-century e-commerce. *Journal of Economic Perspectives, 14*(3), 183–192.

McAfee, R., & McMillan, J. (1992). Bidding rings. *American Economic Review, 82*(3), 579–599.

Menezes, F., & Monteiro, P. K. (2005). *An introduction to auction theory*. Oxford: Oxford University Press.

Milgrom, P. (2004). *Putting auction theory to work*. Cambridge, MA: Cambridge University Press.

Milgrom, P. (2017). Discovering Prices: Auction Design in Markets with Complex Constraints (Kenneth J. Arrow Lecture Series), New York City, NY: Columbia University Press.

Milgrom, P. (2017). *Discovering Prices: Auction Design in Markets with Complex Constraints (Kenneth J. Arrow Lecture Series)*, New York City, NY: Columbia University Press.

Mochón, A., & Sáez, Y. (2014). *Understanding auctions*. London: Springer Nature.

Munoz-Garcia, F. (2017). *Advanced microeconomic theory: An intuitive approach with examples*. Cambridge, MA: The MIT Press.

Munoz-Garcia, F., & Toro-Gonzalez, D. (2020). *Strategy and game theory, practice exercises with answer keys* (2nd ed.). Springer-Nature.

Myerson, R. (1981). Optimal auction design. *Mathematics of Operation Research, 6*(1), 58–73.

Nash, J. F. Jr. (1950). Equilibrium points in n-person games. *Proceedings of the National Academy of Sciences, 36*(1), 48–49.

Orozco-Aleman, S., & Munoz-Garcia, F. (2011). Risk aversion in auctions with asymmetrically informed bidders: A 'Desensitizer' from uncertainty. *Economics Letters, 112*(1), 38–41.

Robinson, M. S. (1985). Collusion and the choice of auction. *RAND Journal of Economics, 16*(1), 141–145.

Sayman, S., & Akçay, Y. (2020). A transaction utility approach for bidding in second-price auctions. *Journal of Interactive Marketing, 49*, 86–93.

Vickrey, W. (1961). Counters peculation, Auctions, and Competitive Sealed Tenders. *The Journal of Finance, 16*(1), 8–37.

Tadelis, S. (2013). *Game theory, an introduction*. Princeton, NJ: Princeton University Press.

Index

A
Acreage of farmland, 267, 276, 279
Affordable bids, 283
Aggressive bids, vi, 19, 34, 40, 42, 43, 63, 68, 76–77, 80, 82, 85, 93, 121, 132, 160, 161, 170, 174, 175, 217, 221
Allocation rule, 204, 243, 244
All-pay auctions, vi, 125–163, 166, 180, 182, 187, 188, 191, 192, 199
All-pay auction under complete information, 125–128
Ascending bids, 2
Assignment rule, 201, 240–242, 247
Asymmetrically distributed valuations, 83–87
Asymmetrically informed risk-averse bidders, 156–163
Asymmetrically informed risk-neutral bidders, 150–156
Asymmetric bids, 7, 21–23
Asymmetric probability distributions, 204
Auction rules, 227

B
Babylon, 28
Backward induction, 87, 288
Bayesian Nash equilibrium (BNE), 18, 19, 39, 110, 115, 188, 201, 203, 207, 209, 210, 213, 215, 217, 218, 220, 238, 289
Best response, 18, 150, 156, 166, 167, 285–286, 289
Beta distribution, 58, 176
Bid, 1, 27, 63, 123, 165, 187, 205, 226, 236, 249
Bidder, 1, 27, 63, 123, 165, 187, 205, 225, 237, 249
Bid equal to valuation, 2–5, 7, 8, 20, 25, 26, 29, 39, 86, 109, 113, 150, 153, 156, 159, 165, 167, 170, 178, 196

Bid shading, 7, 28, 30, 32, 39, 41, 60, 75, 76, 85, 88, 92, 102, 120, 122, 123, 126, 134, 170, 188, 197, 205, 214, 239
Bing, 1
Biodiversity value, 267
Budget constraints, 3, 25–26, 64, 119–123, 283

C
Chain rule, 38, 91
Clicks, 247–250
Collusion, vi, 2, 23–24, 64, 115–118
Common knowledge, 30, 87, 188, 206, 248, 289
Common-value auctions, vii, 205–224
Complex roots, 233
Concave utility function, 74, 156, 182
Conjugate pair, 233
Conservation project, 267
Continuous strategy space, 287
Continuous valuations, 64
Contract, v, vii, viii, 251, 252, 267, 273, 274, 276, 281
Correlated valuations, 2, 10
Cumulative distribution function, 2, 10–12, 21, 28, 43, 48, 49, 51–52, 55, 60, 64, 65, 71, 78, 81, 83, 86, 88, 90, 96, 98–100, 110, 112, 122, 127, 128, 140, 153, 154, 159, 160, 167, 169, 193, 201, 206, 227, 228, 239, 287

D
Density function, 28, 42, 48, 52, 57, 70–72, 81, 94, 96, 98, 103, 105, 128, 140, 176, 193, 227, 228, 287
Descending auction, 28, 60
Direct approach, 28, 36–39, 45, 137–139

Direct revelation mechanism, vii, 201, 202, 237, 239–241
Discrete strategy space, 286
Discrete valuations, 7–9, 109–114
Discriminant, 233
Discriminatory auction, 225, 226
Distortionary taxes, 252
Divide and conquer, 225, 234
Dutch auction, 11, 28, 60–62

E
ebay, 1
Efficiency, 10, 28, 47–48, 82, 139–140, 166, 180, 185, 237, 251–257, 259–262, 264, 265, 267–274, 276–281
English auction, 2, 10–11
Entry fees, vi, 2, 19–21, 64, 99–108, 288
Envelope theorem approach, 28, 45, 115, 126, 130–137, 139, 142, 256, 271
Equilibrium bidding function, vii, 2, 10, 11, 27–29, 31, 34, 35, 37, 39–46, 61, 64, 66, 75–78, 83–88, 95, 102, 104, 117, 119, 122, 130–137, 143–146, 150, 151, 157, 159, 165–179, 182, 183, 188, 198, 207, 208, 211, 212, 214, 216, 217, 219, 220, 227, 234, 235, 239
Expected payment, 66, 73–75, 95, 97, 104, 190, 192, 197–201, 222–223, 240, 242
Expected revenue, vi, vii, 11–19, 48, 61, 64–71, 88, 89, 94, 96, 97, 103, 122, 126, 140–142, 167, 168, 177, 187, 189–204, 206, 221–225, 228–232, 234–237
Expected transfer, 201–202
Expected utility maximization problem, 32, 33, 36, 75, 77, 78, 83, 87, 90, 100, 101, 142, 184
Expected value, 16, 28, 49, 50, 53, 57, 58, 73, 113, 136, 142, 176, 177, 208, 211, 214, 216, 218–220, 227
Exponential distribution, 14, 16, 39, 40, 46, 68, 70, 94, 135, 141, 142, 146, 165, 171, 173, 230, 233
External effects, vii, 251, 261–267

F
First-order stochastic dominance, 85, 98, 113
First-price all-pay auction, 126, 129–146
First-price auction, vi, vii, 7, 13, 19, 27–61, 63–123, 126, 131–135, 139, 142, 150, 153, 159, 162, 171–172, 176, 178, 187–197, 199, 204, 206, 215–218, 221–223, 226, 234–237, 239–241

First-price auction under complete information, 29–30
First-price auction under incomplete information, 27
First-price auction with asymmetrically informed bidders, 152, 159
Flat rate payment scheme, 281

G
Generalized second-price auction, 238, 247–251
Generic distribution, 36–39, 43–47, 60, 63, 64, 71, 122–124, 165, 187
Google, 1, 247
Government securities, 28

H
Highest bidder, 166, 188, 189, 240, 245–250

L
L'Hopital's rule, 15, 70, 214
Linear utility functions, 202
Lotteries, vii, 76, 165–185

M
Marginal benefit, 61, 148, 198, 253, 254, 258, 262, 263, 269
Marginal cost, 29, 61, 89, 198, 252–253, 260, 264, 268
Marginal gain, 63
Marginal loss, 63
Marginal virtual cost, 258, 262, 263
Mechanism design, 197, 237–253
Migratory birds, 267
Misreporting, 238
Mixed-strategy bidding profile, 64, 110
Mixed-strategy Nash equilibrium (msNE), 286, 287
Monitoring cost, 274, 281
Monitoring probability, 274–276, 279, 280
Monotonicity, 269–273, 277–280
More bidders, vi, vii, viii, 4, 21, 34, 35, 40, 68, 108, 120, 121, 128, 134, 171, 172, 180, 183, 205, 214, 228
Multiple units, 225–233
Myerson's Characterization Theorem, 202, 240, 242, 257, 271

N
Nash equilibrium (NE), 285

Negative externalities, 185, 244, 262, 266
No distortion at the top, 259, 260

O

Observed signal, 206, 209, 215, 217, 220
Order of integration, 96, 105, 140
Other distribution forms, 16, 41, 70, 174
Overbidding, 170, 174
Overestimation, 205, 206
Over-reporting, 237

P

Parametric example, 80, 81, 254, 259, 263
Pareto efficiency, 10
Partial derivative, 252, 258
Participation, 19, 92, 98, 100, 102, 104, 125, 246, 252, 253, 256, 257, 269, 275
Payment rule, 237, 240–242
Payoff from losing, 238
Payoff from winning, 22, 32, 36, 79, 83, 238, 239
Perfect monitoring, 267–273
Players, vii, 3, 4, 7, 18, 37, 87, 88, 90, 110, 125, 126, 129, 134, 138, 156, 162, 166, 182, 185, 186, 188, 203, 206, 215, 219, 237, 243–246, 249, 252, 256, 283–289
Political campaigns, 166, 182
Positive externalities, 244, 251, 262–267
Price clock, 11
Probability distribution, 204, 286, 287
Probability of winning, vii, 30, 32, 33, 36, 37, 60, 66, 75, 77, 79, 83, 90, 95, 104, 129, 137, 182, 185, 215, 218, 221, 222
Probability weight, 16, 41–43, 70, 85, 98, 113, 114, 136, 174, 175, 286, 287
Procurement auctions, vi, 251–281
Procurer, 251–263, 266–277, 281
Profitable deviation, 110, 112, 226, 248, 249
Profit margin, 64
Public funds, 252, 253, 269
Pure strategy, 64, 107–110, 125, 126, 147, 156, 159, 286, 287

Q

Quadratic equation, 233
Quasilinear utility function, 252, 267
Quibids.com, 125

R

Radio spectrum, 1
Raising hands, 11, 61
Rate parameter in exponential distribution, 16
R&D race, 125
Reported valuation, 198, 238, 239, 244, 245
Reservation price, v, 2, 10, 11, 17–19, 64, 88–99, 105
Reservation utility, 252, 267, 271
Revelation principle, 201
Revenue equivalence principle, vii, 141, 176, 177, 187–204, 236
Revenue equivalence theorem, 122, 167, 190, 192, 197–204, 223
Risk-averse bidders, 63, 74–82, 156–163, 187, 195–197
Risk-averse seller, 197
Risk aversion, viii, 63, 74, 81, 82, 126, 156, 160–162
Risk loving, 2, 9, 82
Risk neutrality, 195
Risk-neutral seller, 196
Rival, v, vi, 2–4, 8, 18, 22, 25, 29, 33, 37, 43, 60, 85, 91, 109–111, 125–127, 142, 147, 154, 182, 188, 190, 206, 215, 219, 226, 251, 263, 282–286, 289
Rivals' bid profile, 285

S

Sales volume, 64
Sealed envelope, 27
Second highest bidder, 188, 247–249
Second-highest valuation, 12, 26, 51, 189, 195, 237, 240, 241, 243
Second-order statistic, 11–13, 51–55, 58, 168, 196, 228
Second-order stochastic dominance, 196
Second-price all-pay auction, 126, 142–146
Second-price auction, 1–26, 48, 64, 67, 70, 71, 121, 139, 141–142, 165, 170, 172, 173, 176, 177, 187–189, 191–197, 199, 204, 206, 218–223
Second-price auction under complete information, 3, 4
Second price-auction under incomplete information, 4–7
Sequential first-price auction, 64, 87–88
Several units, vi, 225, 237, 245–247
Shadow cost, 252, 253, 269
Single-crossing property, 260, 264, 267

Single unit, 225, 227, 228, 230, 232, 234, 243–245
Spence-Mirrlees sorting condition, 267, 268
Standing bid, 11
Strategy, 2, 64, 125, 181, 188, 219, 226, 238, 283
Strategy profile, 3, 8, 18, 22, 110, 113, 147, 219, 238, 245, 283–288
Strategy set, 283, 284, 286
Strictly dominant strategy, 284
Strictly dominated strategy, 118, 150, 156, 284, 285
Subgame, 87, 288
Subgame perfect equilibrium (SPE), 87, 288
Symmetric bids, 7, 43, 75, 91, 115, 150, 176, 182

T

Third highest bidder, vi, 166, 167, 170, 171, 173, 179, 188, 227, 247–249
Third-highest valuation, 29, 56, 188
Third-order statistic, 227, 228
Third-price auction, vi, 165–185, 188, 189, 191
Town mayor, 252
Truthful revelation, 259
Truthtelling, 238, 239, 248–250
Tulip fever, 28
Types, v, vi, viii, 19, 24, 105, 112, 115, 118, 122, 125, 150, 156, 205, 247, 249, 253, 256, 258, 259, 261

U

Unconstrained bidder, 119, 122–123
Unconstrained problem, 257
Under investment, 274
Under-reporting, 272
Uniform distribution, vi, 14, 16, 23, 27, 30, 32, 41, 42, 45, 49, 52, 53, 57, 60, 63, 70, 72, 74, 85, 87, 88, 93, 102, 115, 131, 133, 141, 145, 171, 174–176, 178, 188, 190, 191, 195, 196, 199, 206, 208, 211, 215, 218, 232, 235
Unilateral incentives, 286, 287, 289

V

Variance, 16, 28, 50, 54, 59, 136, 142, 187, 196, 197, 227
Vickrey–Clarke–Groves (VCG) mechanism, vii, 237, 243–250
Voluntary participation, 253, 256

W

War of attrition, 126, 146–149
Water distribution, vii, 252
Water production, 253, 258
Weakly dominant strategy, 3, 4, 8, 18, 19, 22, 25, 119, 188, 228, 245, 285
Welfare function, 252
Winner's curse, vii, 205–206

Y

Yahoo, 1, 247

The manufacturer's authorised representative in the EU is Springer Nature Customer Service Centre GmbH, Europaplatz 3, 69115 Heidelberg, Germany. If you have any concerns regarding our products, please contact ProductSafety@springernature.com

Printed and bound by CPI Group (UK) Ltd, Croydon, CR0 4YY

25/03/2026

02078170-0003